MW00444385

LENK'S RF HANDBOOK

Other McGraw-Hill Books of Interest

Other books by John Lenk

- LENK'S VIDEO HANDBOOK
- LENK'S LASER HANDBOOK
- LENK'S AUDIO HANDBOOK

Handbooks

Benson • AUDIO ENGINEERING HANDBOOK
Benson • TELEVISION ENGINEERING HANDBOOK
Benson and Whitaker • TELEVISION AND AUDIO HANDBOOK
Coombs • PRINTED CIRCUITS HANDBOOK
Croft and Summers • AMERICAN ELECTRICIANS' HANDBOOK
Di Giacomo • DIGITAL BUS HANDBOOK
Fink and Beaty • STANDARD HANDBOOK FOR ELECTRICAL ENGINEERS
Fink and Christiansen • ELECTRONIC ENGINEERS' HANDBOOK
Hicks • STANDARD HANDBOOK OF ENGINEERING CALCULATIONS
Inglis • ELECTRONIC COMMUNICATIONS HANDBOOK
Kaufman and Seidman • HANDBOOK OF ELECTRONICS CALCULATIONS
Mee and Daniel • MAGNETIC RECORDING HANDBOOK
Tuma • ENGINEERING MATHEMATICS HANDBOOK
Williams and Taylor • ELECTRONIC FILTER DESIGN HANDBOOK

Other

Bartlett • CABLE TELEVISION TECHNOLOGY AND OPERATIONS
Luther • DIGITAL VIDEO IN THE PC ENVIRONMENT
Mee and Daniel • MAGNETIC RECORDING, VOLUMES I-III
Philips International • COMPACT DISC INTERACTIVE
Sherman • CD-ROM HANDBOOK
Whitaker • RF TRANSMISSION SYSTEMS

LENK'S RF HANDBOOK

Operations and Troubleshooting

John D. Lenk

McGRAW-HILL, INC.

New York St. Louis San Francisco Auckland Bogotá Caracas Lisbon London Madrid Mexico Milan Montreal New Delhi Paris San Juan São Paulo Singapore Sydney Tokyo Toronto

Library of Congress Cataloging-in-Publication Data

Lenk, John D.
[RF handbook]
Lenk's RF handbook : operation and troubleshooting / John D. Lenk.
p. cm.
ISBN 0-07-037504-6
1. Radio circuits. 2. Electronic circuit design. 3. Radio circuits—Maintenance and repair. I. Title.
TK6553.L3995 1992
621.384'12—dc20 91-32238
CIP

2 3 4 5 6 7 8 9 0 DOC/DOC 9 8 7 6 5 4 3 2

ISBN 0-07-037504-6

The sponsoring editor for this book was Daniel A. Gonneau, the editing supervisor was Nancy Young, and the production supervisor was Suzanne W. Babeuf. This book was set in Times Roman by McGraw-Hill's Professional Book Group composition unit.

Printed and bound by R. R. Donnelley & Sons Company.

Greetings from the Villa Buttercup!
To my wonderful wife Irene, Thank you
for being by my side all these years!
To my lovely family, Karen,
Tom, Brandon, and Justin.
And to our Lambie and Suzzie,
be happy wherever you are!
To my special readers, may good fortune
find your doorways
to good health and happy things.
Thank you for buying my books and
making me a best seller!
This is book number 73.
Abundance!

ABOUT THE AUTHOR

John D. Lenk is the author of 73 books on electronics which together have sold over a million copies and have been translated into eight languages. His most recent books include *Lenk's Audio Handbook*, *Lenk's Laser Handbook*, and *Lenk's Video Handbook*.

CONTENTS

PREFACE

This is a "something for everyone" RF book. No matter where you are in electronics, this book provides basics, practical considerations, test, and troubleshooting information that can be put to immediate use.

For beginning technicians or students, the book starts with a review and summary of RF basics, from a practical standpoint. This is followed by a chapter describing the practical considerations for RF (such as resonant-circuit design and thermal and mounting problems).

With the basics established, the book provides two chapters devoted to simplified testing and troubleshooting for RF equipment. The RF circuits from a cross section of present-day electronic equipment are chosen as examples. Not only do the chapters cover the operation of the RF circuit, but they describe testing and troubleshooting approaches for that type of RF circuit in step-by-step detail.

For advanced technician or field-service engineer, there are two entire chapters devoted to advanced testing and troubleshooting for sophisticated RF devices such as frequency-synthesis (or digital) tuning and mobile AM/FM/SSB communications equipment. A separate chapter covers RF test equipment, from simple substitute dummy loads for experimenters to measurement of FM deviation with spectrum analyzers.

Chapter 1 is devoted to a review and summary of RF basics, from a practical standpoint.

Chapter 2 covers practical considerations for RF. Such subjects as calculating resonant-circuit values and mounting techniques for RF components are discussed. The main concern is with RF power amplifiers in which thermal problems can arise in design and service.

Chapter 3 is devoted to a review and summary of RF test equipment. Although the emphasis is on communications, a cross section of RF test equipment is discussed.

Chapter 4 covers basic test procedures for RF equipment. These procedures can be applied to a complete piece of equipment (such as a transmitter) or to specific circuits (such as the RF circuits in a receiver). The procedures can also be applied to RF circuits at any time during design or experimentation.

Chapter 5 covers advanced test procedures. Here, the emphasis is on detailed checks for such sophisticated devices as mobile communications sets (CB, amateur, business radio, and cellular telephones).

Chapter 6 is devoted to basic troubleshooting procedures for a cross section of RF circuits. The chapter emphasizes basic problems common to all RF equipment.

Chapter 7 covers advanced troubleshooting. Here, the emphasis is on sophisticated RF devices, such as digital tuning and mobile communications.

John D. Lenk

ACKNOWLEDGMENTS

Many professionals have contributed to this book. I gratefully acknowledge that the tremendous effort needed to make this book such a comprehensive work is impossible for one person and wish to thank all who contributed, both directly and indirectly.

I wish to give special thanks to the following: Dick Harmon, Ross Snyder, Nancy Teater, Mike Arnold, and Joel Salzberg of Hewlett-Packard; Bob Carlson and Martin Plude of B&K-Precision Dynascan Corporation; Joe Cagle and Rinaldo Swayne of Alpine/Luxman; Theodore Zrebiec of Sony; John Taylor and Matthew Mirapaul of Zenith; Thomas Lauterback of Quasar; Donald Woolhouse of Sanyo; J. W. Phipps of Thomson Consumer Electronics (RCA); Tom Roscoe, Dennis Yuoka, and Terrance Miller of Hitachi; and Pat Wilson and Ray Krenzer of Philips Consumer Electronics.

I also wish to thank Joseph A. Labok of Los Angeles Valley College for help and encouragement throughout the years.

And a very special thanks to Daniel A. Gonneau, Editor-in-Chief, Robert McGraw, Nancy Young, Barbara McCann, Charles Love, Peggy Lamb, Thomas G. Kowalczyk, Suzanne W. Babeuf, Stephen Fitzgerald, Kimberly Martin, Kathy Greene, Wayne Smith, and Jeanne Glasser of the McGraw-Hill organization for having that much confidence in the author. I recognize that all books are a team effort and am thankful that I am working with the First Team!

And to my wife Irene, my research analyst and agent, I wish to extend my thanks. Without her help, this book could not have been written.

CHAPTER 1
RF BASICS

This chapter is devoted to a review and summary of RF basics from a practical troubleshooting standpoint. The information here is provided for those readers who are totally unfamiliar with RF (or those who think they know it all).

1.1 RF BANDS

Electric signals (passing through a conductor) generate *electromagnetic waves which are radiated or transmitted* from the conductor when the signal frequency is about 15 kHz and higher. Because of this radiating property, signals of such frequencies are known as *radio-frequency,* or *RF,* signals. It is not practical to design any circuit that covers the entire frequency range or to use all radio frequencies for all purposes. Instead, the RF spectrum is broken down into various *bands,* each used for a specific purpose. In turn, RF circuits are generally designed for use in one particular band.

Table 1.1 shows the most common assignment of RF bands, including both commercial and military bands. Note that radio waves at frequencies above about 1 GHz are known as *microwaves,* and circuits used with microwaves are quite different from those used at lower frequencies. Because of their specialized nature, microwave circuits are not discussed here. Instead, this book concentrates on RF circuits operating at frequencies up to and including the UHF band.

1.2 RF CIRCUIT TYPES

The most common type of RF circuit is an *amplifier,* with *oscillators* running a close second. (Oscillators are discussed in Chap. 2.) RF amplifiers may be divided into two general types: *narrowband amplifiers* (with bandwidths up to several hundred kHz) and *wideband amplifiers* (with bandwidths of several MHz). The following describes the need for both types of amplifiers.

As shown in Table 1.1, the AM broadcast band for the United States is from 535 to 1605 kHz. The frequencies of broadcast stations within this band are spaced from 10 to 15 kHz apart to prevent interference with each other. In the FM broadcast band, the transmitting-station frequencies are spaced 200 kHz apart. In the TV broadcast bands, the stations are about 6 MHz apart.

Within a specific band, each transmitting station is assigned a specific frequency at which the station is to operate. However, each station also transmits

TABLE 1.1 RF-Band Frequency Ranges

Commercial Bands
Very low frequency (VLF) 3–30 kHz
Low frequency (LF) 30–300 kHz
Medium frequency (MF) 300 kHz–3 MHz
High frequency (HF) 3–30 MHz
Very high frequency (VHF) 30–300 MHz
Ultrahigh frequency (UHF) 300 MHz–3 GHz
Superhigh frequency (SHF) 3–30 GHz
Extrahigh frequency (EHF) 30–300 GHz
Military Bands
P-band 225–390 MHz
L-band 390–1550 MHz
S-band 1.5–5.2 GHz
X-band 5.2–10.9 GHz
K-band 10.9–36 GHz
Q-band 36–46 GHz
V-band 46–56 GHz
United States Broadcast Bands
Amplitude modulated (AM) 535–1605 kHz
Frequency modulated (FM) 88–108 MHz
VHF television 54–216 MHz
UHF television 470–890 MHz

across a narrow band of frequencies at either side of the assigned frequency. The band of frequencies is required if the signal is to convey audio intelligence. For example, an AM broadcast station transmits with a band extending 5 to 7.5 kHz on either side of the assigned frequency.

An RF amplifier used in an AM broadcast radio receiver is adjusted to cover a portion of the band about 15 kHz wide, corresponding to the spread of a single station. Under these conditions, the *bandwidth* of the RF amplifier is said to be 15 kHz. The amplifier is adjusted (or tuned) to one station at a time. In the FM broadcast band, where each station is spaced 200 kHz apart, the bandwidth of the RF amplifier is about 150 kHz. Both AM and FM broadcast-band RF amplifiers are essentially narrowband amplifiers. In the television bands, where the stations (or channels) are 6 MHz apart, the RF amplifiers are of the wideband type (or *broadband* type) since the transmitted TV signal is about 4.5 MHz wide.

To sum up, an RF amplifier serves two purposes. One is as a *bandpass filter*, which passes signals from the desired station and rejects all others; the other is to amplify these signals to a suitable voltage (or power) level.

1.3 NARROWBAND RF

The circuit in Fig. 1.1 is a typical narrowband RF amplifier, such as that found in discrete-component radio receivers. In present design, the circuit is usually part

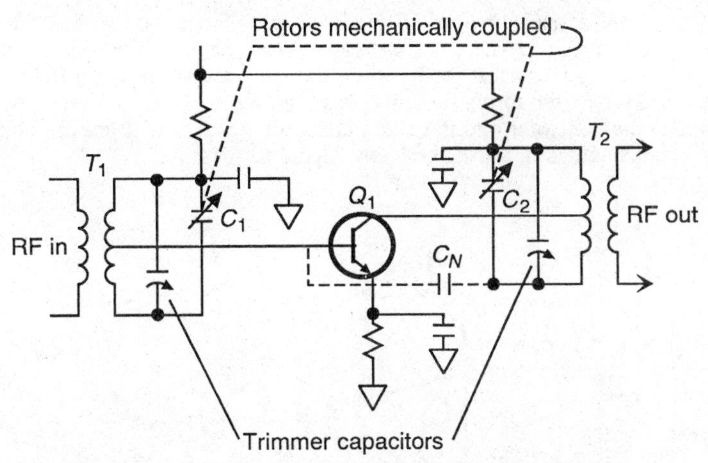

FIGURE 1.1 Typical discrete-component narrowband RF amplifier.

of an IC *tuner package,* as described in Sec. 1.6. For now, let us consider the bandpass filter or tuning function and certain feedback problems.

1.3.1 Narrowband RF Tuning

The circuit in Fig. 1.1 is a single stage of *tuned radio frequency* (TRF) voltage amplification. Input to the stage is by means of transformer T_1, while output is taken from transformer T_2. The secondary of T_1 is tuned to *resonance* at the frequency of the incoming signal by variable capacitor C_1; T_2 is tuned to the same resonant frequency by C_2.

In many cases, the transformers are tuned by adjustable powdered-iron cores. In that case, C_1 and C_2 are fixed capacitors. Since it is impossible to construct two RF circuits that are exactly alike, *trimmer capacitors* (in parallel with C_1 and C_2) vary the overall capacitance of the tuned circuit slightly to compensate for small difference between the circuits.

Although the use of several tuned circuits increases the overall *selectivity* of the amplifier, the need for manipulating a number of variable capacitors can be a problem. This problem is overcome by *ganging,* where the rotors of variable capacitors are mechanically connected so that all rotors move simultaneously when one dial is turned. Any variations in tuning are corrected by the trimmers on each transformer. When the transformers are tuned by adjustable cores, ganging is not used, and each tuning circuit is adjusted separately.

1.3.2 RF Feedback Problems

Undesired feedback (particularly from output to input) is a problem with all RF circuits. There are two types of such feedback: *radiated feedback* and *feedback through the transistor* (or internal feedback). Radiated feedback is prevented by shielding. Fortunately, most modern transistors show little internal feedback at lower frequencies, but feedback can be a problem as frequency increases.

Miller Effect. The most common RF feedback problem is known as *Miller effect,* illustrated in Fig. 1.2. There is some capacitance between the base and emitter of a two-junction transistor (or between the gate and source of a field effect transistor, or FET). This forms the *input capacitance* of an RF circuit. There is also a capacitance between the base and collector (or gate and drain). This capacitance feeds back some of the collector signal to the base.

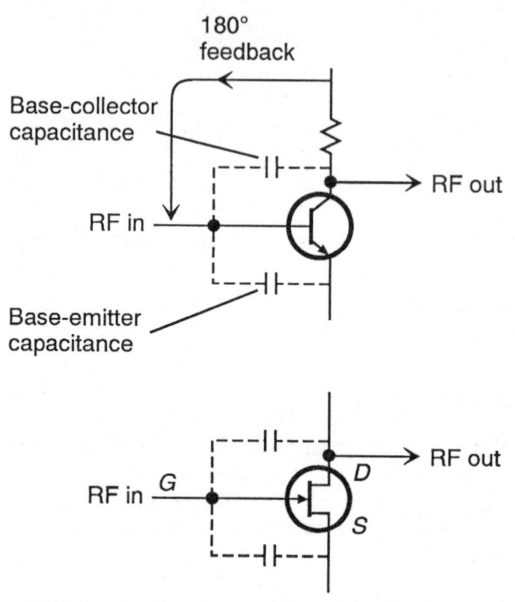

FIGURE 1.2 Transistor input and feedback capacitances.

The collector signal is amplified and is 180° out of phase with the base signal (in a common-emitter amplifier). The collector signal feedback opposes the base signal and tends to distort the input signal. Likewise, the collector-base capacitance is, in effect, in series with the base-emitter capacitance and thus changes the input capacitance.

All of these conditions make for a constantly changing relationship of signals in an RF circuit. For example, if the input-signal amplitude changes, the amount of feedback changes, changing the input capacitance. In turn, the change in input capacitance changes the match between the transistor and the input-tuned circuit, changing the amplitude. Likewise, if the input-signal frequency changes, the feedback changes (since the collector-base capacitive reactance changes), and there is a corresponding change in amplification.

This Miller effect is not necessarily a problem in all RF circuits. FET RF circuits are usually more susceptible to the Miller effect than are two-junction transistor circuits. However, when the Miller effect becomes severe with any RF circuit, the effect can be eliminated (or minimized to a realistic level) by neutralization.

Neutralization. With neutralization, a portion of the signal from the output of the RF circuit is fed back to the input so as to cancel any signal produced by unwanted feedback. This can be used to reduce unwanted feedback resulting from both radiation and internal feedback. Neutralization is done by applying a signal to the input that is equal in magnitude, *but opposite in phase,* to the undesired feedback. The two signals then cancel each other.

The two ends of an output-transformer primary winding (such as T_2 in Fig. 1.1) are of opposite phase. If the opposite-phase signal is fed to the input through a neutralizing capacitor (C_N in Fig. 1.1), the two signals cancel out. As a guideline, the neutralizing capacitor is typically equal to the collector-base capacitance (often a few picofarads).

Common-Base RF Circuits. Instead of neutralization, common-base circuits are sometimes used in RF work to reduce unwanted feedback. Figure 1.3 shows a typical common-base RF amplifier.

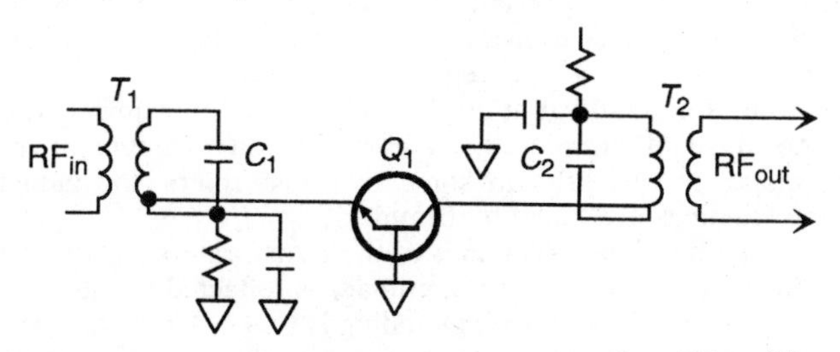

FIGURE 1.3 Typical discrete-component common-base RF amplifier.

Again, the input transformer T_1 is tuned to resonance by variable C_1, while output transformer T_2 is tuned to the same resonant frequency by C_2. The base is grounded, and the input signal is applied to the emitter. The output is taken between the collector and base, which is *common to the input and output circuits.*

The grounded base acts as a shield between the input and output circuits, thus reducing feedback. The circuit in Fig. 1.3 is often found at the antenna input of discrete-component radio receivers. In addition to minimizing undesired feedback, the ground, or common-base, circuit provide a low input impedance (to match a 50-, 75-, or 300-Ω antenna).

1.3.3 IF versus RF

Most radio and TV receivers operate on the *superheterodyne* principle where the frequency of the received radio signal is first converted to a lower predetermined frequency called the *intermediate frequency,* or IF. The circuit (an amplifier) is fixed to operate at this frequency rather than being tunable over the entire band.

The circuit in Fig. 1.4 is a typical IF amplifier, such as that found in discrete-component radio receivers and TV sets. In present design, the circuit is usually part of an IC package, as described in Sec. 1.6. For now, let us consider the tuning function of the IF amplifier.

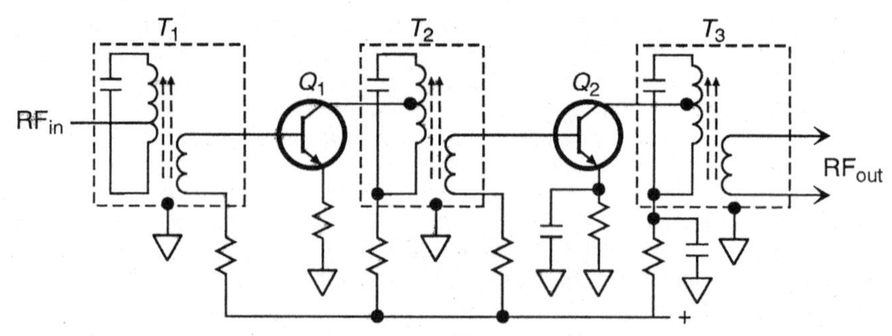

FIGURE 1.4 Typical discrete-component IF amplifier.

The two-stage IF amplifier in Fig. 1.4 is similar to the RF amplifier in Fig. 1.1, except that the IF-amplifier transformers are tuned to the predetermined frequency by means of small fixed capacitors. Since it is not necessary to tune the IF transformers over the entire band, it is practical to tune both the primary and secondary windings of each transformer to the IF. Thus, by adding tuned circuits, the receiver is *selectively* increased.

To compensate for small variations between the IF transformers, each transformer has the adjustable powdered-iron core that can be moved in and out, thus varying the inductance slightly. In some IF transformers, the inductances are fixed, and the windings are tuned by trimmers.

Note that only the primary windings of the transformers are tuned in the Fig. 1.4 circuit. Since the impedance of the primary is reflected to the secondary, the effect is the same regardless of which winding is tuned. However, because there are fewer tuned circuits, the overall selectivity of the amplifier is somewhat reduced. All other factors being equal, the more tuned circuits there are in any amplifier, the greater the selectivity, and vice versa.

The collector of each transistor is connected to a tap on the primary winding of the corresponding output transformer. This is done to match the impedance of the winding to the output impedance of the transistor. Similarly, the secondary of each input transformer has fewer turns than the primary winding so that the winding may match the input impedance of the transistor.

1.3.4 RF Power-Amplifier Circuits

Most radio transmitters (CB, amateur, business radio, cellular telephones) use some form of power amplifier to raise the low-amplitude signal developed by the oscillator to a high-amplitude signal suitable for transmission. For example, most oscillators develop signals of less than 1 W, whereas a solid-state transmitter may require output of several hundred watts (or more).

Figure 1.5 shows two basic RF power-amplifier circuits. In the circuit in Fig. 1.5*a*, the collector load is a parallel-resonant circuit (called a *tank circuit*), consisting of variable capacitor C_1 and coil L_1, tuned to resonance at the desired frequency. The output, which is an amplified version of the input voltage, is from L_2, which together with L_1 forms an output transformer.

The circuit in Fig. 1.5*a* has certain advantages and disadvantages. The winding of L_2 can be made to match the impedance of the load (by selecting the proper

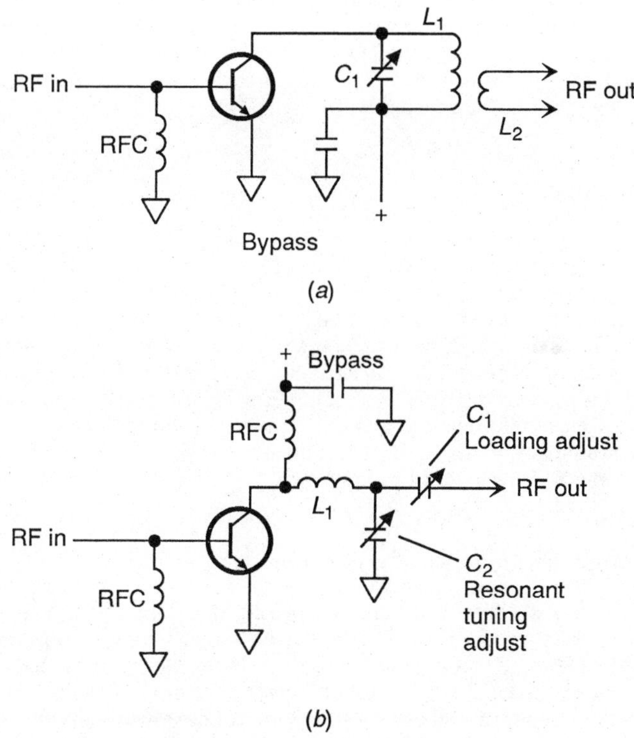

FIGURE 1.5 Basic RF power-amplifier circuits.

number of turns and positioning L_2 in relation to L_1). While that may prove to be an advantage in some cases, it also makes for an *interstage coupling network* that is subject to mismatch and detuning by physical movement or shock. Another disadvantage of the Fig. 1.5*a* circuit is that all current must pass through the tank-circuit coil.

For best transfer of power, the impedance of L_1 should match that of the transistor output. Since two-junction transistor output impedances are generally low, the value of L_1 must be low, often resulting in an impractical size for L_1. The circuit in Fig. 1.5*a* is a carryover from vacuum-tube circuits and, as such, is not often found in present-day RF equipment. A possible exception is in the few low-power FET amplifier circuits.

The circuit in Fig. 1.5*b*, or one of the many variations, is commonly found in RF equipment using two-junction and high-power FET transistors. The collector load is a resonant circuit formed by the network L_1, C_1, and C_2. Note that C_1 is labeled "Loading adjust" whereas C_2 is labeled "Resonant tuning adjust."

As discussed in Chap. 2, these networks provide the dual function of *frequency selection* (equivalent to the tank circuit) and *impedance matching* between transistor and load. To properly match impedances, both the resistive (so-called *real part*) and reactive (so-called *imaginary part*) components of the impedance must be considered (Chap. 2).

Both circuits in Fig. 1.5 are operated as class B, which is typical for RF amplifiers. Class B operation is obtained by connecting the emitter directly to ground and applying no bias to the base-emitter junction. Since any two-junction transistor requires some forward bias to produce current flow, the transistor remains cut off except in the presence of a signal.

1.3.5 RF Multiplier

The circuits in Fig. 1.5 can be used as a frequency multiplier where the collector is tuned to a higher whole-number multiple (harmonic) of the input frequency. Many radio transmitters use some form of multiplier to raise the low-frequency signal developed by the oscillator to a high-frequency signal. For example, most crystals used in oscillators have a fundamental frequency of less than 10 MHz, whereas solid-state transmitters may produce an output in the UHF range.

Although the circuits in RF power multipliers and RF power amplifiers are essentially the same, the efficiency is different. That is, an RF amplifier operating at the same frequency as the input has a higher efficiency than an identical RF circuit operating at a multiple of the input frequency.

1.3.6 RF Amplifier-Multiplier Combinations

The circuits in Fig. 1.5 can be *cascaded* (the output of one circuit applied to the input of the next circuit) to provide increased power amplification and/or frequency multiplication. Typically, no more than three stages are so cascaded. The stages can be mixed. That is, one or two stages can provide frequency multiplication, with the remaining one or two stages providing power amplification. Such arrangements are discussed in Chap. 2.

1.4 WIDEBAND RF

Except for pure sine waves, all signals contain not only the fundamental frequency but harmonic and subharmonic frequencies as well. These harmonics are whole-number multiples of the fundamental frequency. Pulse signals have an especially high harmonic content. For example, the pulses used in TVs contain frequencies ranging from about 30 Hz to 5 MHz (and higher).

A typical RF amplifier with a bandwidth of several hundred kilohertz is not able to uniformly amplify signals with such a broad range of frequencies. For this reason, it is necessary to use special *broadband* or *wideband* RF amplifiers for such applications. These amplifiers are usually known as video frequency (VF) *amplifiers,* or simply *video amplifiers,* in television equipment (TV sets, camcorders, VCRs, and the like) or as *pulse amplifiers* in radar and similar equipment (even though they are RF).

Resistance-coupled (or RC) amplifiers are, in effect, wideband amplifiers. Such circuits amplify uniformly at all frequencies of the audio range, dropping off only at the low- and high-frequency ends. A wideband amplifier (capable of passing RF signals, including pulses) is formed when the uniform amplification is extended to both ends of the frequency range.

A basic RC amplifier circuit is shown in Fig. 1.6*a*. Capacitor C_{out} represents the output capacitance of Q_1, and capacitor C_D represents the *distributed capacitance* of the various components and related wiring. Capacitor C_{in} represents the input capacitance of Q_2.

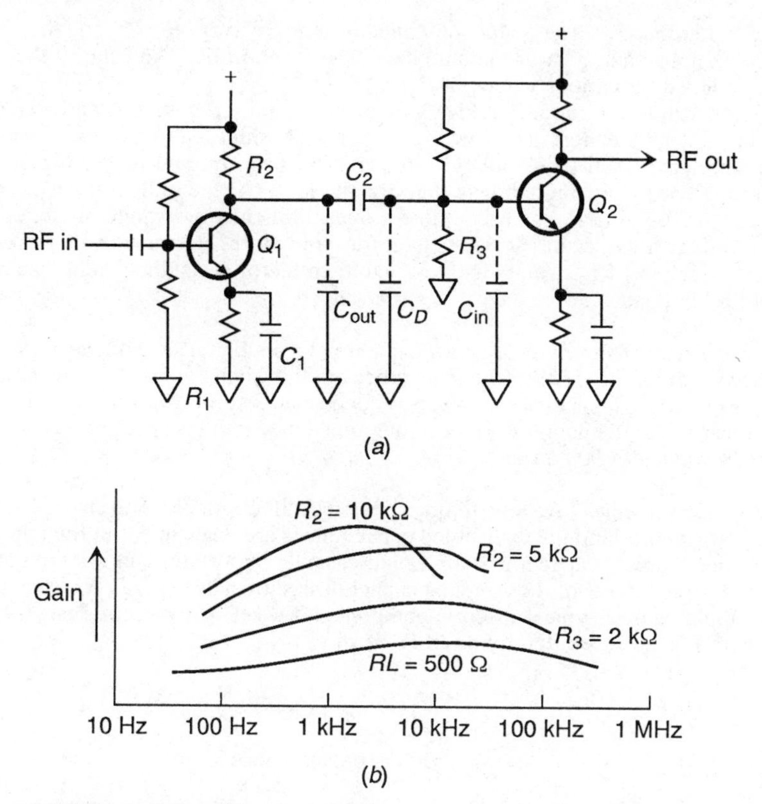

FIGURE 1.6 RC amplifier characteristics.

Coupling C_2 and base resistor R_3 form a voltage divider across the input of Q_2. At low frequencies, the impedance of C_2 is large, and relatively little of the signal is applied to the base of Q_2. Accordingly, the low-frequency response of the circuit is lowered.

Capacitances C_{out}, C_D, and C_{in}, acting in parallel, shunt the load resistor R_2 of Q_1. This lowers the effective resistance of R_2, as well as the high-frequency response of the circuit. (A lower value of R_2 lowers the gain, all other factors being equal.)

1.4.1 Increasing Wideband Response

There are several methods for improving the low- and high-frequency response of RF wideband amplifiers (or RC amplifiers designed for wideband use). In all cases, transistors with small input and output capacitances are used. Likewise, components are carefully placed so that leads and distributed capacitance are kept to a minimum. The following is a summary of additional methods to increase wideband response.

Collector Resistance. The value of collector-load resistor R_2 affects the frequency response and gain of the amplifier. The graph in Fig. 1.6*b* shows the effects produced by various values of R_2.

A large-value R_2 produces a high gain at the middle frequencies and a steep drop in gain at the high and low frequencies. A small-value R_2 produces a much smaller overall gain, but the proportional drop in gain at the high and low frequencies is also much less than for the larger-value collector resistors. With the small-value R_2, gain is uniform over a much wider range of frequencies. In effect, the circuit sacrifices gain for bandwidth. Because of these conditions, wideband RF circuits use low-value collector resistances and transistors with high gain.

Emitter Bypass. The emitter-bypass capacitor C_1 in Fig. 1.6*a* affects the low-frequency gain of the circuit. The impedance of C_1 is higher at lower frequencies. Thus, the circuit gain is lower at lower frequencies. Accordingly, the capacitance of C_1 must be large enough to offer a low impedance (with respect to R_1) at the *lowest frequency to be passed.*

Coupling Capacitance. At low frequencies, the effects of the transistor input-output capacitance and the distributed capacitances are negligible, but the impedance of the coupling capacitor becomes increasingly important. One way to compensate for the effects of the coupling capacitor is shown in Fig. 1.7, which is the video amplifier of a typical discrete-component TV set. (In present design, the circuits in Fig. 1.7 are often found in IC form.)

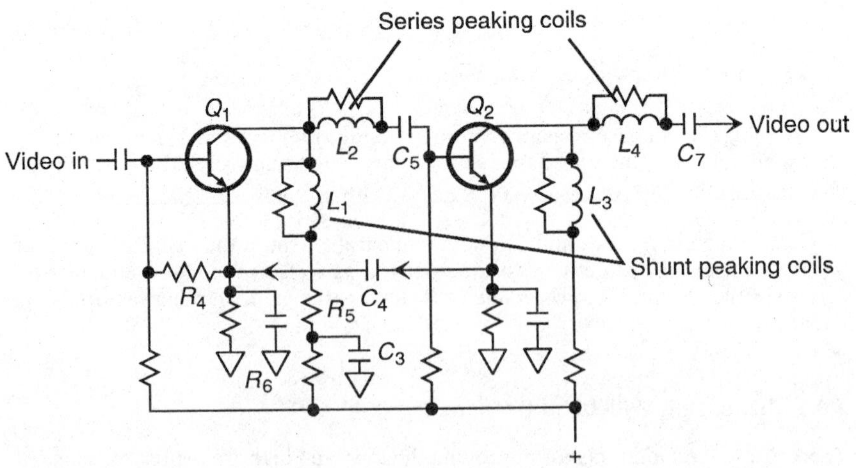

FIGURE 1.7 Typical discrete-component video amplifier.

In the Fig. 1.7 circuit, the load resistance for Q_1 is made up of two parts, R_5 and R_6, connected in series. Capacitor C_3 is the bypass capacitor for R_6. At the higher frequencies, the collector load is effectively R_5 since the small impedance of C_3 at these frequencies permits C_3 to completely bypass R_6 (in effect removing R_6 from the circuit).

At low frequencies, the impedance of C_3 becomes high, and the bypassing effect is greatly reduced. The collector-load resistance then becomes R_5 plus R_6. This greater resistance produces a greater output voltage, thus compensating for any low-frequency drop resulting from the increasing impedance of coupling capacitors (such as C_5 and C_7).

Shunt Peaking Coil. Since the drop in high-frequency response is caused by the shunting effect of the transistor capacitances (and distributed capacitances) upon the load resistor, a small coil L_1 (called a shunt peaking coil) is inserted in series with the load resistances. At low frequencies, L_1 offers very little impedance, and the collector load is, in effect, the resistance of R_5 plus R_6. At high frequencies, the impedance of L_1 is high, and the collector load is the sum of the R_5 plus R_6 resistances and the resistance of L_1. Thus, circuit gain is increased.

Coil L_1 sets up a resonant circuit with the distributed capacitances of the circuit (and the capacitances of the transistor). The value of L_1 is selected so that the circuit is resonant at a frequency where the high-frequency response of the circuit begins to drop. In this way, an additional boost is given to the gain, and the high-frequency end of the response curve is flattened.

Series Peaking Coil. Figure 1.7 shows another similar compensation circuit. Series peaking coil L_2 is connected in series with the coupling capacitor C_5. At high frequencies, L_2 forms a low-impedance series-resonant circuit with the capacitances, causing a larger signal to appear at the base of Q_2.

RF circuits used in video and pulse applications often use both shunt and series peaking coils for high-frequency compensation. Typically, the values of these coils are such that resonance is obtained at the *highest desired frequency.* That is, the capacitances are calculated (or measured), and a corresponding value of inductance is chosen for resonance at the high-frequency end.

Damping Resistances. Note that the coils in Fig. 1.7 are shown with resistances connected in parallel. As discussed in Chap. 2, when resistances are connected across coils, the resonant point of the coils is flattened or broadened. Such resistances are often known as *damping resistors.*

Inverse Feedback. Another method for overcoming the effects of the drop in gain at the low and high ends of the frequency band involves the use of inverse feedback, provided by C_4 and R_4 in Fig. 1.7. This feedback (also known as *negative feedback*) opposes changes in signal level. For example, if the base of Q_1 is swinging positive, the collector of Q_1 swings negative, as do the base and emitter of Q_2. The negative Q_2 emitter signal is fed back to the Q_1 base through C_4 and R_4, thus opposing the original positive swing. This action occurs at both the high and low ends of the frequency band.

Typical Wideband RF Circuit. Figure 1.8 shows a typical wideband RF circuit found in a camcorder. Note that this circuit uses discrete components, even though the camcorder is of present-day design. The circuit illustrates several of the points just discussed for wideband RF.

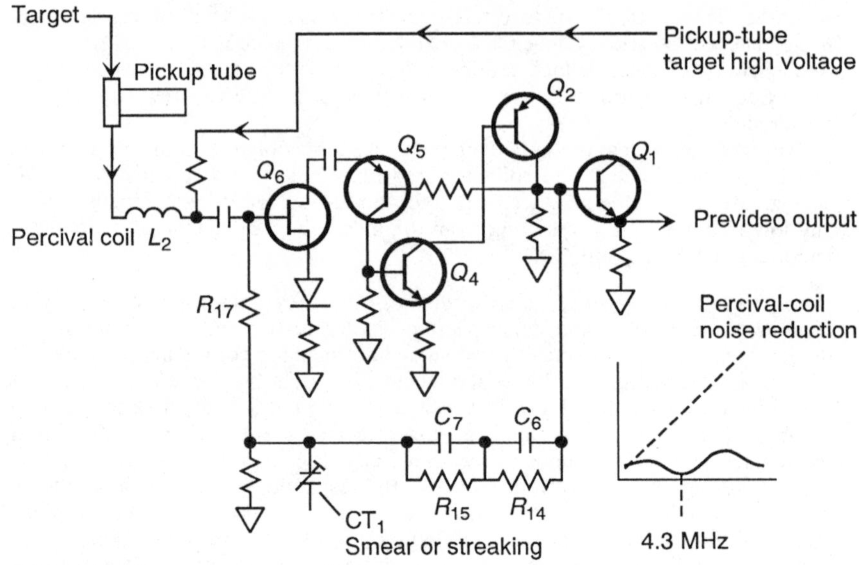

FIGURE 1.8 Camcorder video amplifier circuits.

The function of the Fig. 1.8 circuit is to convert weak signal current from the pickup tube of a camcorder to a voltage signal and then amplify the voltage signal to suitable level. The circuit has one input from the pickup tube and produces one output at the emitter of Q_1 to the prevideo processor circuits. The circuit (technically called a video circuit) also provides for application of the target voltage from the high-voltage power supply to the pickup-tube target.

The preamp has a low output impedance and uses negative feedback in an amplifier with a low-noise junction FET (JFET) input (to improve the signal-to-noise ratio). Negative feedback from output to input is applied through C_6, C_7, R_{15}, R_{14}, R_{16}, and R_{17}. (This is similar to the C_4-R_4 network in Fig. 1.7.) In the Fig. 1.8 circuit, CT_1 (called the *smear* or *streaking* adjustment) varies the low-frequency (400 to 500 kHz) portion of the negative feedback signal to maintain a flat frequency response through the amplifier.

A *Percival compensation* circuit is used at the input to prevent signal-to-noise deterioration because of impedance mismatch. The Percival coil L_2 resonates at about 4.3 MHz (as shown by the curve in Fig. 1.8) and separates the pickup-tube output capacitance from the preamp input capacitance.

1.5 DISCRETE-COMPONENT RF VOLTAGE AMPLIFIERS

The RF voltage amplifiers described in this section are used primarily in receivers. (RF power amplifiers for transmitters are discussed in Chap. 2.) IF amplifiers, or IF limiter-amplifiers, are examples of RF voltage amplifiers. The input, or first stage, of a receiver may include a separate RF voltage amplifier (such as are used with some discrete-component communications receivers). However, most solid-state receivers combine the RF voltage-amplifier function with that of the local oscillator.

In present design, the trend is to incorporate the RF voltage-amplifier circuits described here into integrated circuit (IC) form. For that reason, we do not go into full detail of discrete-component voltage amplifiers. However, the circuits covered in this section provide a basis for understanding the IC RF circuits described in Sec. 1.6.

1.5.1 Basic RF Voltage Amplifier

Figure 1.9 is the schematic of a basic RF voltage amplifier. Such a circuit can be used as an IF amplifier, IF limiter, or separate RF amplifier (with few modifications). Both the input and output are tuned to the desired operating frequency by the resonant circuits. In this case, the resonant circuits are composed of transformers with a capacitor across the primary. The capacitors can be variable but are usually fixed. The resonant circuit is tuned by an adjustable slug between the windings (instead of a powdered-iron core).

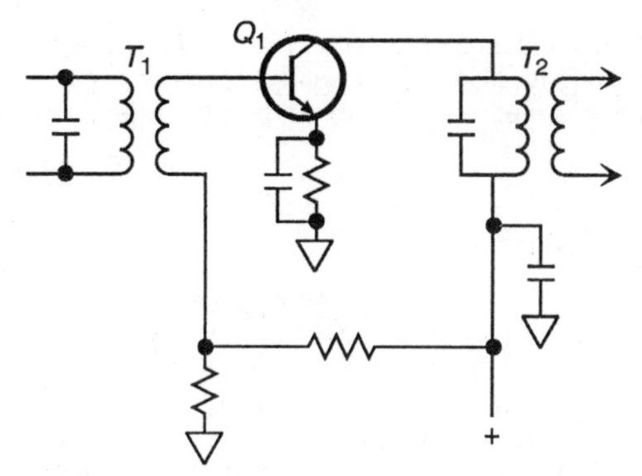

FIGURE 1.9 Typical discrete-component RF voltage amplifier.

1.5.2 Frequency Mixers and Converters

Figure 1.10 shows the schematic of a typical discrete-component frequency mixer and converter. Such a circuit is a combination of an RF voltage amplifier and an RF oscillator. The individual outputs of the two sections are combined to produce an IF output. Usually, the RF oscillator operates at a frequency above the RF amplifier, with the difference in frequency being the IF.

The resonant circuit of T_1 is tuned to the incoming RF signal, T_2 is tuned to the oscillator frequency (RF plus IF), and T_3 is tuned to the intermediate frequency. The resonant circuits of T_1 and T_2 are usually tuned by variable capacitors ganged together (and connected to the station tuning dial) so both the oscillator and RF amplifier remain at the same frequency relationship over the entire tuning range.

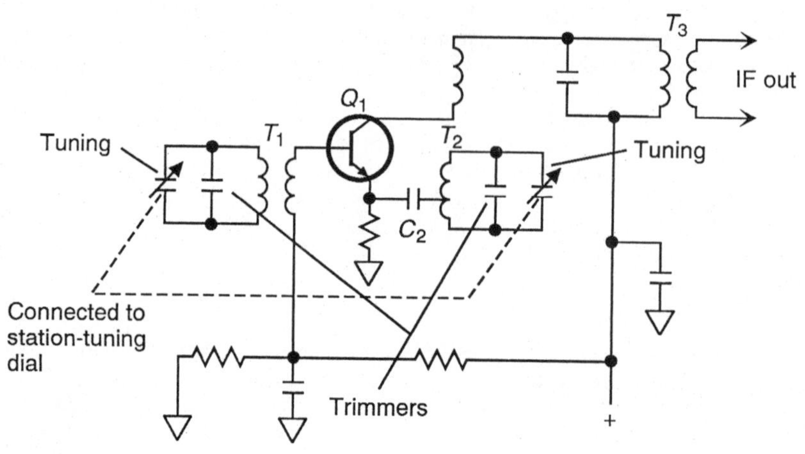

FIGURE 1.10 Typical discrete-component frequency mixer/converter.

For example, if T_1 tunes from 550 to 1600 kHz and T_3 is at a fixed IF of 455 kHz, T_2 must tune from 1005 to 2055 kHz. Usually, trimmer capacitors are connected in parallel with the variable capacitors to permit adjustment over the tuning range.

1.5.3 AVC and AGC for RF Circuits

Most RF circuits used in receivers have some form of automatic volume control and automatic gain control (AVC and AGC) circuit. The terms *AVC* and *AGC* are used interchangeably. AGC is a more accurate term since the circuits involved control the gain of an IF or RF stage (or several stages simultaneously) rather than the volume of an audio signal in an audio stage. However, in a broadcast receiver, the net result is the automatic control of volume. Either way, the circuit provides a constant output despite variations in signal strength. An increased signal reduces stage gain and vice versa.

Figure 1.11 shows the working schematic of two AGC systems common to discrete-component broadcast and communications receivers. Diode CR_1 acts as a variable shunt resistance across the input of the IF stages. Diode CR_2 functions as both the audio detector and AGC bias source.

Under no-signal conditions, or in the presence of a weak signal, diode CR_1 is reverse-biased and has no effect on the circuit. In the presence of a very large signal, CR_1 is forward-biased and acts as a shunt resistance to reduce gain.

The output of CR_2 is developed across R_1 and is applied to the audio stages. Resistor R_1 also forms part of the bias network for the IF-stage transistor. The combined fixed bias (from the network) and variable bias (from the detector) is applied to the IF-stage base-emitter circuit. The detector bias varies with signal strength and is of a polarity that opposes variations in strength. That is, if the signal strength increases, the CR_2 detector bias is more positive (or less negative) for the base of a PNP transistor and vice versa for an NPN transistor.

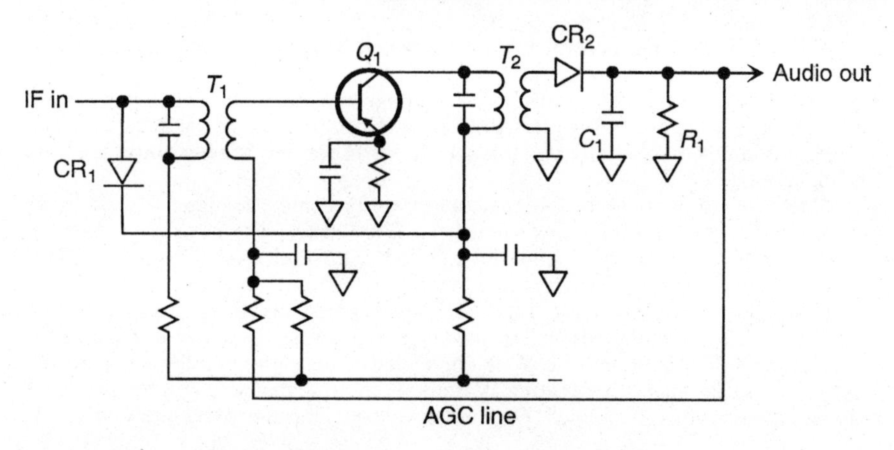

FIGURE 1.11 AGC systems in discrete-component IF/detector circuits.

1.5.4 Television AGC Circuits

Most discrete-component TV sets use a *keyed, saturation-type* AGC circuit such as is shown in Fig. 1.12. The tuner and IF transistors connected to the AGC line are forward-biased at all times. On strong signals, the AGC circuits increase the forward bias, driving the transistors into saturation, thus reducing gain. Under no-signal conditions, the forward bias remains fixed.

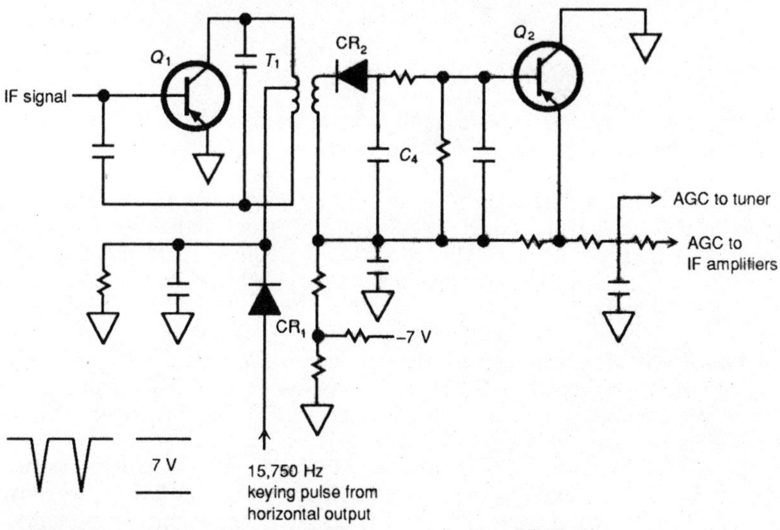

FIGURE 1.12 Signal path through discrete-component AGC circuits.

Although the AGC bias is a dc voltage, the bias is partially developed (or controlled) by bursts of IF signals. A portion of the IF signal is taken from the IF amplifiers and pulsed, or keyed, at the horizontal rate (15,750 Hz). The resultant keyed bursts of signal control the amount of dc voltage produced on the AGC line.

In most present-day sets, the AGC voltage is developed by an IC and applied to the IF amplifier within the same IC, as discussed in Sec. 1.6. The AGC voltage is also made available to the RF tuner and tuner-control circuits as necessary.

Transistor Q_1 is an IF amplifier tuned to the IF center frequency (typically 42 MHz) by T_1. No dc voltage (as such) is supplied to Q_1. The keying pulses from the horizontal flyback transformer are applied to the collector through CR_1. This produces an average collector voltage of about 1 V.

When Q_1 is keyed on, the bursts of IF signals pass through T_1 and are rectified by CR_2. A corresponding dc voltage is developed across C_4 and acts as a bias for AGC amplifier Q_2. Transistor Q_2 is connected as an emitter follower, with the AGC line returned to the emitter. Variations in IF signal strength cause corresponding variations in Q_2 bias, Q_2 emitter voltage, and the AGC line voltage.

1.6 IC RF VOLTAGE AMPLIFIERS

This section describes the RF-amplifier circuits of a typical AM/FM tuner of IC design and the IF and video circuits of an IC-based TV set. Although most of the components for both of the circuits are contained in ICs, there are a number of discrete components between ICs, as well as discrete components used to tune or adjust the ICs. Compare these circuits to the discrete-component circuits described in Sec. 1.5.

1.6.1 Relationship of AM/FM Tuner Circuits

Figure 1.13 shows the relationship of the tuner circuits. Figures 1.14 and 1.15 show further details of the FM and AM sections, respectively.

Amplifier and Multiplex Decoder. IC_{301} is both an amplifier and a multiplex decoder and is used in the audio path for AM and FM. Once audio reaches pin 2 of IC_{301}, whether from the AM or FM section, IC_{301} produces corresponding audio at pins 6 and 7. IC_{301} has only one adjustment control. This is the multiplex VCO-adjust potentiometer R_{305} connected at pin 15.

FM Section. FM broadcast signals are applied to FM tuner package MD_{101}. Note that MD_{101} contains RF voltage amplifiers, local oscillator, mixer/converter, IF amplifiers, and detector. These are FM versions of the circuits shown in Figs. 1.9 through 1.11.

The IF output from MD_{101} is applied to the FM amplifier and detector IC_{201} through amplifiers Q_{101} and Q_{102} and ceramic filters MF_{201} and MF_{202}, as shown in Fig. 1.14. This amplifier-filter combination removes any amplitude modulation and passes only signals of the desired frequency. Note that Q_{101} and Q_{102} are connected as a form of differential amplifier. Also note that IC_{201} has three adjustments: muting-level R_{202}, S-curve null-point T_{201}, and FM distortion T_{202}.

The audio output from the FM section is applied to amplifier-multiplex decoder IC_{301} through audio-select switches Q_{201} and Q_{202}. These circuits select the AM or FM audio for the input to IC_{301}.

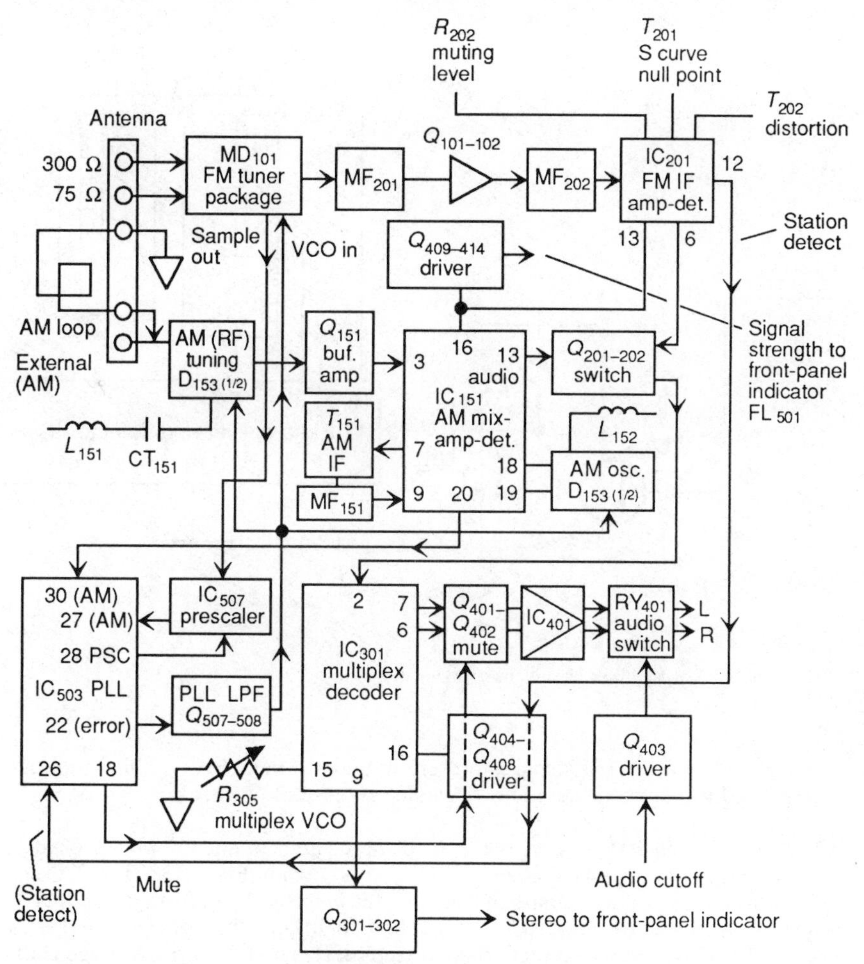

FIGURE 1.13 Relationship of tuner circuits.

Note that IC_{301} has a dual function. When suitable FM is present, IC_{301} is a stereo amplifier (as well as a decoder). IC_{301} functions as a mono amplifier when there is no FM stereo present or when the stereo signal is too weak to produce proper FM.

AM Section. AM broadcast signals are applied to AM IF amplifier-detector IC_{151} through buffer-amplifier Q_{151}, as shown in Fig. 1.15. The RF circuit (L_{151} and CT_{151}) is tuned by one section of D_{153} [a voltage-variable capacitor (VVC) such as described in Chap. 2]. The mixer output of IC_{151} is tuned by AM IF adjust T_{151} and applied to the IF portion of IC_{151} through ceramic filter MF_{151}. Note that IC_{151} contains RF amplifiers, local oscillator, mixer-converter, IF amplifiers, and detector similar to that shown in Figs. 1.9 through 1.11.

The audio output from the detector in IC_{151} is applied to IC_{301} through Q_{201} and Q_{202}. Note that IC_{301} functions as a mono amplifier when the AM mode is

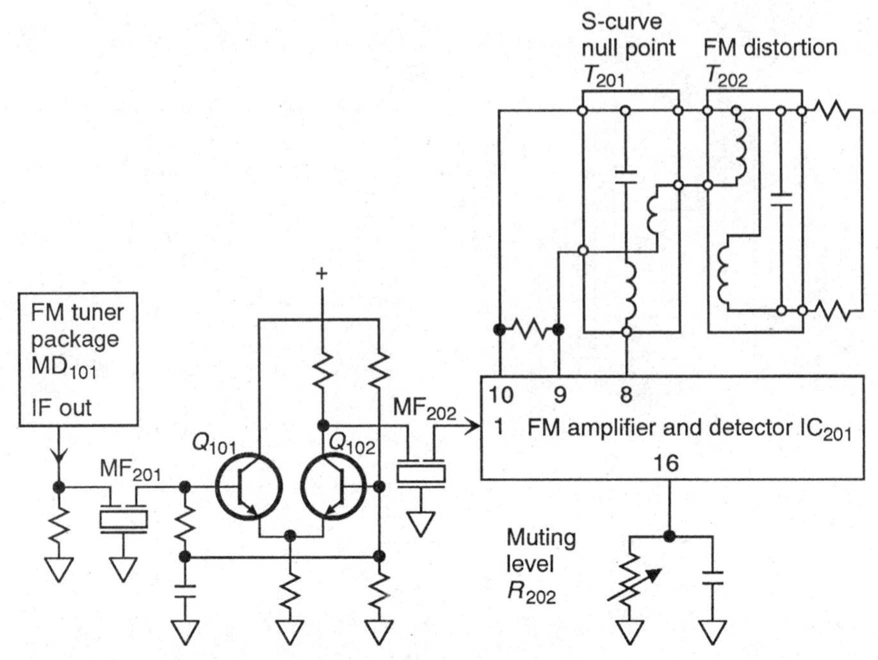

FIGURE 1.14 FM amplifier circuits in IC tuner.

selected. Since no stereo signal is present in the AM mode, IC_{301} shifts to mono operation, just as when there is no FM stereo or stereo is weak.

FM Tuning. The FM section is tuned to the desired frequency (and locked to that frequency) by signals applied to MD_{101}, as shown in Fig. 1.13. The VCO signal (a variable dc voltage, sometimes called the *error voltage*) from phase-locked loop (PLL) microprocessor IC_{503} is applied to MD_{101} through Q_{507} and Q_{508} (which are connected as a direct-coupled amplifier). The IC_{503} error voltage shifts the MD_{101} oscillator as necessary to tune across the FM broadcast band or to fine-tune MD_{101} at a selected station.

The frequency produced by MD_{101} is sampled and applied to the FM input of IC_{503} through prescaler IC_{507}. (Operation of the prescaler is described in Sec. 1.6.2.) The sampled signal serves to complete the FM tuning loop. For example, if MD_{101} drifts from the frequency commanded by IC_{503}, the error voltage from IC_{503} changes the MD_{101} oscillator as necessary to bring MD_{101} back on frequency.

A station-detect signal is produced at pin 12 of IC_{201}. This signal is used to tell PLL microprocessor IC_{503} that an FM station has been located and that the station has sufficient strength to produce good FM operation.

In between FM stations or when the FM station is weak, pin 12 of IC_{201} goes high, turning on Q_{406}. This causes pin 26 of IC_{503} to go low and causes the audio to be muted (so there is no background noise when tuning from station to station).

When there is an FM station of sufficient strength, pin 12 of IC_{201} goes low, turning Q_{406} off and causing pin 26 of IC_{503} to go high. Under these conditions, the audio is unmuted, and IC_{503} fine-tunes the FM section for best reception of the FM station. The muting level is set by R_{202}.

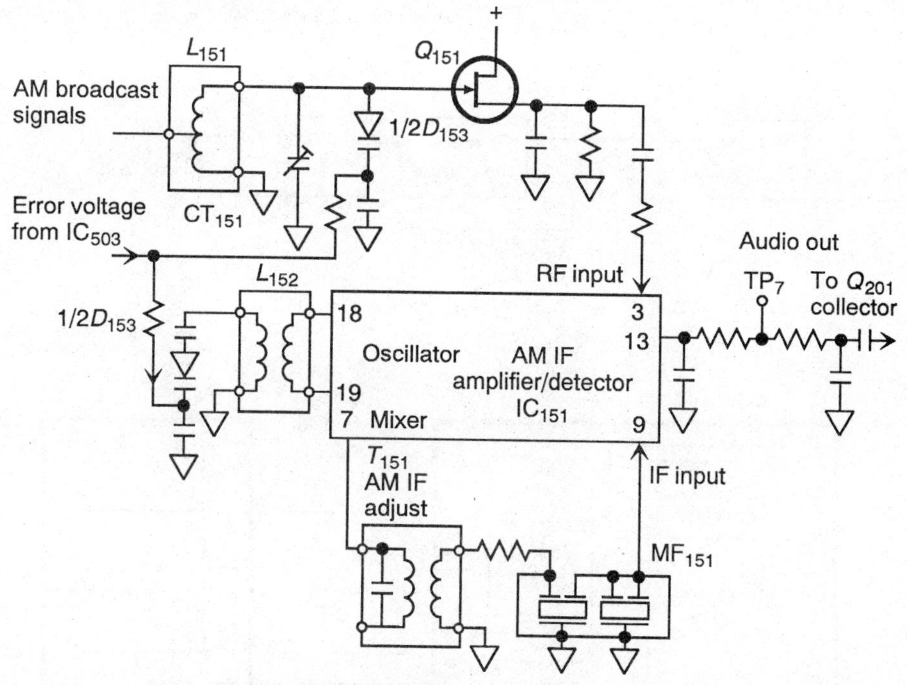

FIGURE 1.15 AM amplifier circuits in IC tuner.

AM Tuning. The AM section is also tuned to the desired frequency (and locked to that frequency) by signals from IC_{503} applied to the two sections of diode D_{153}. This is the same error voltage applied to the VCO of FM tuner MD_{101}. The IC_{503} error voltage shifts the RF and oscillator circuits as necessary to tune across the AM broadcast band.

The oscillator circuit of IC_{151} is adjusted by L_{152}, while the RF input circuit is tuned by tracking adjustments L_{151} and CT_{151}.

The frequency produced by IC_{151} is sampled and applied to the AM input of IC_{503}. The sampled signal serves to complete the AM tuning loop. For example, if the AM section drifts from the frequency commanded by IC_{503}, the error voltage from IC_{503} changes both circuits controlled by D_{153} as necessary to bring the AM section back on frequency.

1.6.2 Frequency Synthesis Tuning of RF Voltage Amplifiers

The RF voltage amplifiers in Fig. 1.13 use some form of frequency synthesis tuning (also known as *quartz tuning* or possibly *digital tuning*). Frequency synthesis (FS) tuning provides for convenient push-button or preset AM and FM station selection (or TV channel selection, as described in Sec. 1.6.3). Most FS systems also provide automatic station search or scan, and automatic fine-tune (AFT) capability.

As shown in Fig. 1.16, the key element in an FS system is the PLL that controls the variable-frequency oscillator (VFO) and/or RF tuning, as required for both station selection and fine tuning. Note that the PLL used in the AM/FM tuner in Fig. 1.13 is essentially the same as the PLL used in the FS tuners of TV sets (Sec. 1.6.3) and VCRs (and can be applied to virtually any RF tuning system).

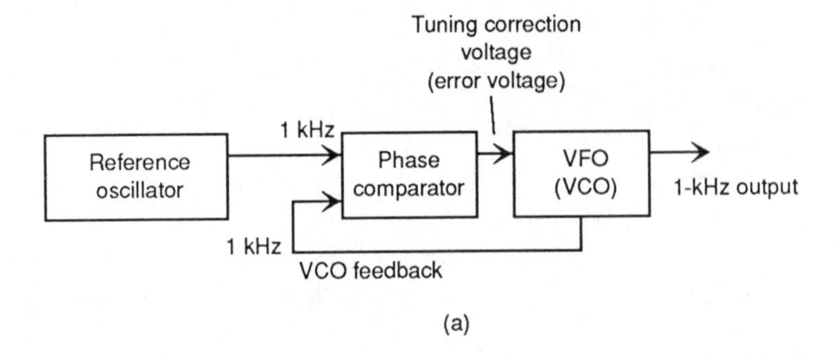

(a)

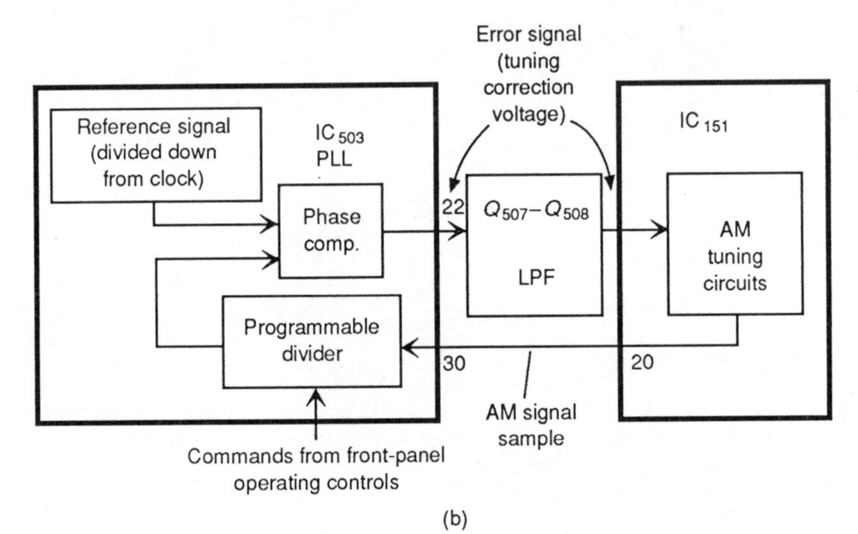

(b)

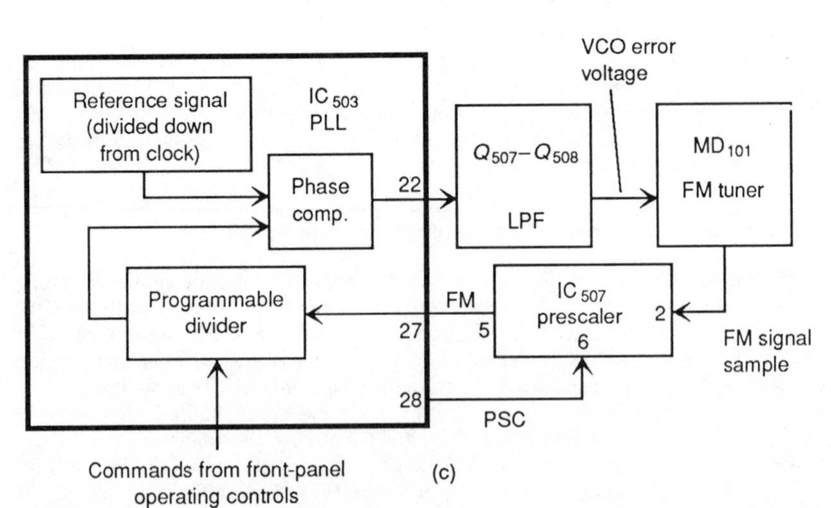

(c)

FIGURE 1.16 Elements of an FS system.

Elements of an FS System. As shown in Fig. 1.16*a,* the basic PLL is a frequency-comparison circuit in which the output of a VFO is compared in frequency and phase to the output of a very stable (usually quartz-crystal controlled) fixed-frequency reference oscillator. Should a deviation occur between the two compared frequencies or should there be any phase differences between the two oscillator signals, the PLL detects the degree of frequency or phase error and automatically compensates by tuning the VFO up or down in frequency or phase until both oscillators are locked to the same frequency and phase.

The accuracy and frequency stability of a PLL circuit depends on the accuracy and frequency stability of the reference oscillator (and on the crystal that controls the reference oscillator). No matter what reference oscillator is used, the VFO of most PLL circuits is a VCO, where frequency is controlled by an error voltage.

AM FS Tuning. Figure 1.16*b* shows how the PLL principles are applied to the AM section of a typical RF circuit (AM/FM tuner). The 1-kHz reference oscillator in Fig. 1.16*a* is replaced by a reference signal, obtained by dividing down the PLL IC_{503} clock (4.5 MHz). This reference signal is applied to a phase comparator within IC_{503}. The other input to the phase comparator is a sample of the AM signal at pin 30 of IC_{503} (taken from pin 20 of IC_{151}).

The output of the phase comparator is an *error signal* or *tuning-correction voltage* applied through low-pass filter Q_{507} and Q_{508} to the AM tuning circuits. The filter acts as a buffer between the comparator and tuning circuits.

Note that the AM sample is applied to the phase comparator through a *programmable divider* or *counter.* The division ratio of the programmable divider is set by commands from the front-panel operating controls (tuning up or down, preset scan, etc.). In effect, the divider is programmed to divide the AM sample by a specific number.

The variable-divider function makes possible many AM local-oscillator frequencies across the RF spectrum. An AM frequency change is made by varying the division ratio with front-panel commands. This produces an error signal that shifts the tuning circuits until the AM signal (after division by the programmable divider) equals the reference-signal frequency, and the tuning loop is locked at the desired frequency.

FM FS Tuning. Figure 1.16*c* shows the PLL circuits for the FM section of our RF tuner. This circuit is similar to the PLLs found in TV sets (Sec. 1.6.3) and VCRs in that a *prescaler* is used. The system in Fig. 1.16*c* is generally called an *extended* PLL and holds the variable oscillator frequency to some harmonic or subharmonic of the reference oscillator (but with a fixed phase relationship between the reference and variable signals).

The PLL in Fig. 1.16*c* uses a form of *pulse-swallow control* (PSC) that allows the division ratio of the programmable divider to be changed in small steps. As in the case of AM, the division ratio of the programmable divider is determined by commands applied to IC_{503} from the front-panel controls.

The PSC system uses a very high-speed prescaler IC_{507} and a variable division ratio. The division ratio of the prescaler is determined by the PSC signal at pin 28 of IC_{503} and can be altered as required to produce subtle changes in frequency needed for optimum station tuning (fine tuning) in the FM mode.

The PSC signal at pin 28 of IC_{503} is a series of pulses. As the number of pulses increases, the division ratio of the prescaler also increases. When a given FM station or frequency is selected by the front-panel controls, the number of pulses on the PSC line is set by circuits in IC_{503} as necessary for each station or frequency.

The overall division ratio for a specific FM station or frequency is the prescaler division ratio, multiplied by the programmable-divider division ratio. The result of division at any FM station or frequency is a fixed output to IC_{503} when the FM tuner is set to the desired frequency.

1.6.3 Frequency-Synthesis Tuning for TV RF Circuits

The RF-tuner circuits of most present-day TV sets also use FS tuning and PLLs. These circuits are similar to, but more complex than, the AM/FM tuners described in Sec. 1.6.2. Figures 1.17 and 1.18 show some typical FS tuning circuits used in TV sets with multiband RF tuners.

The circuit in Fig. 1.17*a* is similar to the basic PLL circuit in Fig. 1.16*a*. Figure 1.17*b* shows a somewhat more sophisticated PLL circuit, one capable of comparing frequencies that are not identical.

The circuit in Fig. 1.17*b* includes a divide-by-10 element, which divides the VCO frequency by 10, and a low-pass filter, which acts as a buffer between the comparator and VCO. Note that while the inputs to the comparator remain at 1 kHz when the loop is locked, the output frequency of the VCO is now 10 kHz because of the divide-by-10 circuit.

Figure 1.17*c* shows a PLL circuit that is similar to that found in modern video equipment (the RF tuners for both TV and VCRs). The system is generally called an *extended* PLL and holds the tuner oscillator frequency to some harmonic (or subharmonic) of the reference oscillator.

The fixed-divide element in Fig. 1.17*b* is replaced by a programmable variable divider (÷N) in Fig. 1.17*c*. A channel change is produced by *varying the division ratio* of the programmable divider with 4-bit data commands from the system-control microprocessor (which, in turn, is operated by front-panel push buttons and/or by remote-control signals).

TV Pulse-Swallow Control. The RF tuners in most present-day TV sets also use PSC with a variable division ratio, such as shown in Fig. 1.17*d*. Here, the variable division ratio depends on the PSC signal from PLL IC_1. As the number of PSC pulses increase, the division ratio increases. A specific number of PSC pulses are produced by IC_1 in response to channel-selection commands.

The overall division ratio for a specific channel is the prescaler division ratio multiplied by the programmable-divider division ratio. The result of division at any channel is a precise 5-kHz output to the comparator when the tuner oscillator is locked to a given channel frequency. In effect, the programmable divider determines the basic channel frequency, and the prescaler performs the fine adjustments (AFT) to the channel frequency.

Note that if you are troubleshooting any RF tuner with PSC and you cannot tune in a channel with AFT or manually, check the PSC line (pin 27 of IC_1 in this case) for pulses and check the tuning voltage to the oscillator. If the pulses are missing, the tuner cannot lock into any channel, even with a tuning voltage present.

Typical TV FS Tuning Circuit. Figure 1.18 shows the circuits of a typical FS (or digital) RF tuning system with PLL and PSC. Note that the PLL IC_1 receives channel commands from a remote control until after the commands have been decoded by a remote-control receiver circuit.

The multiband RF tuner (suitable for TV, VCR, and cable) is controlled by circuits within PLL IC_1 which, in turn, receives commands from the remote-control circuits. IC_1 monitors signals from the IF demodulator circuits to know

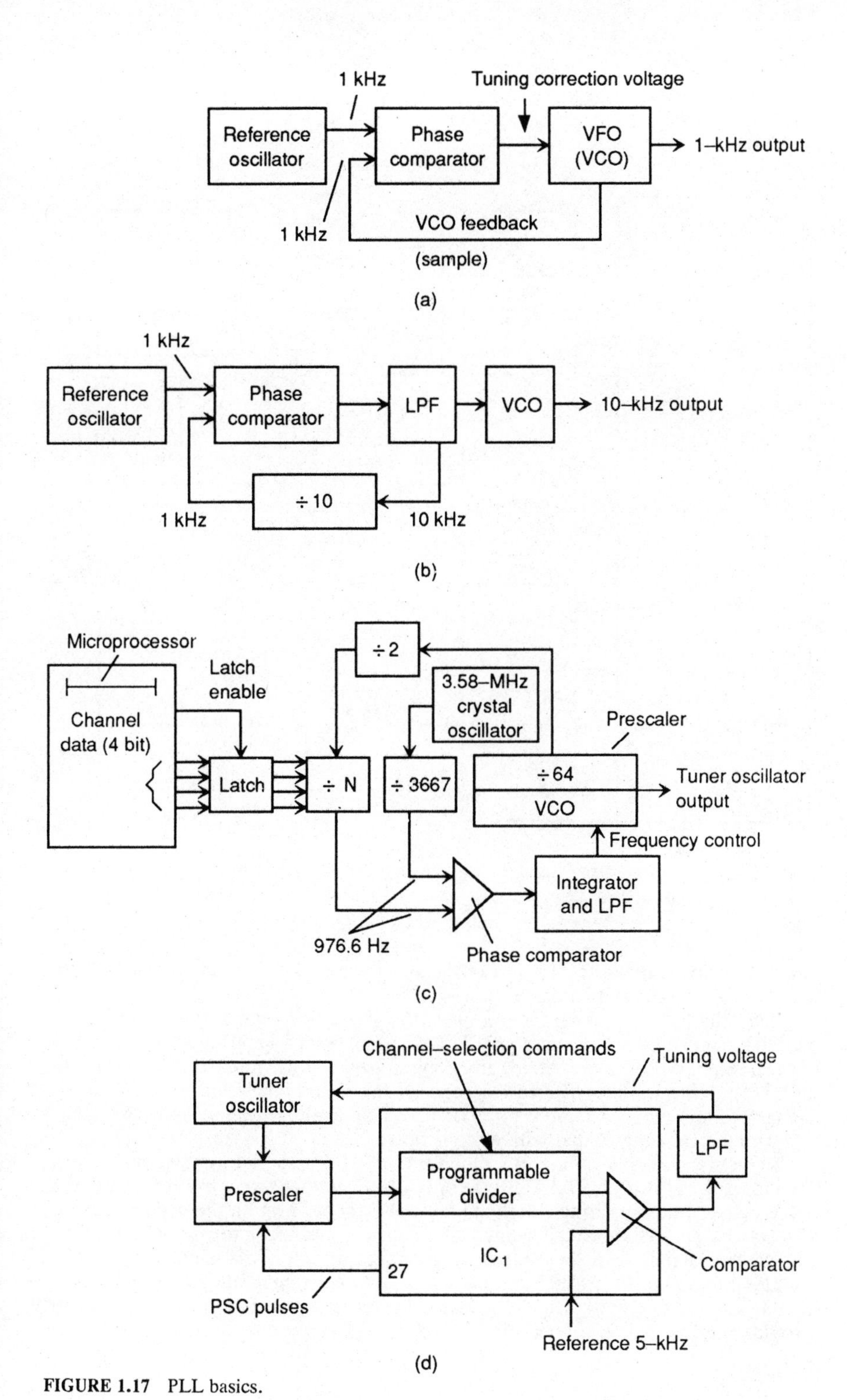

FIGURE 1.17 PLL basics.

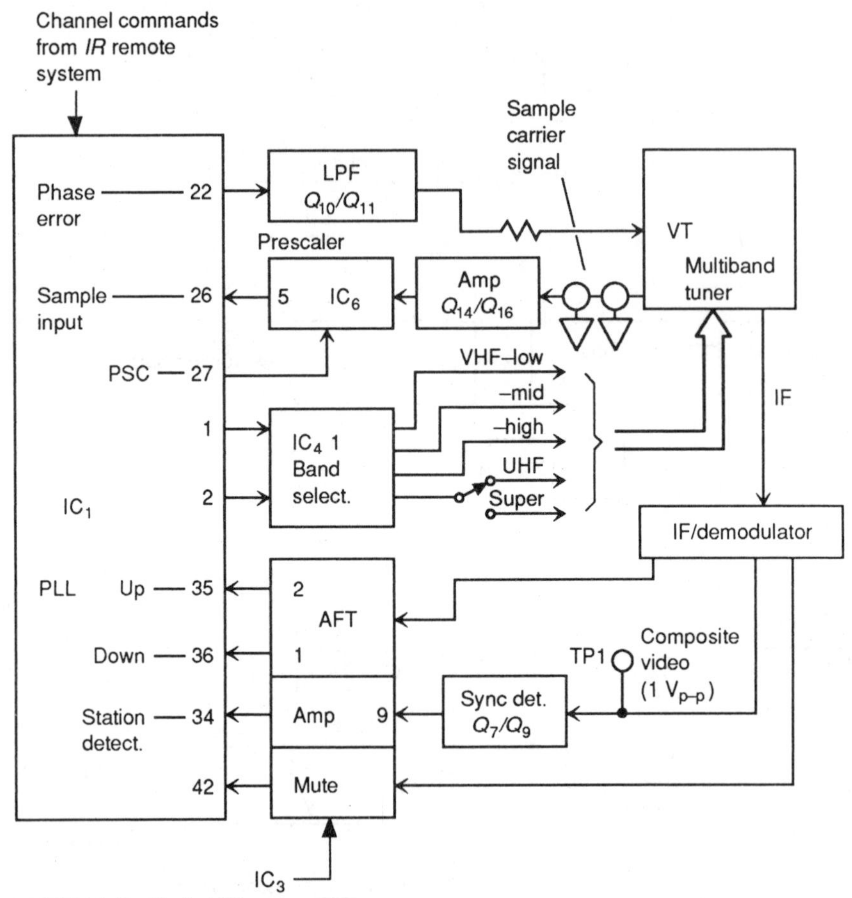

FIGURE 1.18 Typical FS tuning system.

when a station is being received. These signals are the AFT up and down (pins 35 and 36) and station-detect signal (pin 34).

The detected video from the output of the IF demodulator is passed to a sync amplifier and detector Q_7 and Q_9. The detected sync signal (station detect) is passed to pin 9 of IC_3, amplified, and applied to the station-detect input at pin 34 of IC_1. When a station is tuned in properly, the sync is detected from the video signal and applied as a high to pin 9 of IC_3. This applies a high to pin 34 of IC_1, indicating to IC_1 that video with sync is present.

To maintain proper tuning, IC_1 monitors the AFT up and down signals (at pins 35 and 36) from IC_3. The AFT circuit of IC_3 is a *window detector,* monitoring the AFT voltage and outputing a high at pins 1 or 2, depending on the magnitude and direction of the AFT voltage swings (should the tuner oscillator frequency drift).

When a channel is selected, band-switching information is supplied from IC_1 (at pins 1 and 2) to IC_4, which develops four band-switching outputs. (In this particular circuit, one of the four outputs is switched by a normal/cable switch to provide the five bands shown.)

The RF tuner oscillator passes a *sample carrier signal* to oscillator amplifier Q_{14} and Q_{16}. The amplified oscillator signal is then passed to prescaler IC_6. The amount of frequency division is determined by PSC pulses from IC_1. The frequency-divided output of IC_5 (pin 5) is then passed to the sample input of IC_1 (at pin 26). When a channel is selected, circuits within IC_1 produce the appropriate number of PSC pulses at pin 27. The PSC pulses are applied to IC_6 and produce the correct amount of frequency division.

The divided-down oscillator signal (sample input) at pin 26 of IC_1 is divided down again within IC_1 and compared to an internal 5-kHz reference signal. The phase error of these two signals appears at pin 22 of IC_1 and is applied to low-pass filter (LPF) Q_{10} and Q_{11}. The dc output from the LPF is applied to the tuner oscillator. This voltage sets the RF tuner oscillator as necessary to get the proper frequency for the channel selected.

1.6.4 TV IF/Video-Detector Circuits

The IF/video-detector circuits for TV and VCR are far more complex than the simple IF/detector circuit in Fig. 1.11. The basic function of the corresponding TV circuits is to amplify *both* picture and sound signals from the tuner, demodulate *both* signals for application to the video amplifier and sound-IF amplifiers, and trap (or reject) signals from adjacent RF channels.

Discrete-Component TV IF/Video. Figure 1.19 shows the signal path through the IF and video-detector stages of a typical discrete-component TV set. The IF and video-detector functions are usually combined into one or two ICs in present-day sets. In many cases, the IF and video-detector and video-amplifier functions are combined

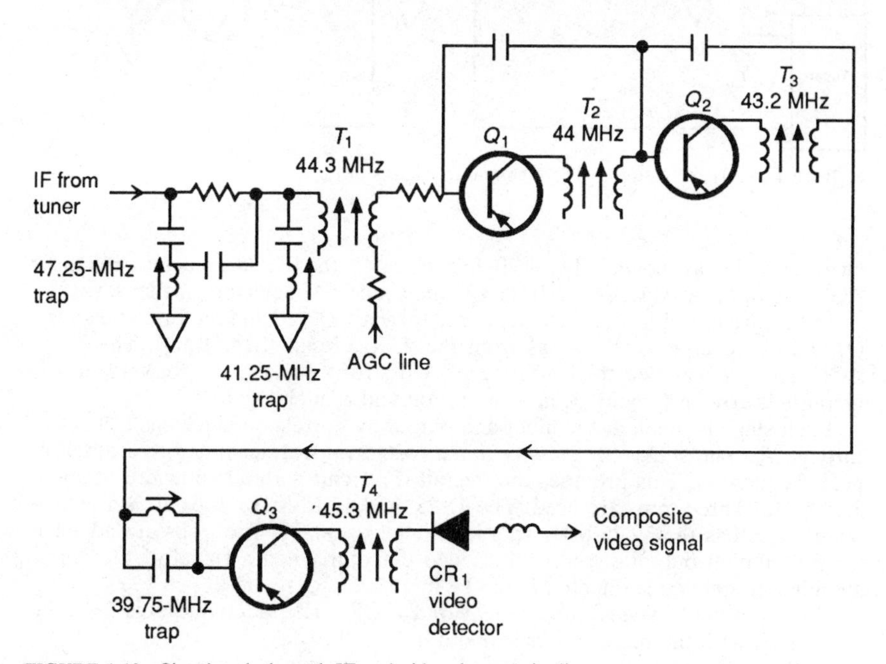

FIGURE 1.19 Signal path through IF and video-detector in discrete-component set.

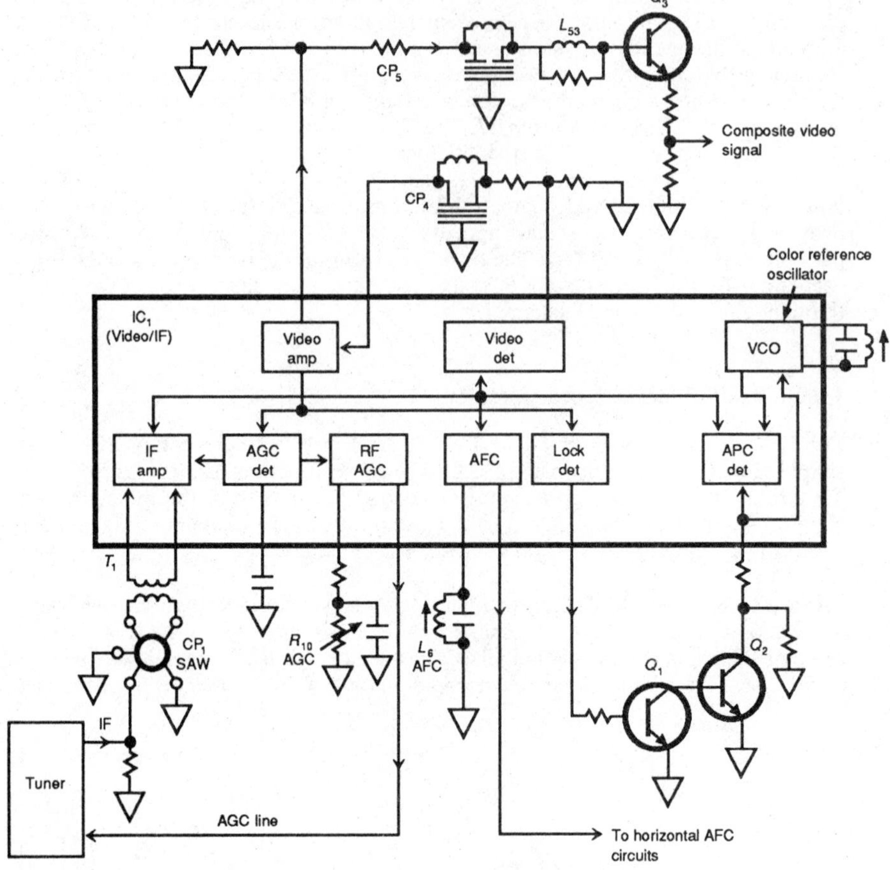

FIGURE 1.20 Signal path through IF and video circuits of IC set.

into a single IC, as shown in Fig. 1.20. In such cases, the IC also contains circuits for AGC, AFC, color APC, sound-IF (SIF), and sound FM-detector functions.

In the circuit in Fig. 1.19, all stages are forward-biased and the first two stages Q_1 and Q_2 receive forward bias from the AGC circuit (Sec. 1.5.4). This same AGC line is connected to the RF tuner. On strong signals, the forward bias is increased, driving Q_1 and Q_2 into saturation and reducing gain.

Each stage is tuned at the input and output by corresponding transformers T_1 through T_4. The stages are *stagger-tuned* (each transformer tuned to a different peak frequency). This provides the overall IF circuit with a bandwidth of about 3.25 MHz. Three traps are used. The 41.25- and 47.25-MHz traps are series resonant, whereas the 39.75-MHz trap is parallel resonant. The traps are adjusted for a minimum output signal at the video detector when a signal of the corresponding frequency is injected at the IF input.

Output of the IF stages can be measured at CR_1. The video output is typically about 1 V, with the picture signals about 0.25 to 0.5 V.

IC TV IF/Video. In the circuit in Fig. 1.20, IF output from the RF tuner is supplied to an IF amplifier within IC_1 through filter CP_1 and T_1. The IF output is applied to a video detector in IC_1. The composite output from the detector is applied through filter CP_4 to a video amplifier within IC_1. (The detector output is also applied to an SIF amplifier in IC_1).

The output of the video amplifier is applied to the picture-tube circuits through filter CP_5, L_{53}, and Q_3. The video-amplifier output is also applied to an AGC detector within IC_1. This AGC circuit controls both the IF amplifier within IC_1 and the tuner (through the RF amplifier in IC_1). The AGC circuit is adjusted by R_{10}.

The output of the IC_1 video detector is applied to an AFC circuit in IC_1. This circuit is adjusted by L_6 and provides pulses to the horizontal AFC circuits. The IC_1 output is also applied to an APC circuit which controls operation of the color-reference oscillator in IC_1 (in conjunction with an IC_1 lock detector, which receives signals from the IC_1 video amplifier).

CHAPTER 2
PRACTICAL CONSIDERATIONS FOR RF

This chapter covers practical considerations for RF. Such subjects as calculating resonant-circuit values and mounting techniques for RF components are discussed. The main concern is with RF power amplifiers where thermal problems can arise in design and service.

2.1 RESONANT CIRCUITS

Both RF amplifiers and RF oscillators use resonant circuits (or *tank* circuits) consisting of a capacitor and coil (inductance) connected in series or parallel, as shown in Fig. 2.1. Such resonant circuits are used to *tune* the RF amplifier/oscillator network. At the resonant frequency, the inductive and capacitive reactances are equal, and the circuit acts as a high impedance (in a parallel circuit) or a low impedance (in a series circuit). In either case, any combination of capacitance and inductance has some resonant frequency.

Either (or both) the capacitance or inductance can be variable to permit tuning of the resonant circuit over a given frequency range. When the inductance is variable, tuning is usually done by a metal (powdered-iron) slug inside the coil. The slug is screwdriver-adjusted to change the inductance (and thus the inductive reactance) as required. Typical RF circuits may include two resonant circuits in the form of a transformer (RF or IF transformer and the like). Again, either the capacitance or inductance can be variable.

2.1.1 Resonant Frequency versus Q or Selectivity

All resonant circuits have a resonant frequency and a Q factor. The circuit Q depends on the *ratio of reactance to resistance.* If a resonant circuit has pure reactance, the Q would be high (actually infinite). However, this is not practical since any coil has some resistance, as do the leads of a capacitor.

The resonant-circuit Q depends on the individual Q factors of inductance and capacitance used in the circuit. For example, if both the inductance and capacitance have a high Q, the circuit has a high Q, provided that a minimum of resistance is produced when the inductance and capacitance are connected to form a resonant circuit.

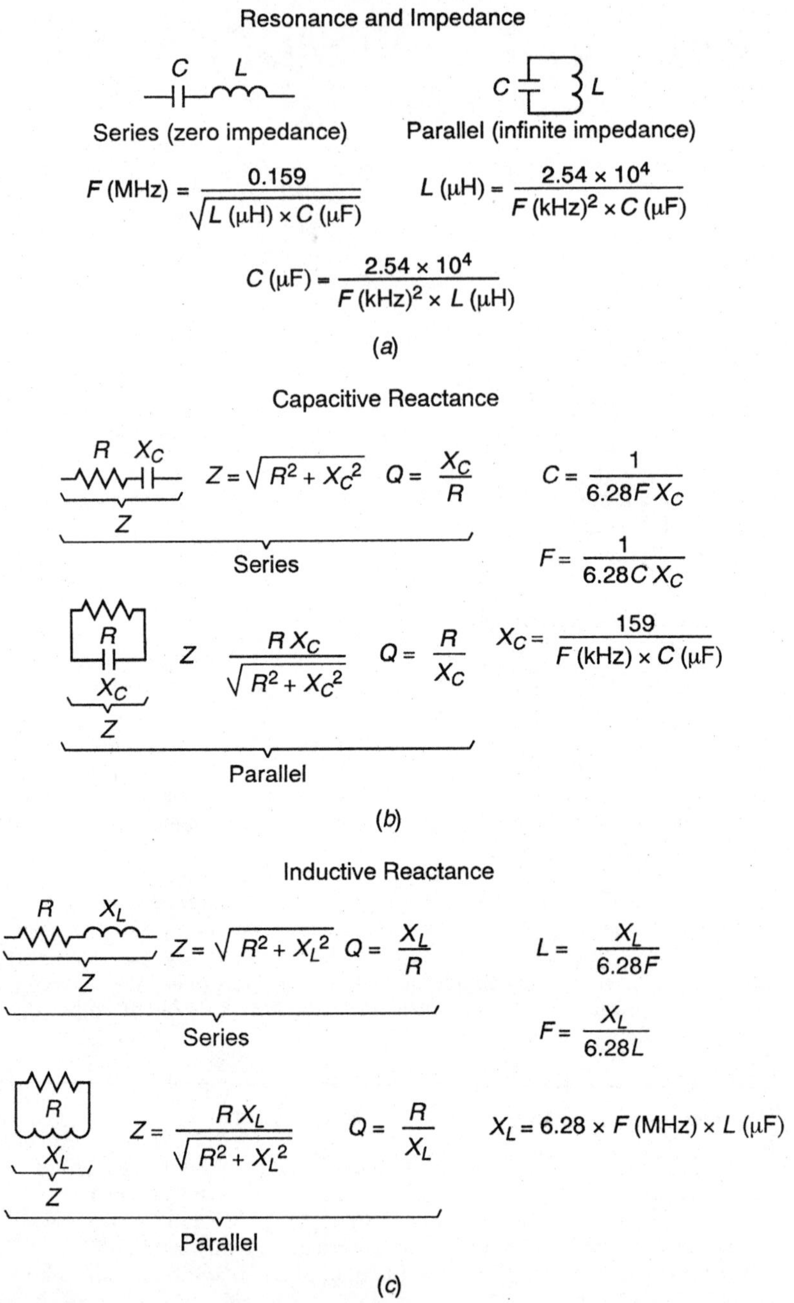
Resonance and Impedance
C L
Series (zero impedance)
C L
Parallel (infinite impedance)
$F\,(\text{MHz}) = \dfrac{0.159}{\sqrt{L\,(\mu\text{H}) \times C\,(\mu\text{F})}}$
$L\,(\mu\text{H}) = \dfrac{2.54 \times 10^4}{F\,(\text{kHz})^2 \times C\,(\mu\text{F})}$
$C\,(\mu\text{F}) = \dfrac{2.54 \times 10^4}{F\,(\text{kHz})^2 \times L\,(\mu\text{H})}$
(a)
Capacitive Reactance
R X_C
Z
$Z = \sqrt{R^2 + X_C^2}$
$Q = \dfrac{X_C}{R}$
Series
$C = \dfrac{1}{6.28 F X_C}$
$F = \dfrac{1}{6.28 C X_C}$
R
X_C
Z
$Z \quad \dfrac{R X_C}{\sqrt{R^2 + X_C^2}}$
$Q = \dfrac{R}{X_C}$
$X_C = \dfrac{159}{F\,(\text{kHz}) \times C\,(\mu\text{F})}$
Parallel
(b)
Inductive Reactance
R X_L
Z
$Z = \sqrt{R^2 + X_L^2}$
$Q = \dfrac{X_L}{R}$
Series
$L = \dfrac{X_L}{6.28F}$
$F = \dfrac{X_L}{6.28L}$
R
X_L
Z
$Z = \dfrac{R X_L}{\sqrt{R^2 + X_L^2}}$
$Q = \dfrac{R}{X_L}$
$X_L = 6.28 \times F\,(\text{MHz}) \times L\,(\mu\text{F})$
Parallel
(c)

FIGURE 2.1 Resonant-circuit equations. (*a*) Resonance and impedance; (*b*) capacitive reactance; (*c*) inductive reactance.

Figure 2.1 has equations that show the relationships among capacitance, inductance, reactance, and frequency, as these factors relate to resonant circuits. Note that there are two sets of equations. One set includes reactance (inductive and capacitive); the other omits it. The reason for the two sets of equations is that some design approaches require the reactance to be calculated for resonant networks (as discussed in Secs. 2.2 through 2.4).

From a practical standpoint, a resonant circuit with a high Q produces a *sharp* resonance curve (narrow bandwidth), whereas a low Q produces a *broad* resonance curve (wide bandwidth). For example, a high-Q resonant circuit provides good harmonic rejection and efficiency in comparison with a low-Q circuit, all other factors being equal. (Efficiency and rejection are discussed in Chaps. 4 and 5.)

The *selectivity* of a resonant circuit is related directly to Q. A very high Q (or high selectivity) is not always desired. Sometimes it is necessary to add resistance to a resonant circuit to broaden the response (increase the bandwidth, decrease the selectivity). The resistances across the coils in Fig. 1.7 is an example.

Typically, resonant-circuit Q is measured at the point on either side of the resonant frequency where the signal amplitude is down 0.707 of the peak resonant value, as shown in Fig. 2.2*a*. (Resonant-circuit Q measurements are included in Chap. 4.)

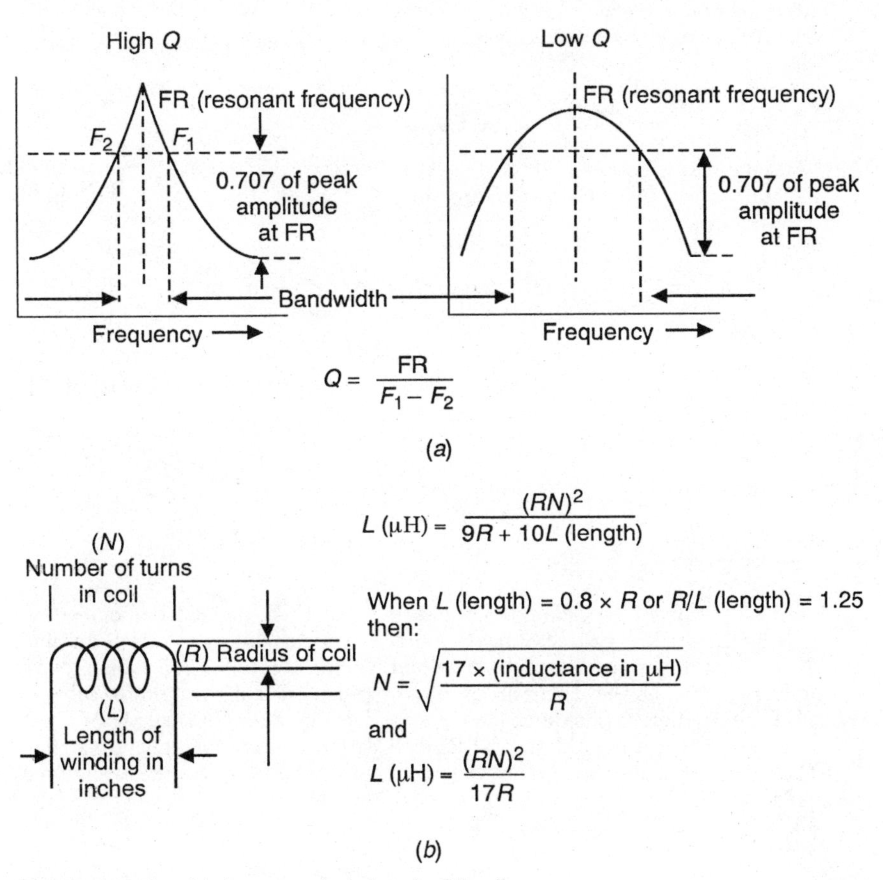

FIGURE 2.2 Q and inductance calculations for RF coils.

Note that Q must be increased for increases in resonant frequency if the *same bandwidth* is to be maintained. For example, if the resonant frequency is 10 MHz, with a bandwidth of 2 MHz, the required circuit Q is 5. If the resonant frequency is increased to 50 MHz, with the same 2-MHz bandwidth, the required Q is 25.

Circuit Q must be decreased for increases in bandwidth if the *same resonant frequency* is to be maintained. For example, if the resonant frequency is 30 kHz, with a bandwidth of 2 kHz, the required circuit Q is 15. If the bandwidth is increased to 10 kHz, with the same 30-kHz resonant frequency, the required Q is 3.

2.1.2 Calculating Resonant Values

Assume that you want a circuit that resonates at 400 kHz with an inductance of 10 μH. What value of capacitor is necessary? Using the equations in Fig. 2.1,

$$C = \frac{2.54 \times 10^4}{400^2 \times 10} = 0.0158$$

Use the nearest standard value of 0.016 μF.

Assume that you want a circuit that resonates at 2.65 MHz with a capacitance of 360 pF. What value of inductance is necessary? Using the equations in Fig. 2.1,

$$L = \frac{2.54 \times 10^4}{2650^2 \times (360 \times 10^{-6})} = 10\ \mu\text{H}$$

Assume that you must find the resonant frequency of a 0.002-μF capacitor and a 0.02-mH inductance. Using the equations in Fig. 2.1, first convert 0.02 mH to 20 μH, then:

$$F = \frac{0.159}{\sqrt{20\ \mu\text{H} \times 0.002\ \mu\text{F}}} = \frac{0.159}{\sqrt{0.4}} = \frac{0.159}{0.2}$$
$$= 0.795\ \text{MHz, or } 795\ \text{kHz}$$

Assume that an RF circuit must operate at 40 MHz with a bandwidth of 8 MHz. What circuit Q is required? Using the equations in Fig. 2.2*a*,

$$\text{FR} = 40 \qquad F_1 - F_2 = 8 \qquad Q = \frac{40}{8} = 5$$

2.1.3 RF Coils

Figure 2.2*b* shows the equations necessary to calculate the self-inductance of a single-layer, air-core coil (the most common type of coil used in RF circuits). Note that maximum inductance is obtained when the ratio of coil-radius to coil-length is 1.25 (when the length is 0.8 of the radius). RF coils wound for this ratio are the most efficient (maximum inductance for minimum physical size).

Assume that you must design an RF coil with 0.5-μH inductance on a 0.25-in radius (air core, single layer). Using the equations in Fig. 2.2*b,* for maximum efficiency, the coil length must be 0.8*R,* or 0.2 in. Then,

$$N = \sqrt{\frac{17 \times 0.5}{0.25}} = \sqrt{34} = 5.8\ \text{turns}$$

For practical purposes, use six turns and spread the turns slightly. The additional part of a turn increases inductance, but the spreading decreases inductance. After the coil is made, the inductance should be checked with an inductance bridge or as described in Chap. 4.

2.1.4 Voltage-Variable Capacitors

Many present-day RF circuits are tuned with VVCs. (Note that a VVC is sometimes called a *voltage-variable diode* because the device is constructed more like a diode than a capacitor.) Figures 1.13 and 1.15 show how VVCs (D_{153}) are used to tune circuits in RF tuners.

The capacitance of a VVC is controlled by external voltage and is varied when the external voltage is varied (such as the error voltage applied to D_{153} in Fig. 1.15). If a VVC is used in an RF circuit, it is possible to vary the circuit capacitance and thus vary the resonant frequency (or tune the circuit) with a variable voltage. From a design standpoint, the main concern with VVCs in resonant circuits is the tuning range of the circuit. All other factors being equal, the tuning range depends on the capacitance range of the VVC.

2.2 RF CIRCUITS FROM SCRATCH

This section provides a review and summary of RF design from a very practical standpoint. The information is presented primarily for the serious experimenter or hobbyist who wants to design RF circuits from scratch. While this book makes no pretense at being an engineering design text, the information presented here provides the background necessary to understand the test and troubleshooting described in the remaining chapters. If you understand what is required in design of RF circuits, you have a starting point for troubleshooting such circuits (when the circuits fail to perform as outlined).

2.2.1 Basic RF-Circuit Design Approaches

There are two basic approaches for the design of RF circuits: y-*parameters* and *large-scale characteristics.* With either approach, it is difficult at best to provide simple, step-by-step procedures for designing RF circuits to meet all possible circuit conditions. In practice, there are several reasons why this procedure often results in considerable trial and error. Here are some typical problems.

First, not all required design characteristics are always available in datasheet form. For example, input and output admittances may be given at some low frequency but not at the desired operating frequency.

Often, manufacturers do not agree on terminology. A good example of this is in y-parameters, where one manufacturer uses letter subscripts and another uses number subscripts (y_{fs} or y_{21}). Of course, this type of variation can be eliminated by conversion.

In some cases, manufacturers give the required information on datasheets but not in the required form. For example, some manufacturers may give the input capacitance in farads rather than listing the input admittance in mhos. (The input admittance is then found when the input capacitance is multiplied by $6.28F$, where F is the frequency of interest.)

This conversion is based on the assumption that the input admittance is primarily capacitive and thus depends on frequency. The assumption is not always true for the frequency of interest, so it may be necessary to use complex admittance-measuring equipment to make actual tests of the transistor.

The input and output tuning circuit of an RF circuit (amplifier, multiplier, etc.) must perform three functions. First, the circuits (capacitors and coils) must tune the amplifier to the desired frequency. Second, the circuits must match the input and output impedances of the transistor to the impedances of the source and load (to minimize signal loss). Third, as in the case with any amplifier, there is some feedback between output and input. If the admittance factors are just right, the feedback may be of sufficient amplitude and proper phase to cause *oscillation* in the circuit. (The circuit is considered *unstable* when this occurs.)

Circuit instability in any form is always undesirable and can be corrected by feedback (called *neutralization*) or by changes in the input-output tuning networks. Generally, the changes involve *introducing some slight mismatch* to improve stability. Although the neutralization and tuning circuits are relatively simple, the equations for determining stability (or instability) and impedance matching are long and complex. As a general rule, *such equations are best solved by computer-aided design methods.*

In an effort to cut through the maze of information and complex equations, we discuss simplified, practical, noncomputer RF-circuit design. Armed with this information you should be able to interpret datasheets, or test information, and use the data to design tuning networks that provide stable RF-circuit operation at the frequencies of interest.

With each step we discuss the various alternative procedures and types of information available. Specific design examples are used to summarize the information. On the assumption that you may not be familiar with two-port networks, we start with a summary of the *y*-parameter system.

2.2.2 *y*-Parameters

Impedance (Z) is a combination of resistance (R, the real part) and reactance (X, the imaginary part). Admittance (y) is the reciprocal of impedance is composed of conductance (g, the real part) and susceptance (jb, the imaginary part). Thus, g is the reciprocal of R, and jb is the reciprocal of X.

To find g, divide R into 1; to find R, divide g into 1. Z is expressed in ohms (Ω). y, being a reciprocal, is expressed in mhos or millimhos (mmhos). For example, an impedance Z of 50 Ω equals 20 mmhos (1/50 = 0.02; 0.02 mho = 20 mmhos).

A *y*-parameter is an expression for admittance in the form

$$y_i = g_i + jb_i$$

where g_i = real (conductive) part of input admittance
jb_i = imaginary (susceptive) part of input admittance
y_i = input admittance (the reciprocal of Z_i)

The term $y_i = g_i + jb_i$ expresses the *y*-parameter in *rectangular form.* Some manufacturers describe the *y*-parameter in *polar form.* For example, they give the *magnitude* of the input as $|y_i|$ and the angle of the input admittance as $\angle y_i$. Quite often, manufacturers mix the two systems of *vector algebra* on data sheets.

Conversion of Vector-Algebra Forms. In case you are not familiar with the basics of vector algebra, the following notes summarize the steps necessary to manipulate vector-algebra terms. With this background, you should be able to perform all calculations required for simplified RF-circuit design (using y-parameters).

To convert from rectangular to polar form:

1. Find the magnitude from the square root of the sum of the squares of the components:

$$\text{Polar magnitude} = \sqrt{g^2 + jb^2}$$

2. Find the angle from the ratio of the component values:

$$\text{Polar angle} = \arctan\frac{jb}{g}$$

The polar angle is leading if the jb term is positive and is lagging if the jb term is negative. For example, assume that the y_{fs} is given as $g_{fs} = 30$ and $jb_{fs} = 70$. This is converted to polar form by

$$|y_{fs}| \text{ polar magnitude} = \sqrt{30^2 + 70^2} = 76$$

$$\angle y_{fs} \text{ polar angle} = \arctan\frac{70}{30} = 67°$$

Converting from polar to rectangular form:

1. Find the real (conductive, or g) part when polar magnitude is multiplied by the cosine of the polar angle.
2. Find the imaginary (susceptance, or jb part) when polar magnitude is multiplied by the sine of the polar angle.

If the angle is positive, the jb component is also positive. When the angle is negative, the jb component is also negative.

For example, assume that the y_{fs} is given as $|y_{fs}| = 20$ and $\angle y_{fs} = -33°$. This is converted to rectangular form by:

$$20 \times \cos 33° = g_{fs} = 16.8$$

$$20 \times \sin 33° = jb_{fs} = 11$$

The Four Basic y-Parameters. Figure 2.3 shows the Y-equivalent circuit for a FET. A similar circuit can be drawn for a two-junction transistor when analyzing small-signal characteristics.

Note that y-parameters can be expressed with number or letter subscripts. The number subscripts are universal and can apply to two-junction transistors, FETs, and IC RF circuits. The letter subscripts are most popular on FET datasheets.

The following notes can be used to standardize y-parameter nomenclature. Note that the letter s in the letter subscript refers to common-source operation of a FET circuit and is equivalent to a common-emitter two-junction circuit.

y_{11} is *input admittance* and can be expressed as y_{is}.

y_{12} is *reverse transadmittance* and can be expressed as y_{rs}.

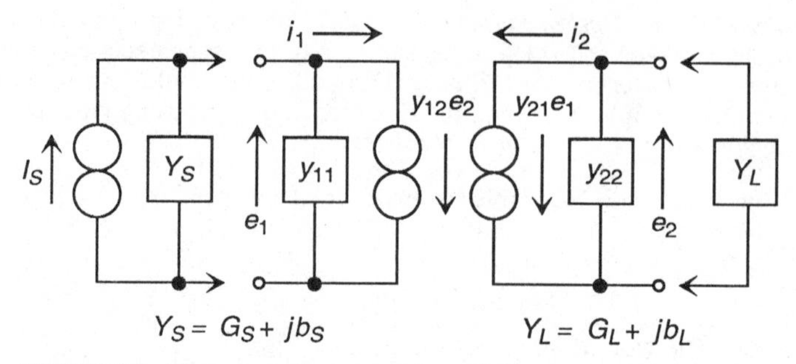

FIGURE 2.3 The *y*-equivalent circuit for a FET.

y_{21} is *forward transadmittance* and can be expressed as y_{fs}.

y_{22} is *output admittance* and can be expressed as y_{os}.

Input admittance, with Y_L = infinity (a short circuit of the load), is expressed as

$$y_{11} = g_{11} + jb_{11} = \frac{di_1}{de_1} \qquad \text{(with } e_2 = 0\text{)}$$

This means that y_{11} is equal to the difference in current i_1, divided by the difference in voltage e_1, with voltage e_2 at 0. The voltages and currents involved are shown in Fig. 2.3.

Some datasheets do not show y_{11} at any frequency but give *input capacitance* instead. If one assumes that the input admittance is entirely (or mostly) capacitive, the input impedance can be found when the input capacitance is multiplied by $6.28F$ (F = frequency in Hz) and the reciprocal is taken.

Because admittance is the reciprocal of impedance, admittance is found when input capacitance is multiplied by $6.28F$ (where admittance is capacitive). For example, if the frequency is 100 MHz and the input capacitance is 9 pF, the input admittance is: $6.28 \times (100 \times 10^6) \times (8 \times 10^{-12})$, or about 5 mmhos.

This assumption is accurate only if the real part of y_{11} (or g_{11}) is negligible. Such an assumption is reasonable for most FETs but not necessarily for all two-junction transistors. The real part of two-junction transistor input admittance can be quite large in relation to the imaginary jb_{11} part.

Forward transadmittance, with Y_L = infinity (a short circuit of the load), is expressed as

$$y_{21} = g_{21} + jb_{21} = \frac{di_2}{de_i} \qquad \text{(with } e_2 = 0\text{)}$$

This means that y_{21} is equal to the difference in output current i_2 divided by the difference in input voltage e_1, with voltage e_2 at 0. In other words, y_{21} represents the *difference in output current for a difference in input voltage.*

Two-junction transistor datasheets often do not give any value for y_{21}. Instead, forward transadmittance is shown by a *hybrid system* of notation using h_{fe} or h_{21} (which means hybrid forward transadmittance with common emitter). No matter what system is used, it is essential that the values of forward transadmittance be considered at the frequency of interest.

Output admittance, with Y_s = infinity (a short circuit of the source or input), is expressed as

$$y_{22} = g_{22} + jb_{22} = \frac{di_s}{de_2} \qquad \text{(with } e_1 = 0\text{)}$$

Reverse transadmittance, with y_2 = infinity (a short circuit of the source or input, is expressed as

$$y_{12} = g_{12} + jb_{12} = \frac{di_1}{di_2} \qquad \text{(with } e_1 = 0\text{)}$$

y_{12} is usually not considered an important two-junction transistor parameter. However, y_{12} may appear in equations related to RF design.

2.2.3 *y*-Parameter Measurement

It is obvious that *y*-parameter information is not always available or in a convenient form. In practical design, it may be necessary to measure the *y*-parameter using test equipment. The main concern in measuring *y*-parameters is that the measurements are made under conditions simulating those of the final circuit. For example, if supply voltages, bias voltages, and operating frequency are not identical (or close) to the final circuit, the tests may be misleading.

Although the datasheets for transistors to be used in RF circuits usually contain input and output admittance data, it may be helpful to know how this information is obtained. There are two basic methods for measuring the *y*-parameters of transistors.

One method involves *direct measurement* of the parameter (such as measuring changes in outputs for corresponding changes in input). The other uses *tuning substitution* (where the transistor is tuned for maximum transfer of power, and the admittances of the tuning circuits are measured). We summarize both methods in the following paragraphs

Direct Measurement of $\mathbf{y_{fs}}$ **(**$\mathbf{y_{21}}$**).** Figure 2.4 shows a typical test circuit for direct measurement of y_{fs} (which may be listed as y_{21}, g_m, or even g_{fs}). Although a FET is shown, the same circuit can apply to any single-input device, such as a two-junction transistor.

The value of *RL* must be such that the drop produced by the FET current drain is negligible, and the operating-voltage point (V_{DS} in the case of a FET) is correct for a given power supply voltage (V_{DD}) and operating current (I_D). For example, if I_D is 10 ma, V_{DD} is 20 V, and V_{DS} is 15 V, *RL* must drop 5 V at 10 mA. Thus the value of *RL* is 5 V/0.01 A = 500 Ω.

During testing, the signal source is adjusted to the frequency of interest. The amplitude of the signal source V_{in} is set to some convenient number such as 1 V or 100 mV. The value of y_{fs} (y_{21}) is calculated from the equation in Fig. 2.4 and is expressed in mhos (or millimhos and μmhos, in practical circuits). As an example, assume that the value of *RL* is 1000 Ω, V_{in} is 1 V, and V_{out} is 8 V. The value of y_{fs} is 8/(1 × 1000) = 0.008 mho = 8 mmho = 8000 μmho.

Direct Measurement of $\mathbf{y_{os}}$ **(**$\mathbf{y_{22}}$**).** Figure 2.5 shows a typical test circuit for direct measurement of y_{os} (which may be listed as y_{22}, g_{os}, g_{22}, or even r_d, where

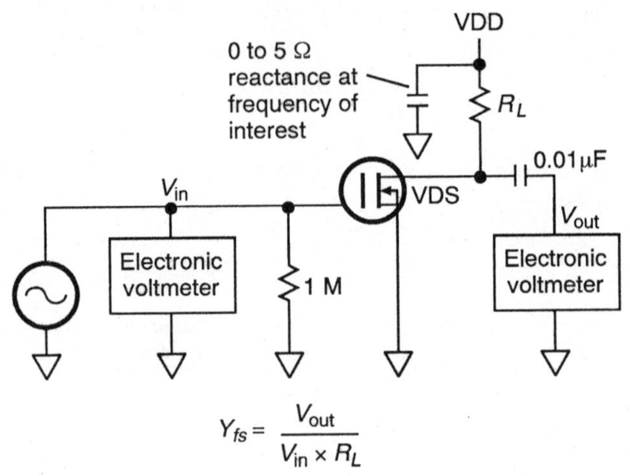

FIGURE 2.4 Direct measurement of y_{fs}.

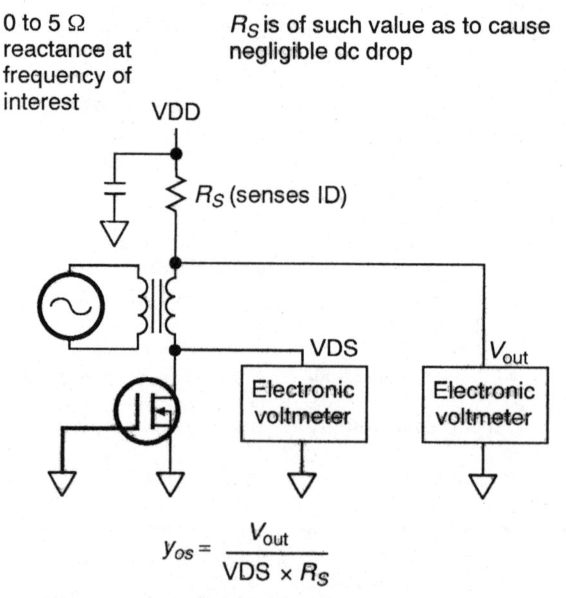

FIGURE 2.5 Direct measurement of y_{os}.

$r_d = 1/y_{os}$). Some datasheets give y_{os} as a complex number with both the real (g_{os}) and imaginary (b_{os}) values shown by means of curves.

The value of RS must be such as to cause a negligible drop (so that V_{DS} can be maintained at the desired level, with given V_{DD} and I_D). During testing, the signal source is adjusted to the frequency of interest. Both V_{out} and V_{DS} are measured, and the value of y_{os} is calculated from the equation in Fig. 2.5.

Direct Measurement of $\mathbf{y}_{is}$ ($\mathbf{y}_{11}$). Although y_{is} is not generally a critical factor for FETs, it is necessary to know the values of y_{is} to calculate impedance-matching networks for FET RF amplifiers (as discussed in Sec. 2.2.9). If it is necessary to establish the imaginary part (b_{is}), use an admittance meter or RX meter.

Direct Measurement of $\mathbf{y}_{is}$ ($\mathbf{y}_{12}$). Again, although y_{rs} is not generally a critical factor for FETs, it is necessary to know the values of y_{is} to calculate impedance-matching networks. Although the real part of g_{rs} remains at zero for all conditions and at all frequencies, the imaginary part b_{rs} does vary with voltage, current, and frequency. (The reverse susceptance varies and, under the right conditions, can produce undesired feedback from output to input.) This condition must be accounted for in the design of RF amplifiers to prevent feedback from causing oscillation. If it is necessary to establish the imaginary part (b_{rs}), use an admittance meter or RX meter.

Tuning-Substitution Measurement of* y*-Parameters. Figure 2.6 is a typical test circuit for measurement of y-parameters using the tuning-substitution method. Although a two-junction transistor is shown, the circuit can be adapted for use with FETs.

During testing, the transistor is placed in a test circuit designed with variable components to provide wide tuning capabilities. This is necessary to ensure cor-

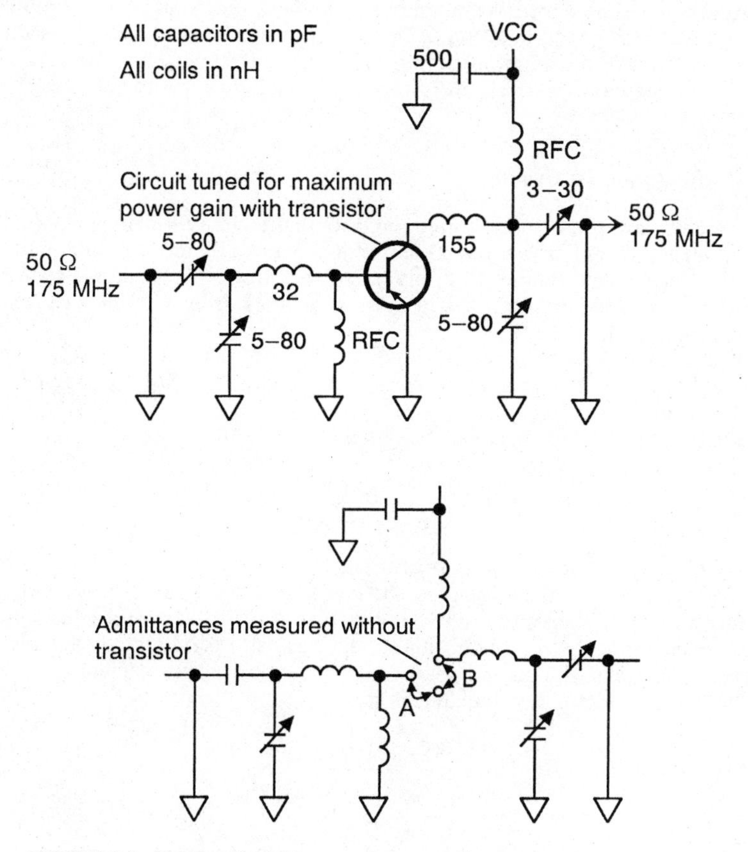

FIGURE 2.6 Tuning-substitution measurement of y-parameters.

rect matching at various power levels. The circuit is tuned for *maximum power gain* at each power level for which admittance information is desired.

After the test amplifier is tuned for maximum power gain, the dc power, signal source, circuit load, and test transistors are disconnected from the circuit. For total circuit impedance to remain the same, the signal-source and output-load circuit connections are terminated at the characteristic impedances (typically 50 Ω).

After the substitutions are complete, complex admittances are measured at the base- and collector-circuit connections of the test transistor (points A and B, respectively, in Fig. 2.6), using a precision admittance meter.

Note that the transistor input and output admittances are the *conjugates* of the base-circuit connection and the collector-circuit connection admittances, respectively. For example, if the base-circuit connection (point A) admittance is $8 + j_3$, the input admittance of the transistor is $8 - j_3$.

In some systems of two-junction transistor RF-circuit design, the networks are calculated on the basis of input-output resistance and capacitance, instead of admittance (although admittances are often used to determine stability before going into design of the RF networks). Such a system (often called the *large-signal design approach*) is described in Sec. 2.2.9. With this approach, the admittances measured in the circuit in Fig. 2.6 are converted to resistance and capacitance.

Admittances are expressed in mhos (or mmhos). Resistance is found by dividing the real part of the admittance into 1. Capacitance is found by dividing the imaginary part of the admittance into 1 (to find reactance); then the reactance is used in the equation $C = 1/6.28F_{XC}$ to find the actual capacitance.

2.2.4 RF-Amplifier Stability

Two factors are used to determine the potential stability (or instability) of transistors in RF-amplifier circuits. One factor is known as the Linvill C factor; the other is the Stern k factor. Both are calculated from the equations requiring y-parameter information (to be taken from datasheets or by actual measurement at the frequency of interest).

The main difference between the two factors is that the Linvill C factor assumes that the transistor is not connected to a load. The Stern k factor includes the effect of a specific load.

The Linvill C factor is calculated from

$$C = \frac{y_{12}y_{21}}{2g_{11}g_{22} - \text{Re}(y_{12}y_{21})}$$

where Re $(y_{12}y_{21})$ is the real part of $y_{12}y_{21}$.

If C is less than 1, the transistor is unconditionally stable. That is, using a conventional (unmodified) circuit, no combination of load and source admittance can cause oscillation. If C is greater than 1, the transistor is potentially unstable (certain combinations of load and source admittance can cause oscillation).

The Stern k factor is calculated from

$$k = \frac{2(g_{11} + G_S)(g_{22} + G_L)}{y_{12}y_{21} + Re(y_{12}y_{21})}$$

where G_S and G_L are source and load conductance, respectively; (G_S = 1/source resistance; G_L = 1/load resistance).

If k is greater than 1, the amplifier circuit is stable (the opposite of Linvill). If k is less than 1, the amplifier is unstable. In practical design, a Stern k factor of 3 or 4 should be used, rather than 1, to provide a safety margin. This accommodates parameter and component variations (particularly with regard to bandpass response).

Note that both equations are fairly complex and require considerable time for solution. In practical work, *computer-aided design techniques are used for stability equations.*

Some manufacturers provide alternative solutions to the stability and load-matching problems, usually in the form of datasheet graphs, such as shown in Fig. 2.7. Note that the device is unconditionally stable at frequencies above 250 MHz. At frequencies below about 50 MHz, the device becomes highly unstable.

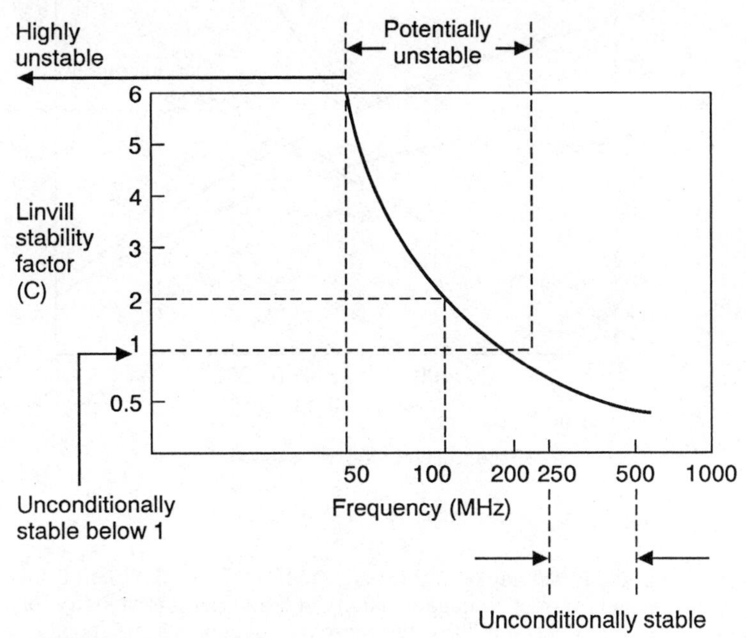

FIGURE 2.7 Typical stability graphs.

2.2.5 RF-Amplifier Stability Solutions

There are two basic design solutions to the problem of unstable RF-amplifier circuits: neutralization and mismatch.

Neutralization. When an RF amplifier is neutralized, part of the output is fed back (after the output is shifted in phase) to the input, canceling oscillation. With neutralization, an RF amplifier can be matched perfectly to the source and load.

Neutralization is sometimes known as a *conjugate match.* In a perfect conjugate match the transistor input and source, as well as the transistor output and load, are matched resistively, and all reactance is tuned out. Neutralization requires extra components and can *create a problem when frequency is changed.*

Mismatch Solution. Mismatch solution involves introducing a specific amount of mismatch into either the source or load tuning networks so that any feedback is not sufficient to produce instability or oscillation. The mismatch solution, sometimes called the *Stern solution* (since the Stern k factor is involved), requires no extra components *but does reduce gain.*

Neutralization versus Mismatch. Figure 2.8 shows a comparison of the neutralized and mismatch solutions. The higher-gain curve represents neutralized operation (also called *unilateralized gain* in some literature). The lower-gain curve represents the power gain when the Stern k factor is 3.

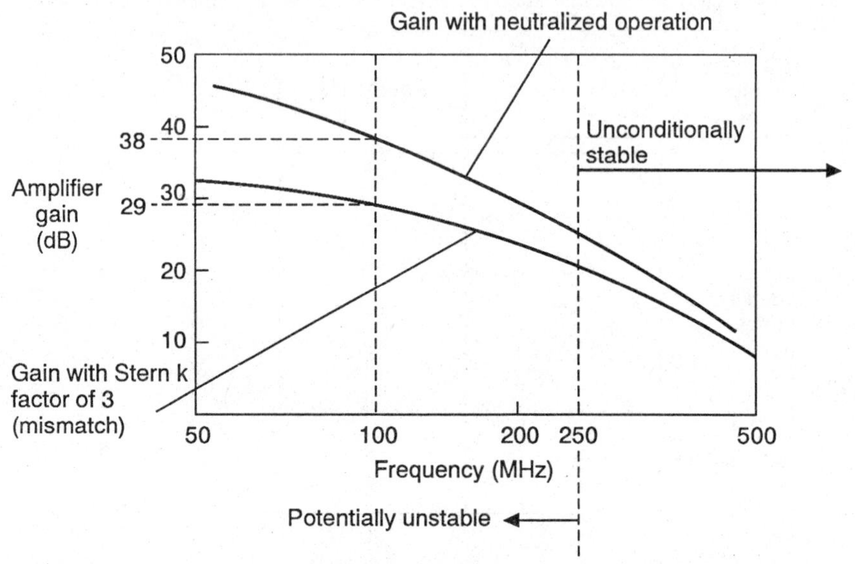

FIGURE 2.8 Comparison of neutralized and mismatch solutions for stability.

Assume that the frequency of interest is 100 MHz. If the RF circuit is matched directly to the load (perfect conjugate match) without regard to stability (or using neutralization to produce stability), the top curve applies and the power gain is about 38 dB. If the amplifier is matched to a load and source where the Stern k factor is 3 (resulting in a mismatch with the actual load and source), the lower curve applies, and the power gain is about 29 dB.

RF-Amplifier Gain. Although there are many expressions used to show RF-amplifier gain, maximum available gain (MAG) and maximum usable gain (MUG) are most useful.

MAG is the gain in a conjugately matched, neutralized RF circuit and is expressed as

$$\text{MAG} = \frac{(y_{21})^2 R_{\text{in}} \times R_{\text{out}}}{4}$$

where R_{in} and R_{out} are the input and output resistances, respectively, of the transistor.

An alternative MAG expression is

$$\text{MAG} = \frac{(y_{21})^2}{4\text{Re}(y_{11})\text{Re}(y_{22})}$$

where $\text{Re}(y_{11})$ is the real part (g_{11}) of the input admittance, and $\text{Re}(y_{22})$ is the real part (g_{22}) of the output admittance.

MUG is usually applied as the *stable gain* that may be realized in a *practical* (neutralized or unneutralized) RF amplifier. In a typical unneutralized circuit, MUG is expressed as

$$\text{MUG} = \frac{0.4y_{21}}{6.28F \times \text{reverse transfer capacitance}}$$

MAG and MUG are often omitted on datasheets for two-junction transistors. Instead, gain is listed as h_{fe} at a given frequency. This is supplemented with graphs that show available power output at given frequencies with a given input.

2.2.6 RF Circuits with Neutralization

There are several methods for neutralization of RF amplifiers. The most common method is the *capacitance-bridge* technique shown in Fig. 2.9. Capacitance-bridge neutralization becomes more apparent when the circuit is redrawn, as shown in Fig. 2.9*b*. The condition for neutralization is that IF = IN (the neutralization current IN must be equal to the feedback current IF in amplitude, but of opposite phase).

The equations normally used to find the value of the feedback neutralization capacitor are long and complex. However, for practical work, if the value of C_1 is made quite large in relation to C_2 (at least 4 times), the value of C_N can be found by

$$C_N = C_F\left(\frac{C_1}{C_2}\right)$$

where C_F is the reverse capacitance of the transistor.

In simple terms, the value of C_N is approximately equal to the value of reverse capacitance times the ratio of C_1/C_2. For example, if reverse capacitance (sometimes listed as collector-to-base capacitance) is 7 pF, C_1 is 30 pF, and C_2 is 3 pF, the C_1/C_2 ratio is 10, and $C_N = 10 \times 7$ pF = 70 pF.

2.2.7 RF Circuits without Neutralization (Mismatching)

A stable design with a potentially unstable transistor is possible without external feedback (neutralization) by proper choice of source and load values. This can be seen by inspection of the Stern k factor equation (Sec. 2.2.4). G_S and G_L can be made large enough (mismatched) to yield a stable circuit, regardless of the potential instability. Using this approach, a circuit-stability factor (typically $k = 3$) is selected, and the Stern k factor equation is used to arrive at values of G_S and G_L that produce the desired k.

Of course, the actual G of the source and load cannot be changed. Instead, the *input and output tuning circuits are designed as if the actual values are changed.*

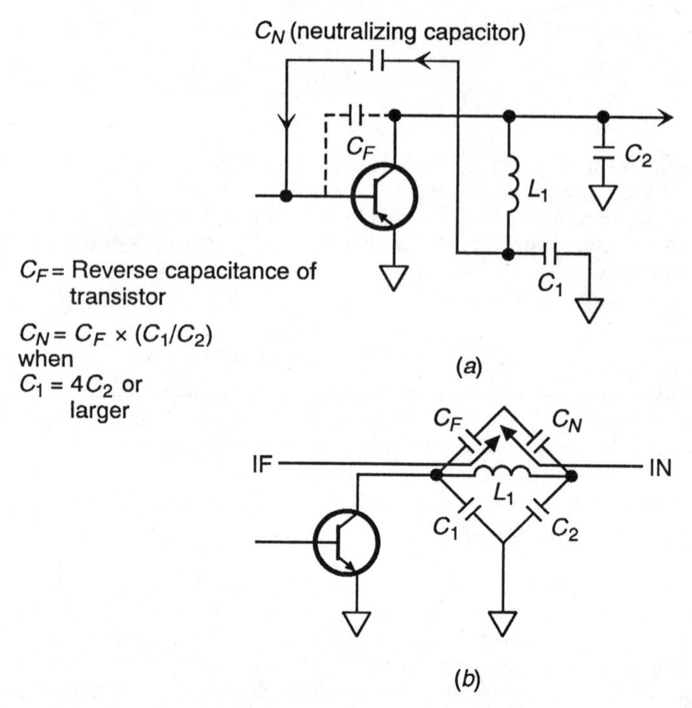

FIGURE 2.9 Basic neutralization technique for RF circuits.

This results in a mismatch and a reduction in power gain but does produce the desired stability, as discussed next.

2.2.8 *y*-Parameter Design Example

This section describes the design of tuning networks for RF amplifiers using *y*-parameters and a *simplified Stern approach.* Note that a full set of *y*-parameters must be available (either in the form of datasheet graphs or obtained by actual test of the transistor, as described in Sec. 2.2.3).

Assume that the circuit in Fig. 2.10 is to be operated at 100 MHz. (Although a FET is shown in Fig. 2.10, the design procedures also apply to two-junction transistors.) The source and load impedance are both 50 Ω. The characteristics of the FET are given in the datasheet. The problem is to find the optimum values of C_1 through C_4, as well as L_1 and L_2.

FET Characteristics. At 100 MHz, *y*-parameters for our FET are:

$$y_{is} = y_{11} = 0.15 + j3.0$$

$$y_{fs} = y_{21} = 10 - j5.5$$

$$y_{os} = y_{22} = 0.04 + j1.7$$

$$y_{rs} = y_{12} = 0 - j0.012$$

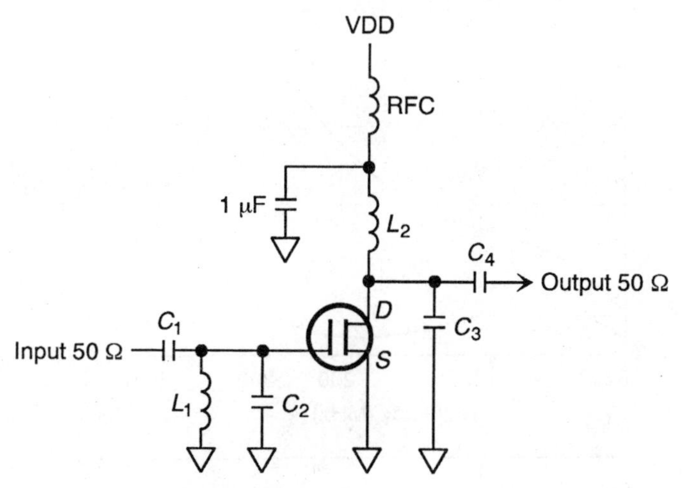

FIGURE 2.10 Basic RF-amplifier circuit.

From Fig. 2.7, the Linvill stability factor C is 2.0. As discussed, a Linvill C factor greater than 1 indicates that the device (a FET in this case) is potentially unstable. Neutralization or mismatching is necessary to prevent oscillation. Mismatching is used in this example (to minimize components and eliminate problems when operating frequency is changed).

Figure 2.8 shows that for a circuit stability (Stern k) factor of 3.0, the gain is about 29 dB. The load and source admittances for the required mismatch and gain (found in Fig. 2.11) are:

$$Y_L = 0.35 - j2.1 \text{ mmho}$$

$$Y_S = 1.30 - j4.4 \text{ mmho}$$

Output Matching. The first step is to design a network that will match the FET to a load and source of 50 Ω. The calculations are easier if the admittances in Fig. 2.11 are converted to impedance. Start by matching the output impedance to the load.

The 50-Ω load impedance must be transformed to the optimum load for the FET ($Y_L = 0.35 - j2.1$). This transformation is performed by the network shown in Fig. 2.12*a*. In effect, R_L is in series with C_4. The 50 Ω must be transformed to

$$R_L = \frac{1}{G_L} = \frac{1}{0.35 \times 10^{-3}} \approx 2860\ \Omega$$

As discussed, the true value of R_L is not changed, but the network is designed as if the value of R_L is changed to the optimum load for the FET. The series capacitive reactance (for C_4) required for this matching is

$$X_{C4,\text{series}} = R_S\left(\sqrt{\frac{R_P}{R_S}} - 1\right)$$

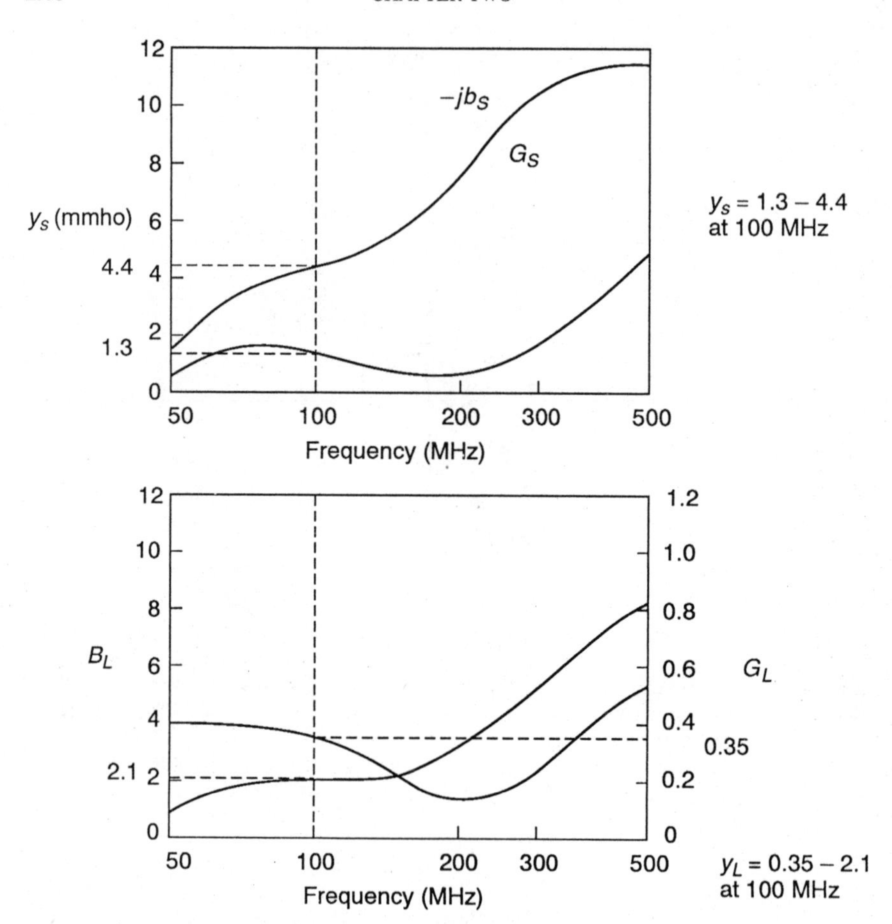

FIGURE 2.11 Typical load and source admittance graphs.

where R_P is the transformed parallel resistance and R_S is the true series resistance. Therefore,

$$X_{C4,\text{series}} = 50\left[\sqrt{\frac{2860}{50}} - 1\right] \approx 372\ \Omega$$

The capacitance for C_4 that provides this reactance at 100 MHz is

$$C_4 = \frac{1}{6.28\text{F} \times X_{C4}} = \frac{1}{6.28(10^8)(327)} \approx 4.3\ \text{pF}$$

The parallel equivalent of this capacitance is needed for determining the bandwidth and resonance later in design.

Series $R_L = 50\ \Omega$ $R_L = 50\ \Omega$ C_4 Parallel

(a) Output impedance transformation

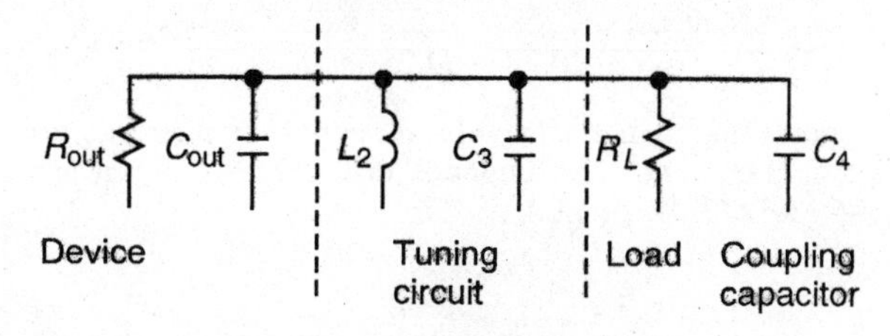

(b) Equivalent output circuit

FIGURE 2.12 Basic RF-network impedance transformation. (*a*) Output impedance transformation; (*b*) Basic RF-network impedance transformation.

$$X_{C4,\text{parallel}} = X_{C4,\text{series}}\left[1 + \left(\frac{R_S}{X_{C4,\text{series}}}\right)^2\right]$$

$$= \frac{R_P}{X_{C4,\text{series}}/R_S}$$

$$= \frac{2860}{372/50} \approx 382\ \Omega$$

Output Network Design. Figure 2.12*b* shows an equivalent circuit for the output circuit after transformation of the load. Since the resistance across the output circuit is fixed by the parallel combination of R_{out} and R_L (after transformation), the desired bandwidth of the output circuit is determined by C_3.

As discussed, the output admittance Y_{out} of the FET is not equal to y_{os} under all conditions. Only when the input is terminated in a short circuit or the feedback admittance is zero does Y_{out} equal y_{os}. When y_{os} is not zero and the input is terminated with a practical source admittance, the true output admittance is found from

$$Y_{\text{out}} = y_{os} - \frac{y_{fs}y_{rs}}{y_{is} + Y_S}$$

$$= 0.04 - j1.7 - \frac{(10 - j5.5)(0 - j0.012)}{(0.15 + j3.0) + (1.3 - j4.4)}$$

$$= -\ 0.66 + j2.05 \text{ mmho}$$

Therefore,

$$R_{\text{out}} = \frac{1}{G_{\text{out}}} = \frac{1}{-\ 0.066 \times 10^{-3}} = -\ 15.2\ \text{k}\Omega$$

$$C_{\text{out}} = \frac{B_{\text{out}}}{6.28\text{F}} = \frac{2.05 \times 10^{-3}}{6.28(10^8)} \approx 3.2\ \text{pF}$$

Note that the negative output impedance indicates the instability of the unloaded amplifier.

Now the total impedance across the circuit can be calculated:

$$R_{total} = \frac{1}{G_{out} + G_L}$$
$$= \frac{1}{(-0.66 \times 10^{-3}) + (0.35 \times 10^{-3})} \approx 3.52\ \text{k}\Omega$$

Since the output impedance is several times higher than the input impedance of the FET, amplifier bandwidth depends primarily on output loaded Q. For a bandwidth of 5 MHz (3-dB points),

$$C_{total} = \frac{1}{6.28R_{total}(BW)}$$
$$= \frac{1}{6.28(3.52 \times 10^3)(5 \times 10^6)} \approx 9\ \text{pF}$$

Therefore,

$$C_3 = C_{total} - C_{out} - C_{4,parallel} = 9.0 - 3.2 - 4.2 = 1.6\ \text{pF}$$

The output inductance (L_2) that resonates with C_{total} at 100 MHz is 280 nH. This completes output network design.

Input Network Design. The input network can be calculated in the same manner and yields the following results:

$$Y_S = 1.30 - j4.4$$
$$X_{C1,series} = 190\ \Omega$$

Therefore,

$$C_1 = 8.4\ \text{pF}$$
$$X_{C1,parallel} = 203\ \Omega$$

Therefore,

$$C_{1,parallel} = 7.8\ \text{pF}$$
$$Y_{in} = y_{is} = \frac{y_{fs}y_{rs}}{y_{os} + Y_L} = 0.25 + j4.52\ \text{mmho}$$

Therefore,

$$R_{in} = 14\ \text{k}\Omega \qquad C_{in} = 7.2\ \text{pF} \qquad T_{total} = 950\ \Omega$$

The bandwidth of the input tuned circuit is chosen to be 10 MHz. Therefore,

$$C_{total} = 17\ \text{pF} \qquad \text{and} \qquad L_1 = 150\ \text{nH}$$
$$C_2 = 17 - 7.2 - 7.8 = 2\ \text{pF}$$

Bypass Capacitor. This completes the design of the tuned circuits (resonant networks). It is important that the circuit be well bypassed to ground at the signal frequency *since only a small impedance to ground may cause instability or loss of gain.* To be on the safe side, select a bypass capacitor value where the reactance is about 1 or 2 Ω at the operating frequency. A 1-μF capacitor provides less than 1-Ω reactance at 100 MHz.

2.2.9 Large-Signal Design Example

As discussed in Sec. 2.2.3, it is possible to design the tuning networks for RF circuits without using a full set of *y*-parameters or admittances. Instead, the networks are designed using the input-output capacitances and resistances of the transistor (the so-called large-signal characteristics). Often, the capacitance and resistance information is available on datasheets in the form of graphs. This is especially true for transistors designed for use as RF power amplifiers. Again, although this book makes no pretense at being an engineering design text, this section describes the design of tuning networks for RF amplifiers using the large-signal approach.

Assume that the circuit in Fig. 2.13*a* is to be used as an RF power amplifier with a 50-Ω antenna and a 28-V power supply. The circuit is to operate at 50 MHz and produce 50-W output. Before we get into design calculations, let us review some design considerations.

RF Amplifier versus Multiplier. The same basic circuits in Fig. 2.13 can be used as RF amplifiers or frequency multipliers. However, in a multiplier circuit, the output must be tuned to a multiple of the input.

RF-frequency-multipliers may or may not provide amplification. Usually, most of the amplification is supplied by the final amplifier stage, which is not operated as a multiplier. That is, the input and output of the final stage are at the same frequency. A typical RF transmitter has three stages: an oscillator to generate the basic signal frequency (Sec. 2.3), an intermediate stage that provides amplification and/or frequency multiplication, and a final stage for power amplification.

Tuning Controls. Note that the circuit in Fig. 2.13*a* has two tuning controls (variable capacitors in this case) in the output network, while the network circuit in Fig. 2.13*b* has only one adjustment control. The circuit in Fig. 2.13*a* is typical for power amplifiers, where the output is tuned to the resonant frequency by one control and adjusted for proper impedance match by the other control (often called the *loading control*). In practice, both controls affect tuning and loading (impedance matching). The circuit in Fig. 2.13*b* is typical for multipliers or intermediate amplifiers where the main concern is tuning to the resonant frequency.

Parallel Capacitors. Note that the variable capacitors are connected in parallel with fixed capacitors in both networks. This parallel arrangement serves two purposes. First, there is a minimum fixed capacitance in case the variable capacitor is adjusted to minimum value. In some cases, if minimum capacitance is not included in the network, a severe mismatch can occur when the variable capacitor is at minimum (possibly resulting in damage to the transistor). The second purpose for a parallel capacitor is to reduce the required capacitance rating (and thus the physical size) of the variable capacitor.

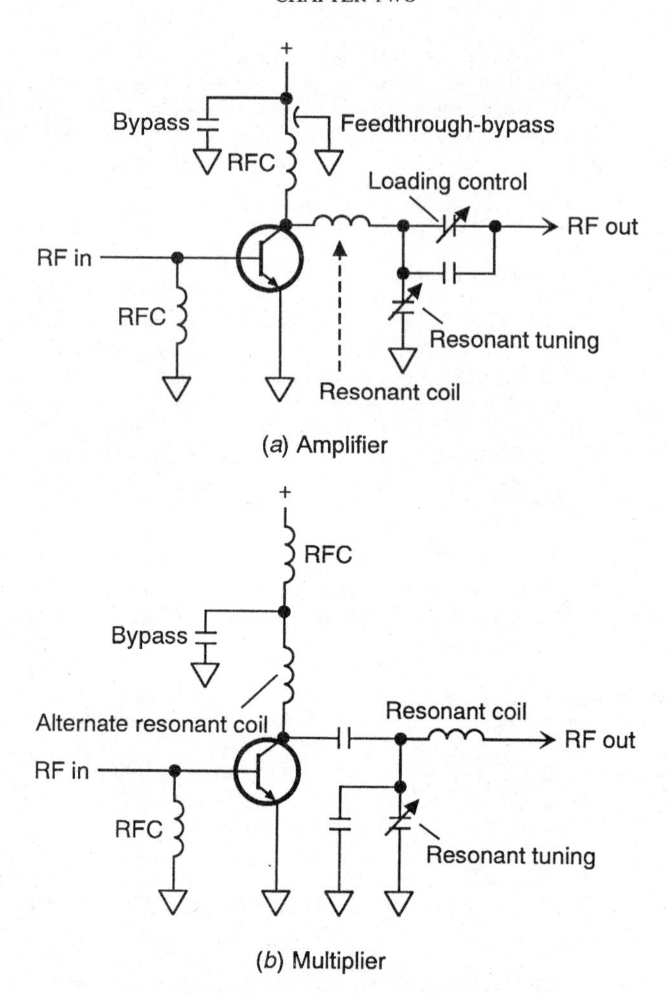

FIGURE 2.13 Basic RF amplifier (*a*) multiplier and (*b*) circuits.

Midrange Capacitors. When designing networks such as shown in Fig. 2.13, use a capacitor with a midrange capacitance equal to the desired capacitance. For example, if the desired capacitance is 25 pF (to produce resonance at the normal operating frequency), use a variable capacitor with a range of 1 to 50 pF. If such a capacitor is not readily available, use a fixed capacitor of 15 pF in parallel with a 15-pF variable capacitor. This provides a capacitance range of 16 to 30 pF, with a midrange of about 23 pF. Of course, the maximum capacitance range depends on the required tuning range of the circuit. (A wide frequency range requires a wide capacitance range.)

Bias. Generally, RF-power transistors remain cut off until a signal is applied. Therefore, the transistors are never conducting for more than 180° (half a cycle) of the 360° input signal cycle. In practice, the transistors conduct for about 140°

of the input cycle, either on the positive or negative half, depending on the transistor type (NPN or PNP). No bias, as such, is required for this class of operation (class C).

Grounded Emitter. The emitter is connected directly to ground. In those transistors where the emitter is connected to the case (typical in many RF power transistors), the case can be mounted on a chassis that is connected to the ground side of the power-supply voltage. A direct connection between emitter and ground is of particular importance in high-frequency operation. If the emitter is connected to ground through a resistance (or a long lead), an inductive or capacitive reactance can develop at high frequencies, resulting in undesired changes in the network.

RFC Connections. The transistor base is connected to ground through an RF choke (RFC). This provides a dc return for the base, as well as RF signal isolation between base and emitter or ground. The transistor collector is connected to the power-supply voltage through an RFC and (in some cases) through the coil portion of the resonant network. The RFC provides dc return, but RF signal isolation, between collector and power supply.

When the collector is connected to the power supply through the resonant network, the coil must be capable of handling the full collector current. For this reason, final (power) amplifier networks should be chosen so that collector current *does not pass through the coil.* The circuit in Fig. 2.13*a* is therefore preferable to the circuit in Fig. 2.13*b* for power amplifiers. The circuit in Fig. 2.13*b* should be used where the current is low (such as an intermediate amplifier).

RFC Ratings. The ratings for RFCs are sometimes confusing. Some manufacturers list a full set of characteristics: inductance, dc resistance, ac resistance, Q, current capability, and normal frequency range. Other manufacturers give only one or two of these characteristics. AC resistance and Q usually depend on frequency. A nominal frequency-range characteristic is a helpful, but usually not critical, design parameter.

All other factors being equal, the dc resistance should be at a minimum for any circuit carrying a large amount of current. For example, a large dc resistance in the collector of a final power amplifier can result in a large voltage drop between the power supply and collector.

Usually, the selection of a trial value for an RFC is based on a tradeoff between inductance and current capability. The minimum current capacity should be greater (by at least 10 percent) than the maximum anticipated direct current. The inductance depends on operating frequency. *As a trial value,* use an inductance that produces a reactance between 1000 and 3000 Ω at the operating frequency.

Bypass Capacitors. The power-supply circuits of power amplifiers and multipliers must be bypassed, as shown in Fig. 2.13. The feed-through bypass capacitors are used at higher frequencies where the RF circuits are physically shielded from the power supply and other circuits. A feed-through capacitor permits direct current to be applied through a shield but prevents RF from passing outside the shield (RF is bypassed to the ground return). *As a trial value,* use a total bypass capacitance range of 0.001 to 0.1 μF.

Checking Bypass Capacitors and RFCs. From a practical standpoint, the best test for adequate bypass capacitance and RFC inductance is the presence of RF signals on the power-supply side of the dc voltage line. If RF signals are present on the power-supply side of the line, the bypass capacitance and/or the RFC inductance are not adequate. (A possible exception is where the RF signals are being picked up because of inadequate shielding.)

If the shielding is good and RF signals are present in the power supply, increase the bypass capacitance value. As a second step, increase the RFC inductance. Of course, circuit performance must be checked with each increase in capacitance or inductance value. For example, *too much bypass capacitance can cause undesired feedback and oscillation*; too much RFC inductance can reduce amplifier output and efficiency. The procedures for measurement of RF signals are described in Chaps. 4 and 5.

Amplifier Efficiency. A class C RF amplifier has a typical efficiency of about 65 to 70 percent. That is, the RF power output is 65 to 70 percent of the dc input power. To find the required dc input power, divide the desired RF power output by 0.65 to 0.7. For example, if the desired RF output is 50 W, the dc input power is 50/0.7, or about 70 W.

Since the collector of an RF amplifier is at a dc potential approximately equal to the power supply (slightly less because of a drop across the RFC and/or coil), divide the input power by the power-supply voltage to find the collector current. For example, with a dc input of 70 W and a 28-V power supply, the collector current is about 2.7 A.

Transistor Characteristics. It is obvious that the transistors must be capable of handling the full power-supply voltage at the collectors and that the current and/or power rating is greater than the maximum calculated values. Likewise, the transistor must be capable of producing the necessary power output at the operating frequency.

It is also obvious that the transistors must provide the necessary power gain at the operating frequency. Likewise, the input power to an amplifier must match the desired output and gain. For example, assume that a 50-W, 50-MHz transmitter is to be designed and that transistors with a power gain of 10 are available. Generally, a transistor oscillator produces less than a 1-W output. Therefore, an intermediate amplifier is required to deliver an output of 5 W to the final amplifier. The intermediate amplifier requires about a 7-W dc input (50/7 = 7). Assuming a gain of 10 for the intermediate amplifier, an input of 0.5 W is required from the oscillator.

Intermediate-Amplifier Efficiency. When an intermediate amplifier is also used as a frequency multiplier, the efficiency drops from the 65 to 70 percent value. As a guideline, the efficiency of a second-harmonic amplifier (output at twice the input frequency) is 42 percent, the third harmonic is 28 percent, the fourth harmonic is 21 percent, and the fifth harmonic is 18 percent. Therefore, if an intermediate amplifier is to be operated at the second harmonic and produce a 5-W RF power output, the required dc input power is about 12 W (5/0.42 = 12).

Power Gain with Frequency Multiplication. Another problem to be considered in frequency multiplication is that power gain (as listed on the datasheet) may not remain the same as when amplifier input and output are at the same frequency. Some datasheets specify power gain at the basic frequency and then derate the power gain for second-harmonic operation. As a guideline, always use the minimum power-gain factor when calculating power input and output values.

Resonant-Network Design. The resonant network must be designed such that the network is resonant at the desired frequency (inductive and capacitive reactance must be equal at the selected frequency), and the network must match the transistor output impedance to the load. Here are the main considerations.

Generally, an antenna load impedance is about 50 Ω, while the output impedance of a typical transistor is a few ohms (at radio frequencies). When one circuit feeds into another, the network must match the output impedance of one transistor to the input impedance of another. Any mismatch can result in a loss of power between stages or to the final load.

Transistor impedance (both input and output) has both resistive and reactive components and therefore varies with frequency. To design a resonant network for the output of a transistor, it is necessary to know the *output reactance* (usually capacitive), the *output resistance* at the operating frequency, and the *output power* (the so-called *large-signal parameters*). It is also necessary to know the input resistance and reactance (at a given frequency and power) when designing the resonant network of the stage feeding into a transistor.

Generally, the large-signal parameters are shown by graphs similar to Fig. 2.14. The reactance is found using the corresponding frequency and capacitance. For example, the output capacitance shown on the graph in Fig. 2.14 is about 15 pF at 80 MHz. This produces a capacitive reactance of about 130 Ω at 80 MHz. The reactance and resistance can then be combined to find impedance, as shown in Fig. 2.14.

R_S = Series resistance $\quad$ R_P = Parallel resistance

X_S = Series reactance $\quad$ X_P = Parallel reactance

$X_C = 1/(6.28FC)$ $\quad$ $X_L = 6.28FL$

$$\text{Parallel output resistance} = \frac{\text{Collector voltage}^2}{2 \times \text{power output}}$$

To convert X_S and R_S to parallel:

$$R_P = R_S\left[1 + \left(\frac{X_S}{R_S}\right)^2\right] \qquad X_P = \frac{R_P}{(X_S/R_S)}$$

To convert X_P and R_P to series:

$$R_S = \frac{R_P}{1 + (R_P/X_P)^2} \qquad X_S = R_S\frac{R_P}{X_P}$$

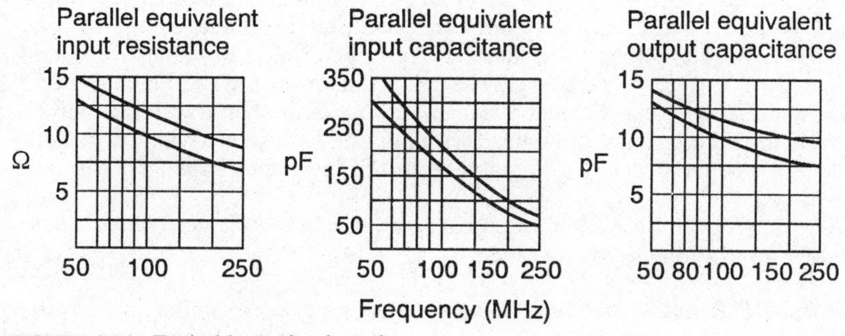

FIGURE 2.14 Typical large-signal graphs.

Input and output transistor impedances are generally listed on datasheets in parallel form. That is, the datasheets assume that the resistance is in parallel with the capacitance. However, some networks require that the impedance be calculated in series form, so it is necessary to convert using the equations in Fig. 2.14.

Resonant Network Calculations. Figure 2.15 shows five typical resonant networks, together with the calculations necessary to find the component values. Any of the resonant networks can be used as the tuning networks for RF amplifiers and/or multipliers. Note that the network in Fig. 2.15*a* is similar to that in Fig. 2.13*a,* while Fig. 2.15*c* is similar to Fig. 2.13*b* (except for the power connection).

The resistor and capacitor shown in the box labeled "transistor to be matched" represent the *complex output impedance* of a transistor. When the network is to be used with a final amplifier,the resistor labeled R_L is the antenna impedance or other load. When the network is used with an intermediate amplifier, R_L represents the input impedance of the following transistor. It is therefore necessary to calculate the input impedance of the transistors being fed by the network, using the data and equations in Fig. 2.14.

The complex impedances are represented in series form in some cases and parallel form in others, depending on which form is the most convenient for network calculation. The resultant network impedance, when terminated with a given load, must be equal to the conjugate of the impedance in the box.

For example, assume that the transistor has a series output impedance of $7.33 - j3.87$. That is, the resistance (real part of impedance) is 7.33 Ω, while the capacitive reactance (imaginary part of impedance) is 3.87. For a maximum power transfer from the transistor to the load, the load impedance must be the conjugate of the output impedance, or $7.33 + j3.87$.

If the circuit is designed to operate into the typical 50-Ω load (antenna), the network must match the $(50 + j0)$-Ω load to the $7.33 - j3.87$ transistor value. In addition to matching, the network provides *harmonic rejection* to prevent transmission on more than one frequency (unless a harmonic is needed in a multiplier stage), low loss, and provisions for *adjustment of both loading and tuning.*

Each network has advantages and disadvantages. The following is a summary of the five resonant networks shown in Fig. 2.15.

The network in Fig. 2.15*a* applies to most RF power amplifiers and is especially useful where the *series real part of the transistor output impedance* (R_1) *is less than 50 Ω.* With a typical 50-Ω load, the required reactance for C_1 rises to an impractical value when R_1 is close to 50 Ω.

The network in Fig. 2.15*b* (often called a *pi network*) is best suited where the parallel *resistor* (R_1) *is high* (near the value of R_L, typically 50 Ω). If the network in Fig. 2.15*b* is used with a low value of R_1, the inductance of L_1 must be very small, while C_1 and C_2 are very large (beyond practical limits).

The networks in Fig. 2.15*c* and 2.15*d* produce practical values for C and L, especially *where* R_1 *is very low.* The main limitation for the networks of Fig. 2.15*c* and 2.15*d* is that R_1 must be *substantially lower* than R_L. These networks, or variations thereof, are often used with intermediate stages where a low output impedance of one transistor is matched to the low input impedance of another transistor.

The network in Fig. 2.15*e* (often called a *tee network*) is best suited *where* R_1 *is much less* (*or much greater*) *than* R_L.

Practical RF Resonant Network. Assume that a network similar to that of Fig. 2.15*a* must match a transistor to a 50-Ω antenna at an operating frequency of 50

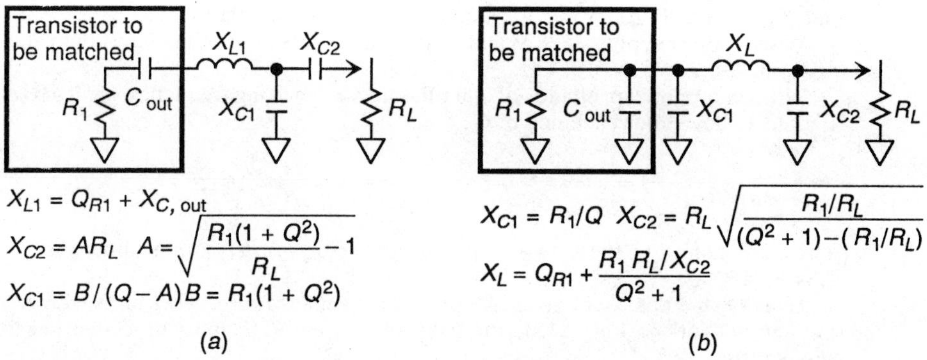

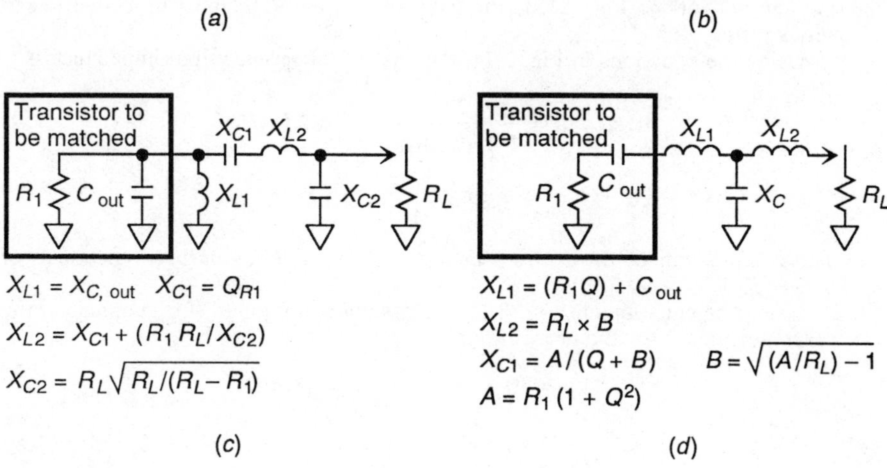

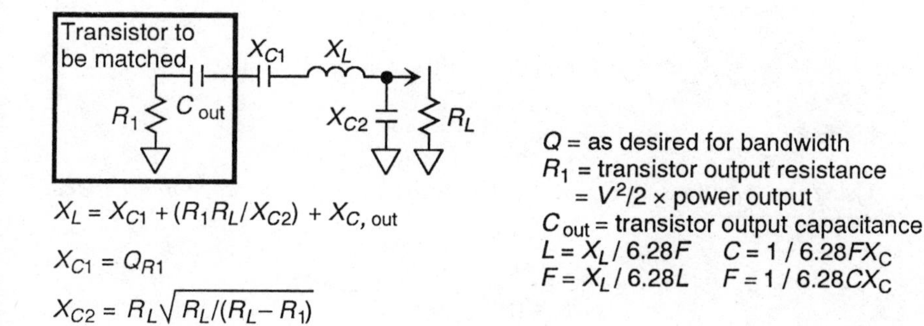

FIGURE 2.15 Typical resonant RF networks. (*a*) R_1 less than R_L; (*b*) R_1 about equal to R_L; (*c*) R_1 very small in relation to R_L; (*d*) R_1 very small in relation to R_L; (*e*) R_1 very large or very small in relation to R_L.

MHz. The required output is 50 W with a 28-V power supply. The transistor output capacitance is 200 pF at 50 MHz (obtained from the datasheet).

With a power supply of 28 V and an output of 50 W, Fig. 2.14 shows the value of R_1 as $28^2/(2 \times 50) = 7.84\ \Omega$.

With an output capacitance of 200 pF, and an operating frequency of 50 MHz, Fig. 2.14 shows the reactance of C_{out} as:

$$\frac{1}{6.28 \times (50 \times 10^6) \times (200 \times 10^{-12})} = 16\ \Omega$$

The combination of these two values results in a *parallel* output impedance of $7.84 - j16$.

Usually, the datasheet gives the output capacitance in parallel form with R_1. For the network in Fig. 2.15*a,* the values of R_1 and C_{out} must be converted to series form.

Using the equations in Fig. 2.14, the equivalent series output impedance is:

$$R_{series} = \frac{7.84}{1 + (7.84/16)^2} = 6.32\ \Omega\ (R_1)$$

$$X_{series} = 6.32 \times \frac{7.84}{16} = 3.1\ \Omega\ (C_{out})$$

The combination of these two values results in a *series output impedance* of $6.32 - j3.1$.

Using the equations in Fig. 2.15 and assuming a Q of 10 (for simplicity), the reactance values for the network are:

$$X_L = 10 \times (6.32) + 3.1 = 66.6\ \Omega$$

$$A = \sqrt{\frac{6.32(1 + 10^2)}{50} - 1} = 3.3$$

$$X_{C2} = 3.3 \times 50 = 165\ \Omega$$

$$B = 6.32(1 \times 10^2) = 638.32$$

$$X_{C1} = \frac{638.32}{10 - 3.3} = 95\ \Omega$$

Using the equations in Fig. 2.15, the corresponding *inductance and capacitance* values are:

$$L_1 = \frac{66.6}{6.28 \times (50 \times 10^6)} = 0.21\ \mu\text{H}$$

$$C_1 = \frac{1}{6.28 \times (50 \times 10^6) \times (95)} = 33\ \text{pF}$$

$$C_2 = \frac{1}{6.28 \times (50 \times 10^6) \times (165)} = 19\ \text{pF}$$

If C_1 and C_2 are variable, the values obtained should be the midrange values.

2.3 RF-OSCILLATOR CIRCUITS FROM SCRATCH

This section provides a review and summary of RF-oscillator design from a very practical standpoint and is presented primarily for the serious experimenter or hobbyist, as is the material in Sec. 2.2.

2.3.1 Basic RF-Oscillator Design Considerations

The main concern in any oscillator design is that the transistor oscillates at the desired frequency and produces the desired voltage or power. Most oscillator circuits operate with power outputs of less than 1 W. Many transistors can handle this power dissipation without heat sinks (Secs. 2.4 and 2.5).

Another problem with oscillators is the class of operation. If an oscillator is biased class A (emitter-base forward-biased at all times), the output is free of distortion, but the circuit is not efficient. That is, the power output is low in relation to power input. Class A oscillators are typically 30 percent efficient. Thus, a 3-W input is required for a 1-W output.

For the purposes of calculation, input power for an oscillator can be considered as the product of collector voltage and collector direct current. The heat dissipation must be calculated on the basis of input power. For these reasons, class A oscillators are usually not used for RF and are generally limited to those applications where a good waveform is the prime consideration (such as audio amplifiers).

A class C oscillator (emitter-base reverse-biased except in the presence of the feedback signal) is far more efficient (usually about 70 percent). Thus a 1.5-W input produces a 1-W output. This cuts the heat-dissipation requirements in half. At radio frequencies, the waveform is usually not critical, so class C is in common use for RF oscillators.

One drawback to class C is that the oscillator may not start in the reverse-bias condition. This can be overcome by forward biasing the emitter-base to start the oscillator (start collector-current flow). The arrangement can be aided by an unbypassed-emitter resistor. Collector-current flow builds up a reverse bias across the emitter resistor.

A particular problem with this bias scheme is that too much reverse bias may cause the transistor to cut off during the on half-cycle. To maintain the correct bias relationships, a variable bias charge is obtained by rectifying part of the oscillator signal and filtering the bias, using a large-value capacitor. In practice, the base-emitter diode serves as the rectifier, with the base (or emitter) coupling (feedback) capacitor serving as the bias filter (to retain the correct bias charge during the on cycle).

Because the bias is variable (it changes with the amplitude of the oscillator signal), the capacitor charge must also change. If the capacitor is too small, the oscillator may not start easily, or there may be distortion. If the capacitor is too large, the charge changes slowly, and the oscillator operates intermittently as a blocking oscillator.

Many RF-oscillators require tuned resonant circuits. All of the design considerations in Sec. 2.2 apply to the resonant circuits for RF oscillators.

To sum up, if the selected transistor is capable of producing the required power at the operating frequency, and the correct component values are selected for the resonant circuits, the only major problem in oscillator design is the *correct bias point.* Often, this must be found by trial-and-error test of the circuit in experimental form.

2.3.2 LC and Crystal-Controlled Oscillator

Figure 2.16*a* and 2.16*b* shows two classic *LC* oscillators (the Hartley and Colpitts, respectively). *LC* oscillators are those that use inductances (coils) and capacitors as the frequency-determining components. Typically, the coils and capacitors are connected in series- or parallel-resonant circuits and adjusted to the desired operating frequency. Either the coil or capacitor can be variable.

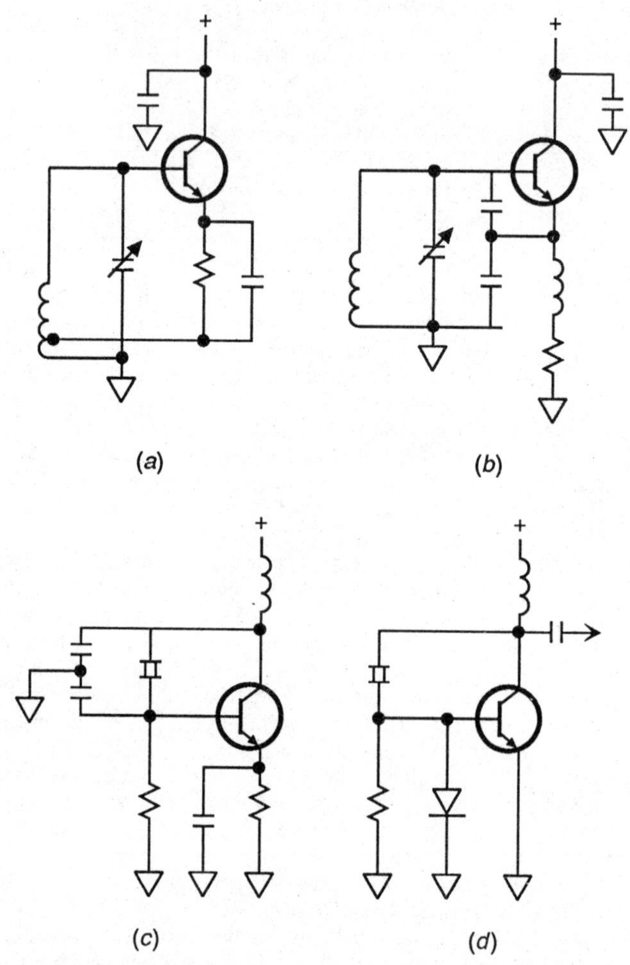

FIGURE 2.16 Classic *LC* RF oscillator circuits. (*a*) Hartley L_C; (*b*) Colpitts L_C; (*c*) Pierce; (*d*) Pierce.

The main problem with a basic *LC* oscillator is that the frequency can drift (with changes in temperature, power-supply voltage, mechanical vibration, etc.). This can be overcome by crystal control. That is, a quartz crystal can be used to set the frequency of operation, with an adjustable *LC* circuit to "trim" the oscillator output to an exact frequency.

Figure 2.17 shows a typical *LC* oscillator with crystal control. This circuit is one of the many variations of the Colpitts oscillator, where the output frequency is fixed and controlled by the crystal. The circuit can be tuned over a narrow RF range by slug-tuned L_1.

	RF power output at a percentage of dc power input			
		Harmonics		
Crystal frequency	Fund	Second	Third	Fourth
Fundamental	30	15	10	5
Third overtone	25	15	10	5
Fifth overtone	20	12	7	3
Seventh overtone	20	12	7	3

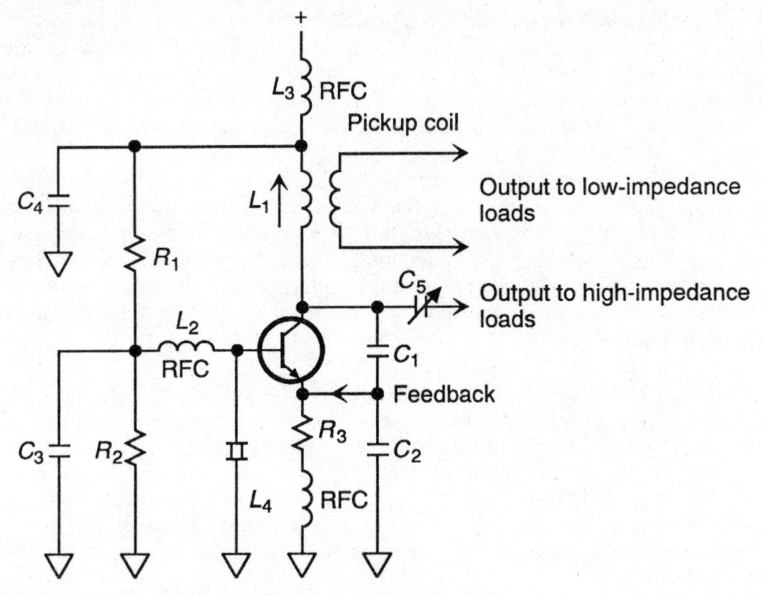

FIGURE 2.17 Typical *LC* oscillator with crystal control.

Figure 2.16*c* and 2.16*d* shows two classic versions of the Pierce oscillator. Such circuits are very popular in RF work because of their simplicity and the minimum number of components. No *LC* circuits are required for frequency control. Instead, the frequency is set by the crystal alone.

2.3.3 RF-Oscillator Design Considerations

Many factors must be considered in the design of RF oscillators. For example, the frequency-determining components must be temperature-stable, and mechanical movement of the individual components should not be possible. The following is a summary of the design considerations for an oscillator such as is shown in Fig. 2.17.

Frequency Stability. Many factors affect oscillator frequency stability. For example, there is usually some optimum bias value and power-supply voltage that produces maximum frequency stability over a given range of operating temperature. However, the one factor that can be controlled by the designer is percentage of feedback. (Note that this percentage refers to feedback versus output voltage.)

Feedback Percentage. The lowest practical feedback level is about 10 percent. The best feedback level is about 15 percent. Rarely is more than 25 to 30 percent ever required, although some oscillators are operated at 40 percent. If the operating frequency is in the VHF or UHF regions, the percentage of feedback must be increased over that of a comparable oscillator operating at low frequencies. Likewise, the percentage of feedback must be increased if the tuning circuits are made high-*C* through substantial increase in tuning capacitor value.

High-**C** ***versus Low-*****C.** The resonant frequency of RF oscillator circuits is usually set by the combination of L and C values. If the value of C is made quite large (with a corresponding lower value of L), the resonant circuit is said to be high-C and usually results in *sharper resonant tuning.* A large value of L (with corresponding lower value of C) introduces more resistance into the resonant circuit, thus lowering the circuit Q to produce *broader resonant tuning* (Sec. 2.1.1).

Crystal Frequency. The resonant circuit (C_1, C_2, L_1, and the transistor output capacitance in Fig. 2.17) should be at the same frequency as the crystal for maximum efficiency (maximum power output for a given power-supply voltage and current). If reduced efficiency is acceptable, the resonant circuit can be at a higher frequency (multiple) of the crystal frequency. However, the resonant circuit should not be operated at a frequency higher than the fourth harmonic of the crystal frequency.

Bias Circuit. The bias-circuit components (R_1, R_2, and R_3 in Fig. 2.17) are selected to produce a given current flow under no-signal conditions. The bias circuit is calculated and tested on the basis of *normal operating point,* even though the circuit is never at the operating point. A feedback signal is always present, and the transistor is always in a state of transition. The collector current should be set at a value to produce the required output power. With the correct bias-feedback relationship, the output power of the oscillator is about 0.3 times the input power.

Typically, the voltage drop across L_1 and L_3 is very small, so the collector voltage equals the power-supply voltage. Thus, to find a correct value of current for a given power output and power-supply voltage, divide the desired output by 0.3 to find the required input power. Then divide the input power by the power-supply voltage to find the collector-current flow.

Feedback Signal. The amount of feedback is set by the ratio of C_1 and C_2 (Fig. 2.17). For example, if C_1 and C_2 are of the same value, the feedback signal is one-half of the output signal. If C_2 is made about 3 times the value of C_1, the feedback signal is about 0.25 of the output signal voltage.

It may be necessary to change the value of C_1 in relation to C_2 for a good bias-feedback relationship. For example, if C_2 is decreased in value, the feedback increases, and the oscillator operates nearer the class C region. An increase in C_2, with C_1 fixed, decreases the feedback and makes the oscillator class A. Remember that any change in C_2 (or C_1) also affects frequency. Thus, if the C_2 and C_1 values are changed, it will probably be necessary to change the value of L_1.

As a first trial value, the amount of feedback should be equal to, or greater than, cutoff. That is, the feedback voltage should be equal to or greater than the voltage necessary to cut off collector current flow. Under normal conditions, such a level of feedback should be sufficient to overcome the fixed bias (set by R_1 and R_2) and the variable bias set by R_3. As discussed, feedback is generally within the limits of 10 and 40 percent, with the best stability in the 15 to 25 percent range.

Frequency. Frequency of the circuit in Fig. 2.17 is determined by the resonant frequency of L_1, C_1, and C_2 and by the crystal frequency. Note that C_1 and C_2 are in series, so the total capacitance is found by the conventional series equation: $C = (C_1 \times C_2)/(C_1 + C_2)$.

Also note that the output capacitance of the transistor must be added to the value of C_1. At low frequencies, the output capacitance can be ignored since the value is usually quite low in relation to a typical value for C_1. At higher frequencies, the value of C_1 is lower, so the output capacitance becomes of greater importance.

For example, if the output capacitance is 5 pF at the frequency of interest, and the value of C_1 is 1000 pF, or larger, the effect of the output capacitance is small. (Transistor output capacitance can be considered as being in parallel with C_1). If the value of C_1 is lowered to 5 pF, however, the parallel output capacitance doubles the value. Thus, the output capacitance must be included in the resonant-frequency calculations.

As discussed, transistor output capacitance is not always listed on datasheets. The capacitance presented by the output of a transistor (collector-to-emitter) is composed of both output capacitance and reverse capacitance. However, reverse capacitance is usually small in relation to output capacitance and can generally be ignored.

When output capacitance is not available on datasheets, it is possible to calculate an *approximate value* of output admittance (the *jb* part, such as jb_{22} represents susceptance, which is the reciprocal of reactance). Thus, to find the reactance presented by the collector-emitter terminals of the transistor at the datasheet frequency, divide the *jb* part into 1. Then find the capacitance that produces such reactance at the datasheet frequency using the equation $C = 1/(6.28\ F_{XC})$, where C is the output capacitance, F is the frequency and XC is the capacitive reactance found as the reciprocal of the *jb* part of the output admittance.

Of course, this method assumes that the *jb* reactance is capacitive, and that the capacitance remains constant at all frequencies (at least the datasheet capacity is the same for the design frequency). Neither of these conditions is always true.

Capacitor C_1 can be made variable. However, it is generally easier to make L_1 variable, since the tuning range of a crystal-controlled oscillator is quite small.

Typically, the value of C_2 is about 3 times the value of C_1 (or the combined values of C_1 and the transistor output capacitance, where applicable). Thus, the signal voltage (fed back to the emitter terminal) is about 0.25 of the total output signal voltage (or about 0.2 of the power-supply voltage, when the usual bias-feedback relationship is established).

Resonant Circuit. Any number of L and C combinations can be used to produce the desired frequency. That is, the coil can be made very large or very small, with corresponding capacitor values. Often, practical limitations are placed on the resonant circuit (such as available variable inductance values).

In the absence of some specific limitations, and as a starting point for resonant-circuit values, the capacitance should be 2 pF/m. For example, if the frequency is 30 MHz, the wavelength is 10 m, and the capacitance should be 20 pF. Wavelength in meters is found by the equation: wavelength (meters) = 300/ frequency (MHz).

At frequencies below about 1 to 5 MHz, the 2-pF/m guideline may result in very large coils to produce the corresponding inductance. If so, the 2 pF/m can be raised to 20 pF/m.

As an alternative method to find realistic values for the resonant circuit, use an inductive reactance value (for L_1) between 80 and 100 Ω at the operating frequency. This guideline is particularly useful at low frequencies (below 1 MHz).

Output Circuit. Output to the following stage can be taken from L_1 by means of a pickup coil (for low-impedance loads) or coupling capacitor (for high-impedance loads). Generally, the most convenient output scheme is to use a coupling capacitor (C_5) and make the capacitor variable. This makes it possible to couple the oscillator to a variable load (a load that changes impedance with changes in frequency).

Crystal. The crystal must, of course, be resonant at the desired operating frequency (or a submultiple thereof, when the circuit is used as a multiplier). Note that the efficiency (power output in relation to power input) of the oscillator is reduced when the oscillator is also used as a multiplier. This is shown in the table in Fig. 2.17, which also illustrates that efficiency is reduced when overtone crystals are used (instead of fundamental crystals).

The crystal must be capable of withstanding the combined dc and signal voltages at the transistor input (base). As a rule, the crystal should be capable of withstanding the full power-supply voltage, even though the crystal is never operated at this level.

Bypass and Coupling Capacitors. The values of bypass capacitors C_3 and C_4 should be such that the reactance is 5 Ω or less at the crystal operating frequency. A higher reactance (200 Ω) could be tolerated. However, because of the low output from most crystals, the lower reactance is preferred.

The value of C_5 should be about equal to the combined parallel output capacitance of the transistor and C_1. Make this the midrange value of C_5 (if C_5 is variable).

RFCs. The values of L_2, L_3, and L_4 should be such that the reactance is between 1000 and 3000 Ω at the operating frequency. The minimum current capacity of the chokes should be greater (by at least 10 percent) than the maximum anticipated direct current. Note that a high reactance is desired at the operating frequency. However, at high frequencies, this can result in very large chokes that produce a large voltage drop (or are too large physically).

Crystal-Oscillator Design Example. Assume that the circuit in Fig. 2.17 is to provide an output at 50 MHz. The circuit is to be tuned by L_1. A 30-V power supply is available. The crystal is not damaged by 30 V and operates at 50 MHz with the desired accuracy. The transistor has an output capacitance of 3 pF and operates without damage at 30 V. The desired output power is 40 to 50 mW.

The collector is operated at 30 V (ignoring the small drop across L_1 and L_3). The values of R_1, R_2, and R_3 should be chosen to provide a current that produces a 40- to 50-mW output with 30 V at the collector. A 45-mW output divided by 0.3 is 150 mW. Thus, the input power (and total dissipation) is 150 mW. Make certain that the transistor permits a 150-mW dissipation at the maximum anticipated temperature. Refer to Secs. 2.4 and 2.5 for temperature and power dissipation calculations.

Assume that the transistor has a 330-mW maximum power dissipation at 25°C, a maximum temperature rating of 175°C, and a 2-mW/°C derating for temperatures above 25°C. If the transistor is operated at 100°C, or 75° above the 25°C level, the transistor must be derated by 150 mW (75 × 2 mW/°C), or 330 − 150 mW = 180 mW. Under these conditions, the 150-mW input power dissipation is safe.

With 30 V at the collector and a desired 150-mW input power, the collector current must be 150 mW/30 V = 5 mA. Bias resistors should be selected to produce this 5-mA collector current using conventional methods. Then, with a 30-V power supply, the output signal is about 24 V (30 × 0.8 = 24). Of course, this depends on the bias-feedback relationship. Also, the collector current does not remain at 5 mA when the circuit is oscillating since the transistor is always in a state of transition.

As a starting point, make C_2 3 times the value of C_1 (plus the transistor output capacitance). With this ratio the feedback is about 25 percent of the output, or 6 V (24 × 0.25 = 6). Considering the amount of fixed and variable bias supplied by the bias network, a feedback of 6 V may be large. However, the 6-V value should be a good starting point.

For realistic values of L and C in the resonant circuit, let C_1 = 2 pF/m, or 12 pF (50 MHz = 6 m; 300/50 = 6). With C_1 at 12 pF, and the transistor output capacitance of 3 pF, the value of C_2 is 45 pF (12 + 3 = 15; 15 × 3 = 45).

The total capacitance across L_1 is: (15 × 45)/(15 + 45) = 13 pF. With a value of 13 pF across L_1, the value of L_1 for resonance at 50 MHz is: $L = (2.54 \times 10^4)/(50^2 \times 13) = 0.8$ μH. For convenience, L_1 should be tunable from about 0.5 to 1.6 μH.

Remember that an incorrect bias-feedback relationship causes distortion of the oscillator waveform or low power or both. The final test of the correct operating point is a good waveform at the operating frequency, together with frequency stability at the desired output power. Also, feedback is set by the relationship of C_1 and C_2. Any change in this relationship requires a corresponding change in the value of L_1. As a guideline, if no realistic combination of C_1, C_2, and L_1 produces the desired waveform and power output, *try a change in the bias*.

The values of C_3 and C_4 should be: 1/(6.28 × 50 MHz × 5) = 630 pF. A slightly larger value (say 1000 pF) assures a reactance of less than 5 at the operating frequency.

The values of L_2, L_3, and L_4 should be: 2000/(6.28 × 50 MHz) = 6.3 μH. Any value between about 3 and 9 μH should be good. The best test for the correct value of an RFC in an RF oscillator is to check for RF at the power-supply side of the line, with the oscillator operating.

As an example, check for RF at the point where L_3 connects to the power supply (not at the L_1 side). There should be no RF (or RF should be no greater than a few microvolts for a typical solid-state oscillator) on the power-supply side of the RFC. Next, check for dc voltage drop across the choke. The drop should be a fraction of 1 V (typically in the microvolt range).

2.4 TEMPERATURE-RELATED PROBLEMS

Three critical parameters for power transistors used in RF circuits are current gain, collector leakage, and power dissipation (in addition to output capacitance). These parameters change with temperature. To compound the problem, a change in parameters can also affect temperature (for example, an increase in current gain or power dissipation results in a temperature increase).

All of these problems can combine to produce *thermal runaway.* Heat is generated when current passes through a transistor junction. If all heat is not dissipated by the case (often an impossibility), the junction temperature rises. This, in turn, causes more current to flow through the junction even though the voltage, circuit values, and so on remain the same. With more current, the junction temperature increases even further, with a corresponding increase in current flow. The transistor burns out if the heat is not dissipated by some means.

When power dissipation is over about 1 W for a single transistor or circuit, *heat sinks* (or special component-mounting provisions) are used to offset thermal runaway. For example, if a transistor (or diode, rectifier, or IC) is used with a heat sink (or is mounted on a metal chassis that acts as a heat sink), an increase in temperature (from any cause) can be dissipated into the surrounding air. Heat sinks are discussed further in Sec. 2.5.

2.4.1 Effects of Temperature

Collector leakage (I_{CBO}) increases with temperature. As a guideline, collector leakage doubles with every 10°C increase in temperature for germanium transistors and doubles with every 15°C increase for silicon transistors. Also, always consider the possible effects of a different collector voltage when approximating collector leakage at temperatures other than those on the datasheet.

Current gain (h_{fe}) increases with temperature. As a guideline, current gain doubles when the temperature is raised from 25 to 100°C for germanium transistors and doubles when the temperature is raised from 25 to 175°C for silicon. If the datasheet does not specify a maximum operating temperature (or there is no datasheet) do not exceed 100°C for germanium or 200°C for silicon.

The power dissipation capabilities of a transistor are directly related to temperature and must be carefully considered when designing or servicing any RF circuit. For example, *do not apply power during service with the heat sinks removed* or you will quickly learn the effects of temperature on transistor power dissipation.

RF components (particularly power transistors) often have some form of *thermal resistance* specified to show the power-dissipation capabilities. Thermal resistance can be defined as the *increase in temperature of the component junction* (with respect to some reference) *divided by the power dissipated,* or degrees centigrade per watt (°C/W).

In RF power transistors, thermal resistance is normally measured from the junction to the case, resulting in the term θ_{JC}. When the case is bolted directly to the mounting surface with a built-in threaded bolt or stud, the terms θ_{MB} (thermal resistance to mounting base) or θ_{MF} (thermal resistance to mounting flange) are used. For RF power components where the junction is mounted directly on a header or pedestal, the total internal thermal resistance from junction to case (or mount) varies from about 50°C/W to less than 1°C/W.

Note that some RF transistor datasheets specify a *maximum case temperature* rather than θ_{JA}. As discussed in Sec. 2.4.3, maximum case temperature can be combined with heat-sink thermal resistance to find maximum power dissipation.

2.4.2 Practical Heat-Sink Considerations

Commercial heat sinks are available for various transistor case sizes and shapes (Sec. 2.5). Such heat sinks are especially useful when the transistors are mounted in sockets which provide no *thermal conduction* to the PC board. Commercial heat sinks are rated in terms of thermal resistance, usually in terms of degrees centigrade per watt.

When heat sinks involve the use of washers, the degree centigrade per watt factor usually includes the thermal resistance between the case and sink, or θ_{CA}. With a washer, only the sink-to-ambient, θ_{SA}, thermal-resistance factor is given. Either way, the thermal-resistance factor represents a temperature increase (in °C) divided by wattage dissipated.

For example, if the heat-sink temperature rises from 25 to 100°C (a 75°C increase) when 25W is dissipated, the thermal resistance is 75/25 = 3. This can be listed on a datasheet as θ_{SA}, or simply as 3°C/W.

All other factors (such as transistor size, mounting provisions, etc.) being equal, the heat sink with the lowest thermal resistance is best (a heat sink with 1°C/W is better than a 3°C/W heat sink).

To operate an RF circuit at full power, there should be no temperature difference between the case and ambient air. This occurs only when the thermal resistance of the heat sink is zero and the only thermal resistance is that between the junction and case. It is not practical to manufacture a heat sink with zero resistance. However, the greater the ratio θ_{JC}/θ_{CA}, the nearer the maximum power limit (set by θ_{JC}) can be approached.

When transistors are mounted on heat sinks, some form of *electrical insulation* is provided between the case and heat sinks (unless a grounded-collector circuit is used). Because good electrical insulators are (usually) good thermal insulators, it is difficult to provide electrical insulation without introducing some thermal resistance between case and heat sink.

The most common materials for electrical insulation of heat sinks are mica, beryllium oxide (Beryllia), and anodized aluminum. The properties of these three materials for the case-to-heat-sink insulation of a TO-3 case are compared as follows:

Material	Thickness (in)	°C/W	Capacitance (pF)
Beryllia	0.063	0.25	15
Anodized aluminum	0.016	0.35	110
Mica	0.002	0.40	90

As shown, any insulation between collector and the mounting surface (such as produced by a washer between the case and heat sink) also results in capacitance between the two metals. This capacitance can be a problem in RF design (Sec. 2.3). Generally, the *capacitance must be added to the output capacitance of the transistor.*

2.4.3 Calculating Power Dissipation

For practical design, the no-signal dc collector voltage and current can be used to calculate power dissipation when a transistor is operated under steady-state conditions (such as an RF amplifier). Other calculations must be used for pulse operating conditions. In theory, there are other currents that produce power dissipation (collector-base leakage current, emitter-base current, etc.). However, these can be ignored, and power dissipation (in watts) can be considered as the dc collector voltage times the collector current.

Once the power dissipation is calculated, the *maximum power dissipation* must be found. Under steady-state conditions, the maximum dissipation capability depends on (1) the sum of the series thermal resistances from the transistor junction to ambient air, (2) the maximum junction temperature, and (3) the ambient temperature.

Assume that it is desired to find the *maximum power dissipation* of a transistor under the following conditions: a maximum junction temperature of 200°C (typical for a silicon power transistor used in RF work), a junction-to-case thermal resistance of 2°C/W, a heat sink with a thermal resistance of 3°C/W, and an ambient temperature of 25°C.

First, find the total junction-to-ambient thermal resistance: 2°C/W + 3°C/W = 5°C/W.

Next, find the maximum permitted power dissipation: (200°C – 25°)/(5°C/W = 35 W (maximum)

If the same transistor is used without a heat sink, but under the same conditions and with a TO-3 case (which is rated at 30°C/W), the maximum power can be calculated as follows:

First, find the total junction-to-ambient thermal resistance: 2°C/W + 30°C/W = 32°C/W

Next, find the maximum permitted power dissipation: (200°C/W – 25°C)/32°C/W = 5 W (approximate)

Some RF power-transistor datasheets specify a *maximum case temperature* rather than a maximum junction temperature. Assume that a maximum case temperature of 130°C is specified instead of a maximum junction temperature of 200°C. In that event, subtract the ambient temperature from the maximum permitted case temperature: 130°C – 25°C = 105°C. Then divide the case temperature by the heat-sink thermal resistance: 105°C/3°C = 35 W maximum power.

2.5 HEAT-SINK AND COMPONENT MOUNTING TECHNIQUES

This section summarizes mounting techniques for metal-packaged power transistors found in RF circuits.

2.5.1 Interface Thermal Resistance

The interface thermal resistance is often referred to as the *case-to-sink resistance* and is sometimes listed as $R\theta_{CS}$. Table 2.1 shows the approximate interface thermal resistance (in °C/W) for a few RF transistor and diode cases. The table also shows recommended hole and drill sizes, as well as torque for the mounting nuts or screws. These values can be used when replacing RF components during service (in the absence of specific service-literature instructions).

As shown in Table 2.1, interface thermal resistance changes quite drastically for different mounting conditions. For example, assume that a T0-3 case is involved. If the case is mounted (on a heat sink or chassis) with an insulator and no thermal compound or lubrication is used, the interface thermal resistance is

TABLE 2.1 Interface Thermal Resistance Characteristics

Package type				Interface thermal resistance (°C/W)				
				Metal to metal		With insulator		
JEDEC outline number	Description	Recommended hole and drill size	Torque in-lb	Dry	Lubed	Dry	Lubed	Type
DO-4	10-32 stud 7/16 hex	0.118, no. 12	15	0.41	0.22	1.24	1.06	3-mil mica
DO-5	1/4-28 stud 11/16 hex	0.25, no. 1	30	0.38	0.20	0.89	0.70	5-mil mica
DO-21	Pressfit, 1/2	See Fig. 2.1*c*	—	0.15	0.10	—	—	—
TO-3	Diamond	0.14, no. 28	—	0.20	0.10	1.45	0.80	3-mil mica
						0.80	0.40	2-mil mica
						0.40	0.35	Anodized aluminum
TO-66	Diamond	0.14, no. 28	—	—	0.50	—	—	—
TO-83	1/2-20 stud	0.50	130	—	0.10	—	—	—

1.45°C/W. If a thermal compound is used, the resistance drops to 0.8°C/W. If circuit conditions make it impossible to eliminate the insulator, the thermal resistance drops to 0.1°C (with thermal compound) or 0.2°C (without thermal compound).

2.5.2 Fastening Techniques

The various types of transistor and diode packages shown in Table 2.1 require different fastening techniques. Mounting details for stud, flat-base, press-fit, and disk-type transistors are shown in Fig. 2.18. Of course, there are many other types of fastening techniques for transistors used in RF amplifiers. The following notes supplement the few examples shown here.

With any of the mounting schemes, the *screw heads must be free of grease* to prevent inconsistent torque readings when tightening nuts. Maximum allowable torque should always be used to reduce thermal resistance. However, *take care not to exceed the torque rating of parts.* Excessive torque applied to disk- or stud-mounted parts can cause damage to the semiconductor.

To prevent galvanic action from occurring when components are used with aluminum heat sinks in a corrosive atmosphere, many devices are nickel or gold plated. Take precautions not to mar the surface.

With press-fit components (Fig. 2.18*c*), the hole edge must be chamfered as shown to prevent shearing off the knurled edge of the component during press-in. The pressing force should be applied evenly on the shoulder ring to avoid tilting or canting the device in the hole during the pressing operation. Also, thermal compound (Sec. 2.5.4) should be used to ease the component into the hole.

With the disk-type mounting (Fig. 2.18*d*), a self-leveling type of mounting clamp is often used to keep the contacts parallel, with even distribution of pressure on each contact area. A swivel-type clamp or a narrow leaf spring in contact with the heat sink is often used.

When reinstalling the component, apply the clamping force smoothly, evenly, and perpendicularly to the disk-type package (to prevent deformation of the device or the sink-mounting surfaces). Use the correct clamping force as specified in the service literature.

2.5.3 Preparing the Heat-Sink Mounting Service

In general, the heat sink should have a flatness and finish comparable to that of the component. For the typical experimenter or hobbyist, the heat-sink surface is satisfactory if the surface *appears flat* against a straightedge and is free of any deep scratches. During the manufacture of commercial RF equipment, it may be necessary to measure the actual flatness with special tools and indicators.

Many commercial or off-the-shelf heat sinks require spot-facing. In general, milled or machined surfaces are satisfactory if prepared with tools in good working order.

The surface must be free from all dirt, film, and oxide (freshly bared aluminum forms an oxide layer in a few seconds). Unless used immediately after machining, it is a good practice to polish the mounting area with no. 000 steel wool followed by an acetone or alcohol rinse. Thermal grease should then be applied *immediately.* The same is true when *reinstalling heat sinks.*

Many aluminum heat sinks are *black anodized* for appearance, durability, performance, and economy. Anodizing is an electrical and thermal insulator that of-

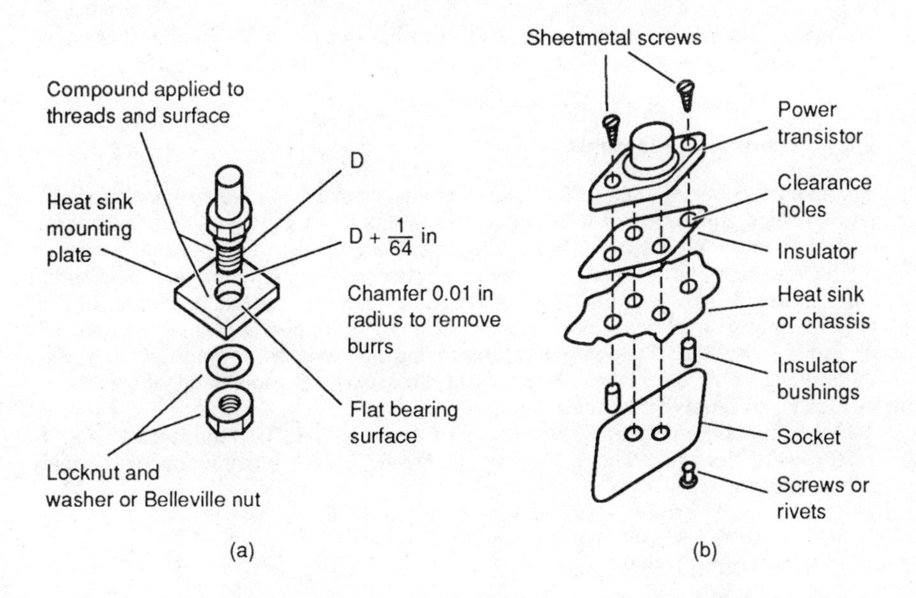

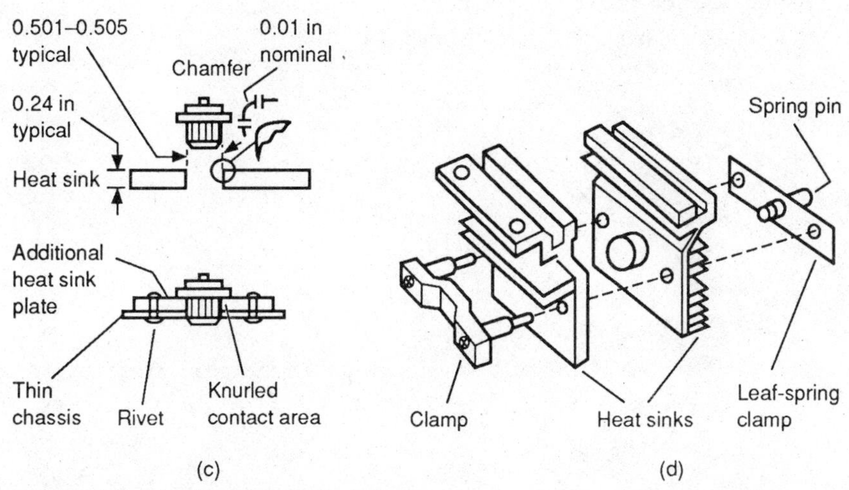

FIGURE 2.18 Transistor fastening techniques.

fers resistance to heat flow. As a result, anodizing should be removed from the mounting area.

Another aluminum finish is *iridite* (chromate acid dip), which offers low resistance because of the thin surface. For best results, the iridite finish must be cleaned of oils and films that collect in the manufacture and storage of the sinks.

Some heat sinks are *painted* after manufacture. Paint of any kind has a high thermal resistance (compared to metal). For that reason, it is essential that paint be removed from the heat-sink surface where the component is attached.

2.5.4 Thermal Compounds

Thermal compounds (also called *joint compounds* or *silicon grease*) are a formulation of fine zinc particles in a silicon oil that maintain a grease-like consistency with time and temperature. These compounds are used to fill air voids between mating surfaces and thus improve contact between the component and heat sink.

Compounds can be applied in a very thin layer with a spatula or lintless brush (wiping lightly to remove excess compound) or by applying a small amount of pressure to spread the compound (removing any excess compound after the mounting is complete). It may be necessary to use a cloth moistened with acetone or alcohol to remove all excess compound.

A typical compound has a resistivity of about 60°C-in/W, compared to about 1200°C-in/W for air. The following are some often recommended thermal compounds:

Dow Corning, Silicon Heat Sink Compound 340

General Electric, Insulgrease

Wakefield, Thermal Compound Type 1201

Astrodyne, Conductive Compound 829

Emerson & Cuming, Inc., Eccotherm TC-4

CHAPTER 3
RF TEST EQUIPMENT

This chapter is devoted to a review and summary of RF test equipment. Although the emphasis is on communications, a cross section of RF test equipment is discussed.

3.1 SIGNAL GENERATORS

The signal generator is an indispensable tool for RF circuit troubleshooting. Without a signal generator you depend entirely on signals transmitted by another piece of RF equipment (such as a communications set or receiver/transmitter), and you are limited to signal tracing only. This means that you have no control over frequency, amplitude, or modulation of the signals and have no means of signal injection.

With a signal generator of the appropriate type, you can duplicate transmitted signals or produce special signals required for alignment and test of all RF circuits. Also, the frequency, amplitude, and modulation characteristics of the signals can be controlled so that you can check operation of the receiver circuits under various signal conditions (weak, strong, normal, or abnormal signals).

3.1.1 Signal Generator Basics

An oscillator (audio, RF, pulse, etc.) is the simplest form of signal generator. At the most elementary level of troubleshooting, a single-stage AF or RF oscillator can provide a signal source. The special test sets described in Sec. 2.13 generally include such basic oscillator circuits. Beyond the simplest troubleshooting, most RF service requires an RF and AF generator and possibly a pulse generator. Another instrument that may be useful in RF work is the probe-type (or pencil-type) generator.

3.1.2 Probe, or Pencil, Generators

Probe generators (also known as pencil-type noise generators, signal injectors, and various other names) are essentially solid-state pulse generators or oscillators with a fast-rise waveform output and no adjustments. The fast-rise output produces simultaneous signals over a wide frequency range. The output signals may be used to troubleshoot the receiver and audio and modulation sections of communications sets, as well as the RF sections of AM/FM tuners. However, except in basic troubleshooting situations, such an instrument has many obvious drawbacks.

For example, to check the selectivity of receiver RF circuits, the signal source must be variable in amplitude (as described in Chap. 5). To check the detector or audio portions of receiver circuits, the signal source must be capable of internal and/or external modulation. These characteristics are not available in the pencil-type unit. As a result, even the least expensive shop-type (or even kit-type) generators have many advantages over the pencil generators.

3.1.3 RF Signal Generators

There are no basic differences between shop- and lab-type generators. That is, both instruments produce RF signals capable of being varied in frequency and amplitude and capable of internal and external modulation. However, the lab-type instruments have several refinements not found in shop equipment, as well as a number of quality features (this accounts for the wide difference in price). The following is a summary of the differences between shop and lab RF generators.

Output Meter. In most shop generators, the amplitude of the RF output is either unknown or approximated by means of dial markings. The lab generator has an output meter. The meter is usually calibrated in microvolts so that the actual RF output may be read directly. Without a built-in output meter, you must monitor the RF signal with an external meter. Remember that the meter must be capable of indicating output signals on the order of a few microvolts to make typical RF circuit checks. Meters are discussed in Sec. 3.3.

Percentage-of-Modulation Meters. Most shop generators have a fixed percentage of modulation (usually about 30 percent). Lab generators provide for a variable percentage of modulation and a meter to indicate this percentage. Some generators have two meters (one for output amplitude and one for modulation percentage). Some generators use the same meter for both functions.

Output Uniformity. Shop generators vary in output amplitude from band to band. Also, shop generators usually cover part of the frequency range with harmonics. Lab generators have a more uniform output over the entire operating range and cover the range with pure fundamental signals.

Wideband Modulation. Often, the oscillator of a shop generator is modulated directly. This can result in undesired frequency modulation. The oscillator of a lab generator is never modulated directly (unless designed to produce an FM output). Instead, the oscillator is fed to a wideband amplifier where the modulation is introduced. Thus, the oscillator is isolated from the modulating signal.

Frequency, or Tuning, Accuracy. The accuracy of the tuning dials for a typical shop generator is about 2 or 3 percent, whereas a lab generator has from about 0.5 to 1 percent accuracy. However, neither instrument can be used as a frequency standard for test or adjustment of RF equipment since the required accuracy is usually much greater. As an example, the FCC requires an accuracy of 0.005 percent or better (preferably 0.0025 percent) for CB equipment.

There are laboratory instruments generally described as *communications monitors,* or frequency meters and signal generators, that provide signals with accuracies of up to 0.00005 percent. These instruments are designed for commercial communications work (radio and TV broadcast, etc.), are quite expensive, and are thus not usually found in a typical service shop.

Combined Frequency Counter and Signal Generator. To overcome the accuracy problem in practical RF work, the simplest approach is to monitor the signal-generator output with a frequency counter. Such counters are discussed in Sec. 3.5. Using this technique, the frequency of the RF signal is determined by the *accuracy of the counter,* not the generator.

Frequency Range. Obviously, any RF generator used in communications work must be capable of producing signals at all frequencies used in the communications set. It is also convenient if the generator can produce signals at both harmonic and subharmonic frequencies.

Frequency Drift. Because a signal generator must provide continuous tuning across a given range, some type of VFO must be used. As a result, the output is subject to drift, instability, modulation (by noise, mechanical shock, or power-supply ripple), and other problems associated with VFOs. Frequency instability does not present too great a problem in practical RF work, provided that you monitor the signal-generator output with a frequency counter. Of course, continuous drift can be annoying. For this reason, lab generators have temperature-compensated capacitors, frequency synthesizers, and PLL circuits to minimize drift. Similarly, the effects of line-voltage variations are offset by regulated power supplies.

Shielding. The better generators have more elaborate shielding, especially for the output-attenuator circuits, where RF signals are most likely to leak. The leakage of RF from signal generators is something of a problem during receiver-circuit sensitivity tests (Chap. 5) or any tests involving low-amplitude (microvolt range) RF signals from the generator.

Band Spread. Shop generators usually have a minimum number of bands for a given frequency range. This makes the tuning-dial or frequency-control adjustments more critical and also more difficult to see. Lab generators usually have a much greater band spread. That is, precision generators cover a smaller part of the frequency range in each band.

3.1.4 Audio, or Function, Generators

Audio generators are useful in testing and troubleshooting the audio and modulation sections of RF circuits (such as the circuits of a communications set). Audio generators may also be used as modulation sources for RF signal generators. For example, if an RF generator has only a 400-Hz internal modulation provision (which is typical), and the test requires 1000 Hz (also typical), you can modulate the RF generator with an audio generator adjusted to 1000 Hz.

Early audio generators produced only sine waves. However, most present-day audio generators also produce square waves at audio frequencies. Some lab audio generators are referred to as function generators because they produce various functions: sine, square, triangular, and/or sawtooth waves. Generally, only the sine waves are of any particular value in RF work. However, almost any audio generator available today has some of the other outputs.

The major differences in audio generators are in quality rather than in special features. For example, the better audio generators are less subject to frequency drift and line-voltage variations. The effects of hum or other line noises are minimized by extensive filtering. Accuracy and dial resolution are generally better for lab generators. This makes the tuning-dial adjustments less critical. Lab generators also have a more uniform output over the entire operating range, whereas shop-type generators may vary in amplitude from band to band.

Remember that if you want accuracy from an audio generator, you must monitor the output signal with a meter (for signal amplitude) and a frequency counter (for signal frequency).

3.1.5 Pulse Generators

The most common use for a pulse generator in RF work is the test and adjustment of *receiver noise blanker circuits* (Chap. 5). Many communications receivers have blanking circuits that detect noise signals or pulses at the receiver input (antenna) and that function to desensitize (or cut off) the receiver in the presence of large noise signals. A pulse generator may be used to simulate noise bursts. However, a pulse generator is not always recommended. In many cases, the noise-blanking circuits are tested and adjusted with an RF generator and modulated by an audio generator.

3.1.6 FM Stereo Generator

The sole purpose of an FM stereo generator is to test and adjust an AM/FM tuner. For that reason, we discuss this special-purpose generator in the related chapter (Sec. 4.10).

3.1.7 Sweep Generator

The main purpose of a sweep generator is to test and adjust the RF and IF circuits of receivers (communications, TV, etc.). For that reason, we discuss this special-purpose generator in the related chapter (Sec. 4.9).

3.2 RF OSCILLOSCOPES

There are two uses for oscilloscopes (or scopes) in RF work: signal tracing and modulation measurement.

3.2.1 RF Signal Tracing with a Scope

RF signals may be traced with a scope if the scope is equipped with the proper probe (probes are discussed further in Sec. 3.4). You can check amplitude, frequency, and waveforms of the signals with a scope. However, many RF service technicians do not use scopes extensively, for the following reasons.

The scope can measure signal amplitude, but a meter is easier to read. The same applies to signal frequency. The frequency counter is easier to read, and it is far simpler to measure frequency with a counter than with a scope, particularly in the typical RF communications range. The scope is a superior instrument for monitoring waveforms. However, most communications signals are sine waves, and waveforms are not critical.

3.2.2 Modulation Checks with a Scope

There are many variations of the basic technique for modulation measurement in RF work. The following is a summary of the most commonly used techniques.

Direct Measurement with High-Frequency Scope. If the vertical channel response of the scope is capable of handling the RF circuit frequency (typically the output frequency of an RF transmitter), the signal can be applied directly through the scope vertical amplifier. The basic test connections are shown in Fig. 3.1. The procedure is as follows:

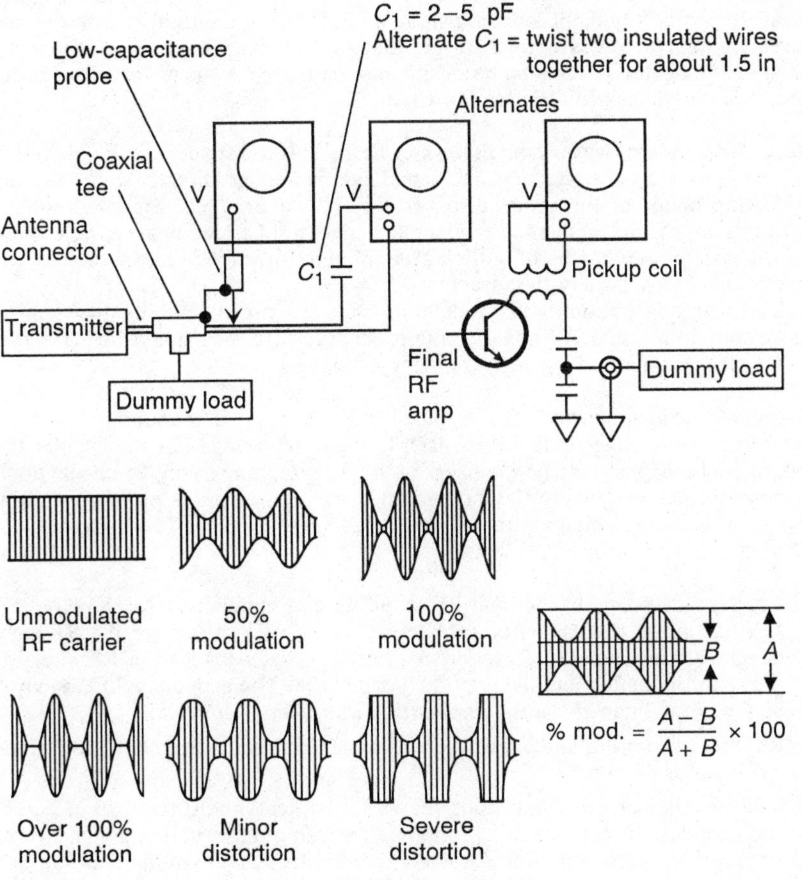

FIGURE 3.1 Direct measurement of RF-circuit modulation.

1. Connect the scope to the antenna jack or the final RF amplifier of the transmitter, as shown in Fig. 3.1. Use one of the three alternatives shown or the modulation measurement described in the service literature.
2. Key the transmitter (press the push-to-talk switch) and adjust the scope controls to produce displays as shown. You can either speak into the microphone (for a rough check of modulation), or you can introduce an audio signal (typically 400 or 1000 Hz) at the microphone jack input (for a precise check of modulation). Note that Fig. 3.1 provides simulations of typical scope displays during modulation tests.
3. Measure the vertical dimensions shown as A and B in Fig. 3.1 (the crest amplitude and the trough amplitude). Calculate the percentage of modulation using the equation in Fig. 3.1. For example, if the crest amplitude (A) is 63 (63 screen divisions, 6.3 V, and so on) and the trough amplitude (B) is 27, the percentage of modulation is

$$\frac{63 - 27}{63 + 27} \times 100 = 40\%$$

Make certain to use the same scope scale for both crest (A) and trough (B) measurements. Remember when making modulation measurements, or any measurement that involves a transmitter, that the RF output (antenna connector) must be connected to an antenna or dummy load. Dummy loads are discussed in Sec. 3.6. Antennas are discussed in Chap. 7.

Direct Measurement with a Low-Frequency Scope. If the scope amplifier is not capable of passing the frequency, the signal can be applied directly to the vertical-deflection plates of the scope display tube. However, there are two problems with this approach. First, the vertical plates may not be readily accessible. Next, the RF signal may not be of sufficient amplitude to produce measurable deflection of the scope display-tube trace.

The test connections and modulation patterns are essentially the same as those shown in Fig. 3.1. Similarly, the procedures are the same as described for direct measurement with a high-frequency scope.

Trapezoidal Measurement. The trapezoidal technique of modulation measurement has an advantage in that it is easier to measure straight-line dimensions than curving dimensions. Any *nonlinearity in modulation* may easily be checked with a trapezoid pattern, where the modulated carrier amplitude is plotted as a function of modulating voltage rather than as a function of time. The basic test connections are shown in Fig. 3.2.

1. Connect the scope to the final RF amplifier and modulator. As shown in Fig. 3.2, use either the capacitor connection or the pickup coil for the RF signal (scope vertical input). However, for best results, connect the RF signal directly to the deflection plates of the scope tube. The scope amplifiers may be nonlinear and can cause the *modulation to appear distorted.*
2. Key the transmitter and adjust the controls (scope controls and R_1) to produce a display as shown.
3. Measure the vertical dimensions shown as A (crest) and B (trough) in Fig. 3.2 and calculate the percentage of modulation using the equation given. For example, if the crest amplitude A is 80, and the trough amplitude B is 40, using the same scale, the percentage of modulation is

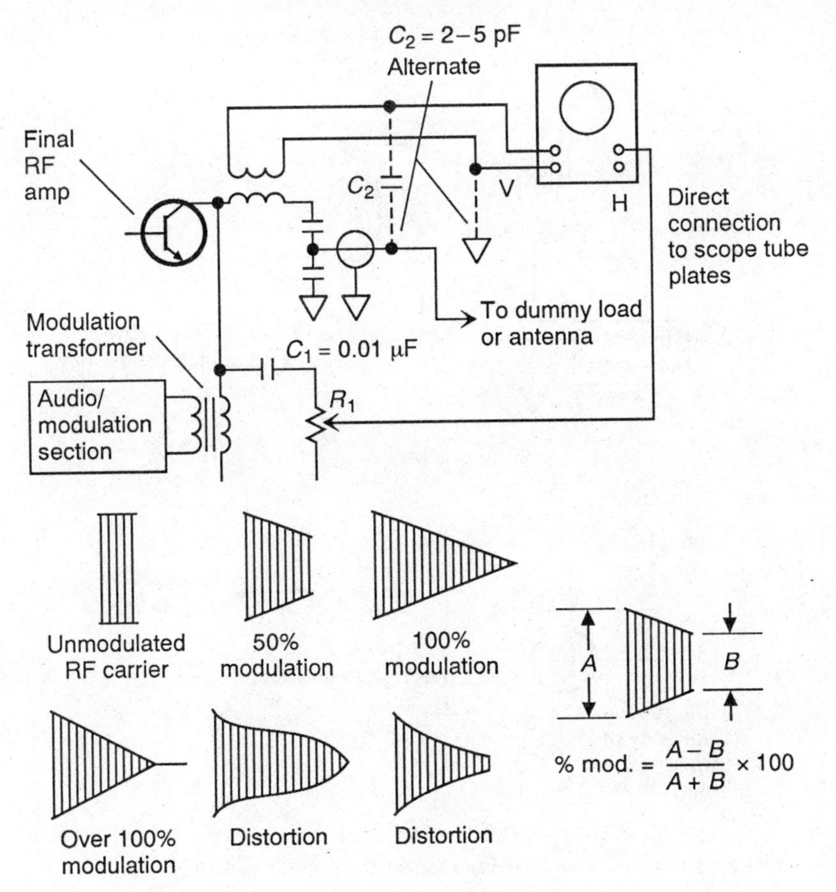

FIGURE 3.2 Trapezoidal measurement of RF-circuit modulation.

$$\frac{80 - 40}{80 + 40} \times 100 = 33\%$$

Again, make sure that the transmitter output is connected to an antenna or dummy load before transmitting.

Down-Conversion Measurement. If the scope is not capable of passing the RF signal, and the signal output is not sufficient to produce a good indication when connected directly to the scope tube, it is possible to use a down-converter. One method requires an external RF generator and an IF transformer. The other uses a receiver capable of monitoring the RF signal frequencies.

The RF-generator method of down-conversion is shown in Fig. 3.3*a*. In this method, the RF generator is tuned to a frequency above or below the transmitter frequency by an amount equal to the transformer frequency. For example, if the transformer is 455 kHz, tune the RF generator to a frequency 455 kHz above (or below) the transmitter frequency.

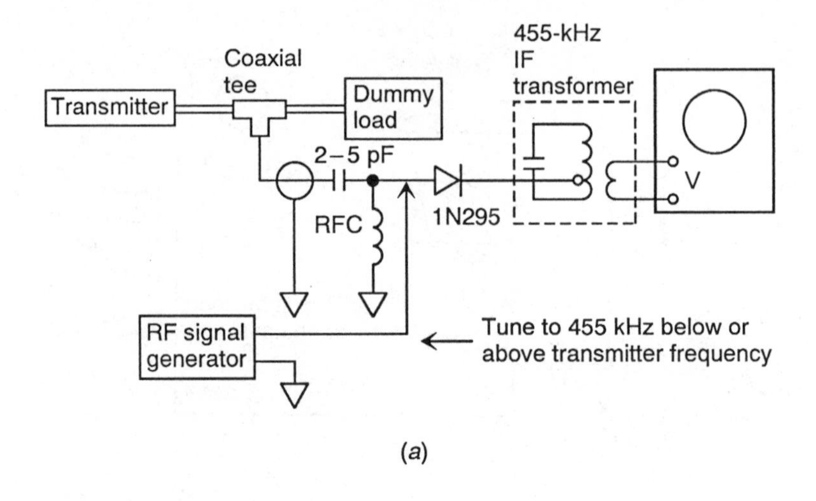

(*a*)

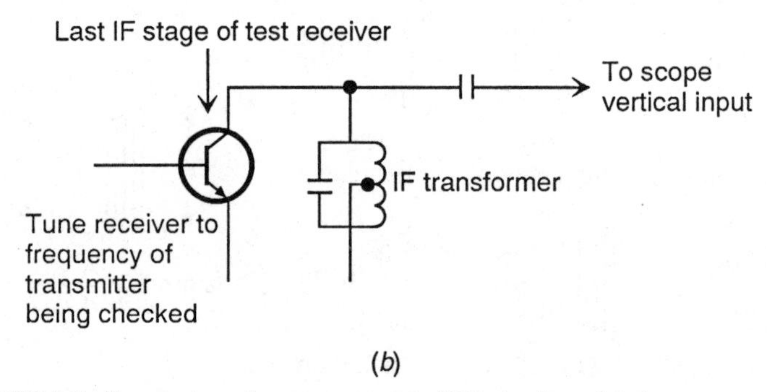

(*b*)

FIGURE 3.3 Down-conversion measurement of RF-circuit modulation.

The receiver method of down-conversion is shown in Fig. 3.3*b*. With this method, the receiver is tuned to the transmitter frequency, and the scope input signal is taken from the last IF-stage output through a 30-pF capacitor.

With either method of down-conversion measurement, the RF generator or receiver is tuned for a maximum indication on the scope screen. Once a good pattern is obtained, the rest of the procedure is the same as described for direct measurement.

The author does not generally recommend the down-conversion methods, except as temporary measures. There are a number of relatively inexpensive scopes available that pass signals up to and well beyond 50 MHz.

Linear-Detector Measurement. If you must use a scope that cannot pass the RF-signal frequency, you can use a linear detector (provided that the scope has a *dc input* where the signal is fed directly to the scope vertical amplifier, not through a capacitor). Most modern scopes have both ac (with capacitor) and dc (no capacitor) inputs. The basic test connections for linear detection of the modulation envelope are shown in Fig. 3.4. The basic test procedure is as follows:

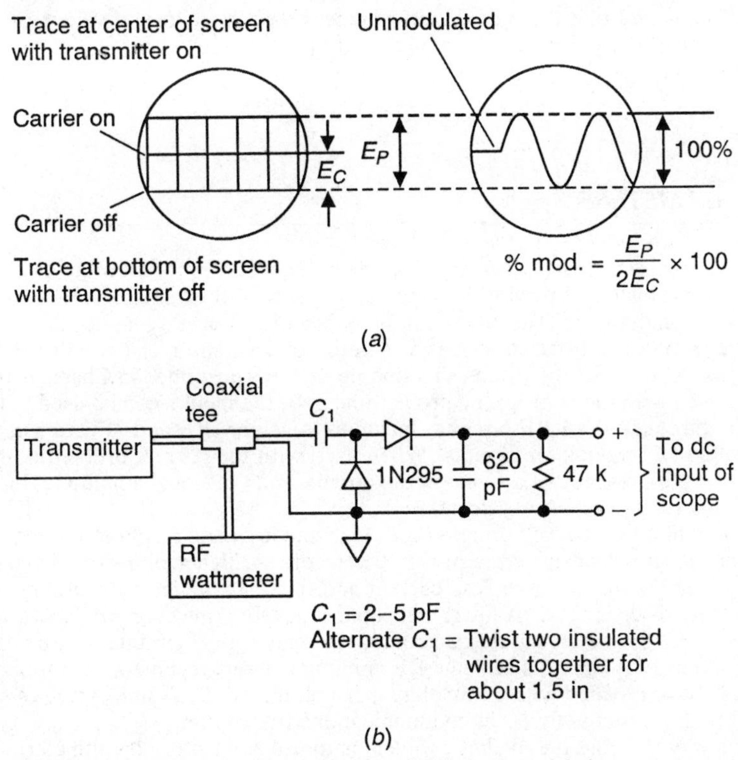

FIGURE 3.4 Linear-detector measurement of RF-circuit modulation.

1. Connect the transmitter output to the scope through the linear-detector circuit, as shown in Fig. 3.4. Make certain to include the dummy load (or wattmeter) as shown.
2. With the transmitter not keyed, adjust the scope position control to place the trace on a reference line near the bottom of the screen, as shown in Fig. 3.4*b* (carrier off).
3. Key the transmitter but do not apply modulation. Adjust the scope gain control to place the top of the trace at the center of the screen, as shown in Fig. 3.4*b* (carrier on). It may be necessary to switch the transmitter off and on several times to adjust the trace properly since the position and gain controls of most scopes interact.
4. Measure the distance (in scale divisions) of the shift between the carrier-on (step 3) and carrier-off (step 2) traces. For example, if the screen has a total of 10 vertical divisions, and the no-carrier trace is at the bottom or zero line, there is a shift of 5 scale divisions to the centerline.
5. Key the transmitter and apply modulation. Do not touch either the position or gain controls of the scope.
6. Find the percentage of modulation using the equation shown in Fig. 3.4. For example, assume that the shift between the carrier and no-carrier trace is five

divisions and that the modulation produces a peak-to-peak envelope of eight divisions. The percentage of modulation is

$$\frac{8}{2 \times 5} \times 100 = 80\%$$

3.3 RF METERS

The meters used for RF work are essentially the same as for all other electronic fields. Most tests and troubleshooting can be done with standard volt-ohmmeter (VOM) or multimeter. The VOM can be either digital or moving-needle.

The meters can be used to measure both voltages and resistances of RF circuits, as required for the troubleshooting procedures described in Chaps. 6 and 7. When used with the appropriate probe (Sec. 3.4), the meters can be used to trace signals throughout all RF circuits, including receiver (RF and IF), transmitter, and audio and modulator circuits. When used with the correct probe, the meter indicates the presence of a signal in the circuit, and the signal amplitude, but not the signal frequency or waveform.

In addition to accuracy, ranges (both high and low) and resolution or readability, meters are rated in terms of ohms per volt. A higher ohms-per-volt rating means that the meter draws less current and thus has the least disturbing effect on the circuit under test. A lower ohms-per-volt rating means more *circuit loading* (Sec. 3.4.8), which should be avoided in some critical circuits. For example, the AVC or AGC circuits of some RF-communications receivers do not operate properly when loaded with a low ohms-per-volt meter. The same is true of some RF oscillator circuits found in communications transmitters.

One way to avoid the loading problem is to use an electronic voltmeter with a high input impedance (which thus draws very little current from the RF circuit under test). The electronic voltmeter can be a vacuum-tube voltmeter (VTVM, yes they still exist), electronic voltmeter (EVM), transistorized voltmeter (TVM), or some similar instrument. Most digital meters are electronic meters and thus draw a minimum of current from the circuit.

One minor problem with some older meters is that the frequency range is not sufficient to cover the entire audio range (up to about 20 kHz). The problem may be overcome with a probe. Also, the range of the audio and modulation circuits of a typical RF communications set is about 3 kHz maximum.

3.4 RF PROBES

In practical RF work, all meters and scopes operate with some type of probe. In addition to providing for electrical contact to the circuit being tested, probes serve to modify the voltage or signal being measured to some condition suitable for display on a scope or readout on a meter.

3.4.1 Basic Probe

In the simplest form, the basic probe is a *test prod.* Such probes work well on circuits carrying dc and audio signals. However, for an RF signal (even at the low end of the RF range), it may be necessary to use a special *low-capacitance* probe. The same is true if the meter or scope gain is high. Hand capacitance can cause hum pickup when simple probes are used for RF. This condition may be offset by shielding in low-capacitance probes. More important, use of a low-capacitance probe prevents meter or scope impedance from being connected directly to the circuit being tested. (Such impedance may disturb circuit conditions.)

3.4.2 Low-Capacitance Probes

Figure 3.5*a* shows the basic circuit of a low-capacitance probe. (You can make up such a probe. However, it is far more practical to use the probe designed for the particular meter or scope.) The series resistance R_1 and capacitance C_1, as well as the parallel or shunt R_2, are surrounded by a shielded handle. The values of R_1 and C_1 are preset at the factory and *should not* be disturbed unless recalibration is required (Sec. 3.4.7).

In many low-capacitance probes, the values of R_1 and R_2 are selected to form a 10:1 *voltage divider* between the circuit being tested and the meter or scope input. Such probes serve the dual purpose of capacitance and voltage reduction. Remember that voltage indications are one-tenth (or whatever value of attenuation is used) of the actual value when a voltage-divider probe is used. The capacitance value of C_1 in combination with the values of R_1 and R_2 also provide a capacitance reduction, usually in the range 3:1 to 11:1.

3.4.3 High-Voltage Probes

High-voltage probes are not generally needed for RF work unless the equipment has a display tube. An obvious exception is in the high-voltage circuits of some high-power transmitters (such as broadcast transmitters). A typical high-voltage probe provides a voltage division of 100:1 or 1000:1.

3.4.4 RF Probes (Supplied with Meter or Scope)

An RF probe is required when the signals to be measured are at radio frequencies and are beyond the capabilities of the meter or scope. Again, always use the probe supplied with the meter or scope. An RF-probe circuit is discussed in Sec. 3.4.9 and shown in Fig. 3.5*b*.

3.4.5 Demodulator Probes

The circuit of a demodulator probe is essentially the same as that of an RF probe, but the circuit values and basic functions are somewhat different. When the high-frequency signals contain modulation (which is typical for the modulated RF-carrier signals of most communications equipment), a demodulator probe is more effective for signal tracing.

Figure 3.5*c* shows a typical demodulator-probe circuit. Note that the circuit is essentially a half-wave rectifier. However, C_1 and R_2 act as a filter. The demodulator probe produces *both an ac and a dc output.* The RF signal is converted

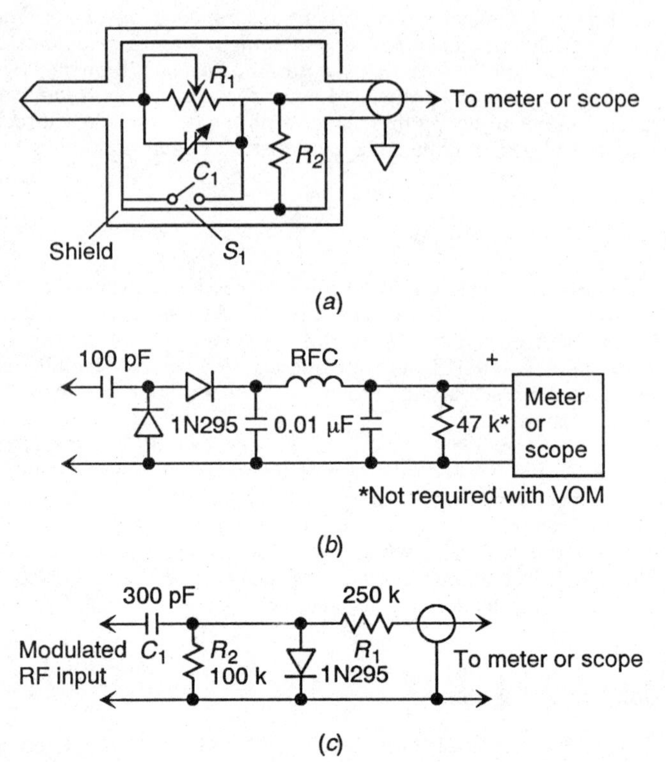

FIGURE 3.5 RF-probe circuits.

into a dc voltage approximately equal to the peak value. The modulation voltage on the RF signals appears as ac at the probe output.

In use, the meter is set to dc and the RF signal is measured. Then the meter is set to ac and the modulating voltage is measured. The calibrating resistor R_1 is adjusted so that the dc scale (of the meter) reads the correct value. If there is no demodulator probe available for a particular meter, you can use the circuit in Fig. 3.5*c* on most meters (for troubleshooting and testing but not for precise voltage measurement). The following steps describe the calibration and fabrication procedure:

1. Connect the probe circuit to a signal generator and meter.
2. Set the meter to measure dc voltage. The meter should be high impedance for best results.
3. Adjust the signal generator voltage amplitude to some precise value, such as 10 V, as measured on the generator output meter.
4. Adjust the calibrating resistor R_1 until the meter indicates the same value (10 V).
5. As an alternative procedure, adjust the signal generator for a 10-V peak output; then adjust R_1 for a reading of 7.07 on the meter readout.

6. Remove the power, disconnect the circuit, measure the value of R_1, and replace the variable resistor with a fixed resistor of the same value.
7. Repeat the test with the fixed resistance in place. If the reading is correct, mount the circuit in a suitable package (such as within a test prod). Repeat the test with the circuit in final form. Also repeat the test over the entire frequency range of the probe. Generally, the probe in Fig. 3.5*c* provides satisfactory response up to about 250 MHz.
8. Remember that the meter must be set to measure direct current since the probe output is dc when there is no modulation.

3.4.6 Solid-State Signal-Tracing Probe

It is possible to increase the sensitivity of a meter or scope with an amplifier. Such amplifiers are particularly useful with a VOM for measuring small-signal voltages during testing or troubleshooting. An amplifier is usually not required for an electronic meter or scope because such instruments contain built-in amplifiers.

Figure 3.5*d* shows a typical probe and amplifier circuit. Such an arrangement increases sensitivity by at least 10:1 and provides good response up to about 500 MHz. The circuit is not normally calibrated to provide a specific voltage indication. Rather, the circuit is used to increase the sensitivity for signal tracing in RF circuits.

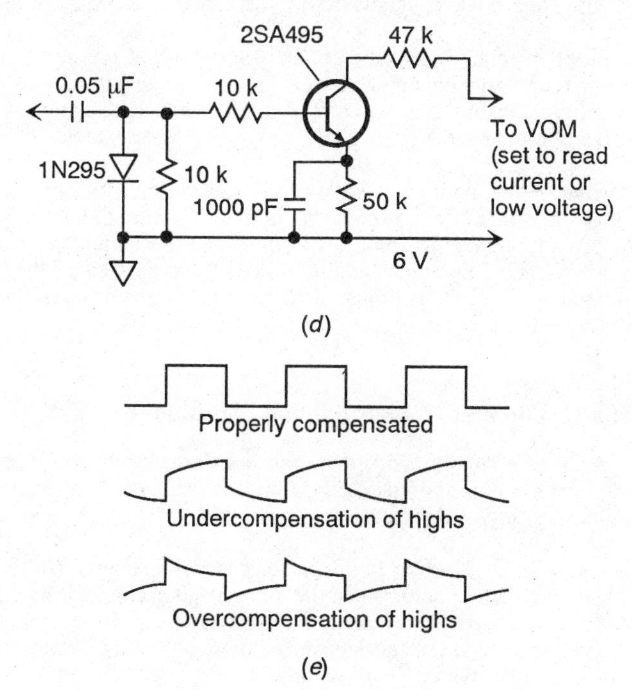

FIGURE 3.5 (*Continued*) RF-probe circuits.

3.4.7 Probe Compensation and Calibration

Probes must be calibrated to provide a proper output to the meter or scope with which they are used. Probe compensation and calibration are best done at the factory and require precision test equipment. The following paragraphs describe the *general procedures* for compensation and calibrating probes. Never attempt to adjust a probe unless you follow the instruction manual and have the proper test equipment. An improperly adjusted probe produces erroneous readings and may cause undesired circuit loading.

Probe Compensation. The capacitors that compensate for excessive attenuation of high-frequency signal components (through the probe resistance dividers) affect the entire frequency range from some midband point upward. Capacitor C_1 in Fig. 3.5*a* is an example of such a compensating capacitor.

Compensating capacitors must be adjusted so that the higher-frequency components are attenuated by the same amount as are low frequency and direct current. It is possible to check the adjustment of the probe-compensating capacitors using a square wave signal source. This is done by applying the square wave signal directly to the scope input and then applying the same signals through the probe and noting any change in pattern. In a properly compensated probe, there should be no change (except for a possible reduction of the amplitude).

Figure 3.5*e* shows typical square wave displays with the probe properly compensated, undercompensated (high frequencies underemphasized), and overcompensated (high frequencies overemphasized). Proper compensation of probes is often neglected, especially when probes are used interchangeably with meters or scopes having different input characteristics. It is recommended that any probe be checked with square wave signals before the probe is used in testing and troubleshooting.

Another problem related to probe compensation is that the input capacitance of the meter or scope may change with age. (This is not generally a problem with present-day equipment but can occur on older vacuum-tube meters and scopes when the tubes are changed.)

Probe Calibration. The main purpose of probe calibration is to provide a specific output for a given input. For example, the value of R_1 in Fig. 3.5*c* is adjusted (or selected) to provide a specific amount of voltage to the meter or scope. During calibration, a voltage of known value and accuracy is applied to the input. The output is monitored, and R_1 is adjusted to produce a given value (0.707 of RF peak value, for example).

3.4.8 Probe Testing and Troubleshooting Techniques

Although a probe is a simple instrument and does not require specific operating procedures, several points should be considered to use a probe effectively in testing and troubleshooting of RF circuits.

Circuit Loading. When a probe is used, the probe impedance (rather than the meter or scope impedance) determines the amount of circuit loading. Connecting a meter or scope to a circuit may alter the signal at the point of connection. To prevent this, the impedance of the measuring device must be large in relation to that of the circuit being tested. Thus, a high-impedance probe offers less circuit loading, even though the meter or scope may have a lower impedance.

Measurement Error. The ratio of the two impedances (of the probe and the circuit being tested) represents the amount of probable error. For example, a ratio of 100:1 (perhaps a 100-MΩ probe used to measure the voltage across a 1-MΩ circuit) accounts for an error of about 1 percent. A ratio of 10:1 produces an error of about 9 percent.

Effects of Frequency. The input impedance of a probe is not the same at all frequencies. Input impedance becomes smaller as frequency increases. (Capacitive reactance and impedance decrease with an increase in frequency.) All probes have some input capacitance. Even an increase at audio frequencies may produce a significant change in impedance.

Shielding Capacitance. When using a shielded cable with a probe to minimize pickup of stray signals and hum, the additional capacitance of the cable should be considered. The capacitance effects of a shielded cable can be minimized by terminating the cable at one end in the cable's characteristic impedance. Unfortunately, this is not always possible with the input circuits of most meters and scopes.

Relationship of Loading and Attenuation Factor. The reduction of loading (either capacitive or resistive) caused by a probe may not be the same as the attenuation factor of the probe. (Capacitive loading is almost never reduced by the same amount as the attenuation factor because of the additional capacitance of the probe cable.) For example, a typical 5:1 attenuator probe may reduce capacitive loading by 2:1.

Checking Effects of a Probe. When testing any circuit (but particularly an RF circuit), it is possible to check the effect of a probe on the circuit with the following test. Attach and detach another connection of similar kind (such as another probe) and observe any difference in meter reading or scope display. If there is little or no change when the additional probe is touched to the circuit, it is safe to assume that the probe has little effect on the circuit.

Probe Length and Connections. Long probes should be restricted to the measurement of dc and low-frequency ac. The same is true for long ground leads. The ground lead should be connected where no hum or high-frequency signal components exist in the ground path between the point and the signal-pickoff point. Keep all test connections as short as possible for RF work.

Measuring High Voltages. Avoid applying more than the rated voltage to a probe. Fortunately, most commercial probes will handle the highest voltages found in RF work (with the possible exception of high-power transmitters).

3.4.9 An RF Probe for Testing and Troubleshooting

Figure 3.5*b* shows the circuit diagram of a probe that is suitable for testing and troubleshooting of most RF circuits. The probe is designed specifically for use with a VOM or digital meter and converts both audio-frequency and RF signals to direct current. The probe is full-wave and thus produces a larger signal than the half-wave demodulator probe in Fig. 3.5*c*.

The probe in Fig. 3.5*b* operates satisfactorily up to about 250 MHz; it is essentially a signal-tracing device but does not provide accurate voltage readings. The meter used with the probe must be set to read direct current because the probe output is dc. However, if the RF-input signal is amplitude-modulated, the probe output is pulsating direct current.

3.5 FREQUENCY METERS AND COUNTERS

There are two basic types of frequency-measuring devices for RF work: the heterodyne, or zero-beat, frequency meter and the digital electronic counter.

3.5.1 Heterodyne, or Zero-Beat, Frequency Meter

In the early days of radio communications, the heterodyne meter (Fig. 3.6*a*) was the only practical device for frequency measurement of transmitter signals. The signals to be measured are applied to a mixer, together with the signals of a known frequency (usually from a variable-frequency oscillator in the meter). The meter oscillator is adjusted until there is a null or "zero beat" on the output device, indicating that the oscillator is at the same frequency as the signals to be measured. The frequency is read from the oscillator frequency-control dial. Precision frequency meters often include charts or graphs to help interpret frequency-dial readings so that exact frequencies can be pinpointed.

3.5.2 Electronic Digital Counter

The electronic counter has all but replaced the heterodyne meter in today's RF work. One reason is that the counter is generally easier to operate and has much greater resolution, or readability. Using the counter, you need only connect the test leads to the circuit or test point, select a time base and attenuator and multiplier range, and read the signal frequency on a convenient digital readout.

Digital-Counter Basics. Although there are many types of digital counters, all counters have several basic functional sections in common. These sections are interconnected in a variety of ways to perform the various counter functions. Figure 3.6*b* shows the basic counter circuit for *frequency measurement* (which is the most common use of a counter in RF work). A typical digital counter also provides a *total operation* (where the instrument adds up events), a *period operation* (where the instrument measures intervals up to a given time), and a *time-interval* operation (where the instrument measures time between two events). However, frequency measurement is the prime function of a counter in RF work.

All counters have some form of *main gate* that controls the count-start and count-stop with respect to time. Counters also have some form of crystal-controlled *time base* that supplies the precise increment of time to control the gate for a frequency measurement. The accuracy of the counter depends on the accuracy of the time base, plus or minus one count. For example, if the time-base accuracy is 0.005 percent, the overall accuracy of the counter is 0.005 percent plus or minus one count. The one-count error occurs because the count may start and stop in the middle of an input pulse, thus omitting the pulse from the total count. Likewise, part of the pulse may pass through the gate before the gate closes, thus adding a pulse to the count.

Most electronic counters have *dividers* that permit variation of gate time.

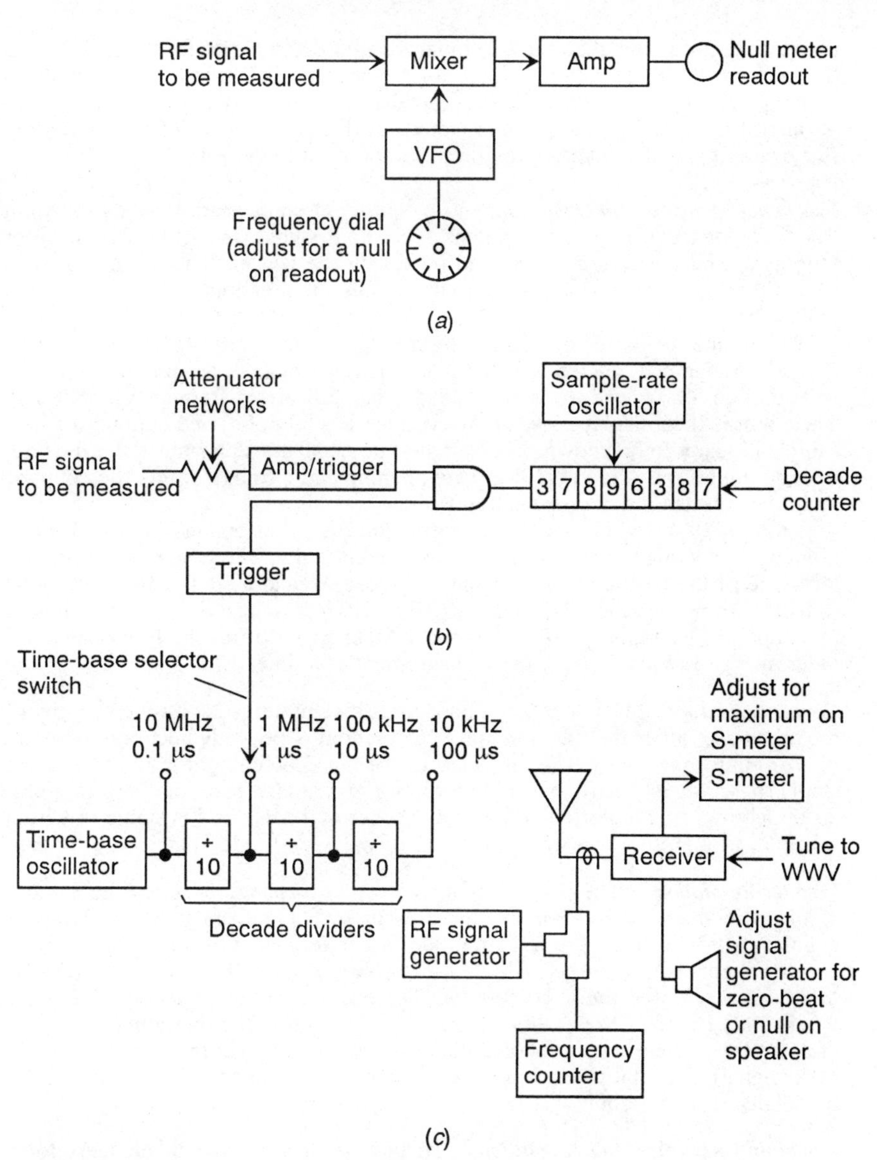

FIGURE 3.6 Frequency meter and counter circuits.

These dividers convert the fixed-frequency time base to several other frequencies. In addition to the basic sections, most counters have *attenuator networks, amplifier* and *trigger* circuits (to shape a variety of input signals to the common form), and *logic* circuits to control operation of the instrument.

Electronic counters have some form of *counter and readout.* Early instruments used binary counters and readout tubes that converted the binary count to a decade readout. Such instruments have long since been replaced by decade counters that convert the count to binary-coded decimal (BCD) form, decoders that convert the BCD data to decade form (generally BCD-to-seven-segment decoders), and readouts that display the decade information (generally seven-segment LCDs, LEDs, etc.). One readout, or display, is provided for each digit. For example, eight readouts provide for a count up to 99,999,999.

Frequency Measurement Operation. The counter circuit are arranged as shown in Fig. 3.6*b* for frequency measurement. The input signal (say from a transmitter output) is first converted to uniform pulses by the trigger. The pulses are then routed through the main gate and into the counter and readout circuits, where the pulses are totaled.

The number of pulses totaled during the "gate-open" interval is a measure of the average input frequency for that interval. For example, assume that the gate is held open for 1 s and the count is 888. This indicates a frequency of 888 Hz. The count is then displayed (with the correct decimal point) and held until a new sample is ready to be shown. The sample-rate oscillator determines the time between samples (not the interval of gate opening and closing), resets the counter, and thus initiates the next measurement cycle.

The time-base switch selects the gating interval, thus positioning the decimal point and selecting the appropriate measurement units. The time-base switch selects one of the frequencies from the time-base oscillator. If the 10-MHz signal (directly from the time-base shown in Fig. 3.6*b*) is selected, the time interval (gate-open to gate-close) is 0.1 s. If the 1-MHz signal (from the first decade divider in Fig. 3.6*b*) is chosen, the measurement time interval is 1 s.

Counter Accuracy. The accuracy of a frequency counter is set by the stability of the time base rather than the readout. The readout is typically accurate to within plus or minus one count. The time base of the Fig. 3.6*b* counter is 10 MHz and is stable to within ±10 parts per million (ppm), or 100 Hz. The time base of a precision laboratory counter could be on the order of 4 MHz and is stable to within ±1 ppm, or 4 Hz.

Counter Resolution. The resolution of an electronic counter is set by the number of digits in the readout. For example, assume that you must use a five-digit counter to troubleshoot the RF circuits of a CB set. The CB operating frequencies, or channels, are in the 27-MHz range. Now assume that you measure a 27-MHz signal with the five-digit counter. The count could be 26.999 or 27.001, or within 1000 Hz of 27 MHz. Since the FCC requires that the operating frequency of a CB set be held within 0.005 percent (or about 1350 Hz in the case of a 27-MHz signal), a digital counter for CB troubleshooting must have a minimum of five digits in the readout.

Combining Accuracy and Resolution. To find out if a counter is adequate for a particular RF test, add the time-base stability (in terms of frequency) to the resolution at the operating frequency. Again using the CB-set example, if the accuracy is 100 Hz and the count can be resolved to 1000 Hz (at the measurement frequency), the maximum possible inaccuracy is 1000 + 100 Hz, or 1100 Hz. This is within the approximate 1350 Hz (0.005 percent of 27 MHz) required.

3.5.3 Calibration Check of Counters

The accuracy of frequency-measuring devices used for RF work should be checked periodically, at least every 6 mo. Always follow the procedures recommended in the counter service literature. Generally, you can send the instrument to a calibration lab or to the factory, or you can maintain your own frequency standard. (The latter is not generally practical for most RF service shops.)No matter what standard is used, remember that the standard must be *more accurate, and have better resolution,* than the frequency-measuring device.

3.5.4 WWV Signals

In the absence of a frequency standard, or factory calibration, you can use the frequency information broadcast by U.S. government radio station WWV. These WWV signals are broadcast on 2.5, 5, 10, 15, 20, and 25 MHz continuously night and day, except for silent periods of about 4 min beginning 45 min after each hour. Broadcast frequencies are held accurate to within 5 parts in 10^{11}. This is far more accurate than required for most RF equipment tests.

The hourly broadcast schedules of WWV are subject to change. For full data on WWV broadcasts, refer to National Bureau of Standards (NBS) *Standard Frequency and Time Services* (Miscellaneous Publication 236), available from the Superintendent of Documents, U.S. Government Printing Office, Washington, D.C. 20402.

It is the continuous wave (CW) signals broadcast by WWV that provide the most accurate means of calibrating (or checking) frequency counters. It is not practical to use the WWV signal directly (except on some special counters) but the test connections for check are not complex.

Figure 3.6*c* shows the basic test connections for checking the accuracy of a frequency counter using WWV. Note that a receiver and a signal generator are required. Their accuracy is not critical, but both instruments must be capable of covering the desired frequency range. The procedure is as follows:

1. Allow the signal generator, receiver, and counter being tested to warm up for at least 15 min.
2. Reduce the signal generator output amplitude to zero. (Turn off the signal generator RF output if this is possible without turning off the entire signal generator.)
3. Tune the receiver to the desired WWV frequency. It is generally best to use a WWV frequency that is near the operating frequency of the RF equipment. For example, if a 27-MHz CB set is being tested, use the 25-MHz WWV signal.
4. Operate the receiver controls until you can hear the WWV signal in the receiver loudspeaker.
5. If the receiver is of the communications type, it will have a beat-frequency oscillator (BFO) and an output signal-strength indicator, or S-meter. Turn on the BFO, if necessary, to locate and identify the WWV signal. Then tune the receiver for maximum signal strength on the S-meter. The receiver is now exactly on 25 MHz, or whatever WWV frequency is selected.
6. Turn on the signal generator and tune the generator until the generator output is at zero beat against the WWV signal. As the signal generator is adjusted so that the generator frequency is close to that of the WWV signal (so that the difference in frequency is within the audio range), a tone, whistle, or "beat note" is heard on the receiver. When the signal generator is adjusted to exactly the WWV frequency, there is no "difference signal," and the tone can no longer be heard. In effect, the tone drops to zero, and the two signals (generator and WWV) are at zero beat.
7. Read the counter. The readout should be equal to the WWV frequency. For

example, with a five-digit counter at 25 MHz, the counter reading should be 24.999 to 25.001.

8. Repeat the procedure at other WWV broadcast frequencies.

3.6 RF DUMMY LOAD

Never adjust an RF circuit without a load connected to the output. This will almost certainly cause damage to the RF circuit. For example, when a transmitter is connected to an antenna or load, power is transferred from the final RF stage of the transmitter to the antenna or load. Without an antenna or load, the final RF stage must dissipate the full power and will probably be damaged (even with the heat sinks described in Chap. 2). Equally important, you should not make any major adjustments (except for a brief final tune-up) to a transmitter that is connected to a radiating antenna. You will probably cause interference.

These two problems can be overcome by a nonradiating load, commonly called a *dummy load.* There are a number of commercial dummy loads for communications equipment troubleshooting. The RF wattmeters described in Sec. 3.7 and the special test sets covered in Sec. 3.13 contain dummy loads. It is also possible to make up dummy loads suitable for most RF-circuit troubleshooting.

There are two generally accepted dummy loads: the fixed resistance and the lamp. Remember that these loads are for routine troubleshooting but are not a substitute for an RF wattmeter or special test set.

3.6.1 Fixed-Resistor Dummy Load

The simplest dummy load is a fixed resistor capable of dissipating the full power output of the RF circuit. The resistor can be connected to the output of the RF circuit (say at the antenna connector of a transmitter) by means of a plug, as shown in Fig. 3.7*a*.

Most communications transmitters operate with a 50-Ω antenna and lead-in and thus require a 50-Ω resistor. The nearest standard resistor is 51Ω. This 1-Ω difference is not critical. However, it is essential that the resistor be *noninductive* (*composition or carbon*), *never wire wound.* Wire-wound resistors have some inductance, which changes with frequency, causing the load impedance to change.

Always use a resistor with a power rating greater than the anticipated maximum output power of the transmitter. For example, an AM CB transmitter can (legally) have a 5-W input, which results in an output of about 4 W. A 7- to 10-W resistor should be used for the dummy load in this case. A single sideband (SSB) CB transmitter should not produce more than 12 W of output with full modulation, so a 15- to 20-W dummy-load resistor can be used.

RF Power Output Measurement with a Dummy-Load Resistor. It is possible to get an *approximate* measurement of RF power output from a radio transmitter with a resistor dummy load and a suitable meter. Again, these procedures are a substitute for power measurement with an accurate RF wattmeter.

The procedure is simple. Measure the voltage across the 50-Ω dummy-load resistor and find the power with the equation: power = voltage^2/50. For example, if the voltage measured is 14 V, the power output is: $14^2/50 = 3.92$ W.

Certain precautions must be observed. First, the meter must be capable of

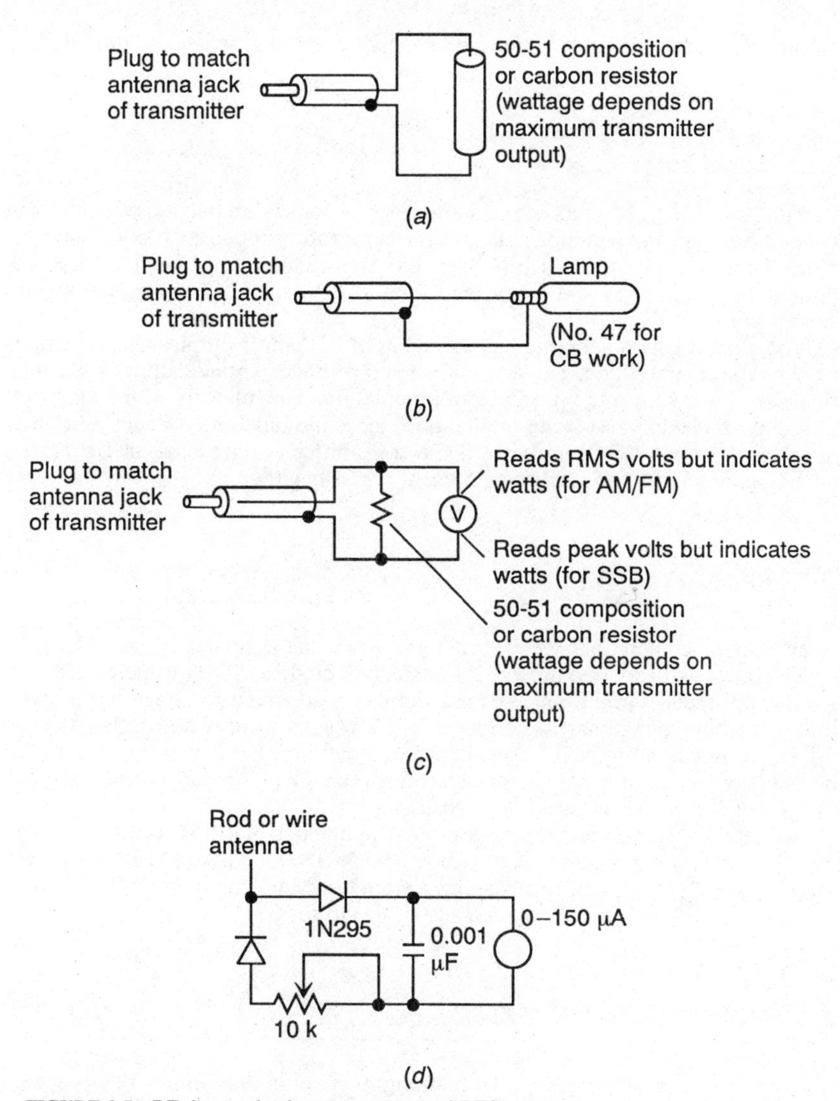

FIGURE 3.7 RF dummy loads, wattmeters, and RFS meters.

producing accurate voltage indications at the transmitter operating frequency. This usually requires a meter with an RF probe (Sec. 3.4), preferably a probe calibrated with the meter. An AM or FM transmitter should be checked with a root-mean-square (rms) voltmeter and with no modulation applied. An SSB transmitter must be checked with a peak-reading voltmeter and with modulation applied (since SSB produces no output without modulation). This usually involves connecting an audio generator to the microphone input of the SSB transmitter circuits. Always follow the service-literature recommendations for all RF power output measurements (frequency, channels, operating voltages, modulation, etc.).

3.6.2 Lamp Dummy Load

Lamps have been the traditional dummy loads for communications equipment troubleshooting. For example, the no. 47 lamp (often found as a pilot lamp for older electronic instruments) provides the approximate impedance and power dissipation required as a dummy load for CB equipment. The connections are shown in Fig. 3.7*b*.

You cannot get an accurate measurement of RF power output when a lamp is used as the dummy load. However, the lamp provides an indication of the relative power and shows the presence of modulation. The intensity of the light produced by the lamp varies with modulation (more modulation produces a brighter glow), so you can tell at a glance if the transmitter is producing an RF carrier (steady glow) and if modulation is present (varying glow).

3.7 RF WATTMETER

A number of commercial RF wattmeters are available for RF work. Also, the special test sets described in Sec. 3.13 usually include an RF wattmeter. The basic RF wattmeter circuit consists of a dummy load (fixed resistor) and a meter that measures voltage across the load, but it reads out in watts (rather than in volts), as shown in Fig. 3.7*c*. You simply connect the RF wattmeter to the circuit output (say the antenna connector of a transmitter), key the transmitter, and read the power output on the wattmeter scale.

Although operation is simple, you must remember that SSB transmitters require a peak-reading wattmeter to indicate peak envelope power (PEP), whereas an AM or FM set uses an rms-reading wattmeter. Most commercial RF wattmeters are rms-reading, unless specifically designed for SSB.

3.8 FIELD-STRENGTH METER

There are two basic types of field-strength meters: the simple relative field strength (RFS) meter and the precision laboratory or broadcast-type instrument. Most communications equipment troubleshooting can be carried out with simple RFS instruments. An exception is where you must make precision measurements of broadcast antenna radiation patterns.

The purpose of a field-strength meter is to measure the strength of the RF signals radiated by an antenna. This simultaneously tests the transmitter output, the antenna, and the lead-in. In the simplest form, a field-strength meter consists of an antenna (a short piece of wire or rod), a potentiometer, diodes, and a microammeter, as shown in Fig. 3.7*d*. More elaborate RFS meters include a tuned circuit and possibly an amplifier.

In use, a field-strength meter is placed near the antenna at some location accessible to the transmitter (where you can see the meter), the transmitter is keyed, and the relative field strength is indicated on the meter. The special test set described in Sec. 3.13 includes an RFS test function. Note that the potentiometer shown in Fig. 3.7*d* provides for calibration of the meter.

3.9 STANDING-WAVE-RATIO MEASUREMENT

The standing-wave ratio of an antenna is actually a measure of match or mismatch for the antenna, transmission line (lead-in), and transmitter or RF output circuit. When the impedances of the antenna, line, and RF circuit are perfectly matched, all of the energy (or RF signal) is transferred to or from the antenna, and there is no loss. If there is a mismatch (as is the case in any practical application), some of the energy, or RF signal, is reflected back into the line. This energy cancels part of the RF signal.

If the voltage (or current) is measured along the line, there are voltage or current maximums (where the reflected signals are in phase with the outgoing signals), and voltage or current minimums (where the reflected RF signal is out of phase, partially canceling the outgoing signal). The maximums and minimums are called *standing waves.* The ratio of the maximum to the minimum is the standing-wave ratio (SWR). The ratio may be related to either voltage or current. Since voltage is easier to measure, voltage is usually used (except in some RF lab work), resulting in the common term *voltage standing-wave* ratio (VSWR). The theoretical calculations for VSWR are shown in Fig. 3.8*a*.

An SWR of 1 to 1, expressed as 1:1, means that there are no maximums or minimums (the voltage is constant at any point along the line) and that there is a perfect match for circuit, line, and antenna. As a practical matter, if this 1:1 ratio should occur on one frequency, the ratio will not occur at any other frequency since impedance changes with frequency. It is not likely that all three elements (circuit, antenna, line) will change impedance by exactly the same amount on all frequencies. Therefore, when checking SWR, always check on all frequencies or channels, where practical. As an alternative, check SWR at the high, low, and middle channels, or frequencies.

In the case of microwave RF signals being measured in the laboratory, a meter is physically moved along the line to measure maximum and minimum voltages. This is not practical at most communications-equipment frequencies because of the physical length of the wave. In communications equipment, it is far more practical to measure forward or outgoing voltage and reflected voltage and then calculate the *reflection coefficient* (reflected voltage/outgoing voltage). The relationship of reflection coefficient to SWR is as follows: reflection coefficient = reflected voltage/forward voltage. For example, using a 10-V forward and a 2-V reflected voltage, the reflection coefficient is 0.2.

Reflection coefficient is converted to SWR by dividing (1 + reflection coefficient) by (1 − reflection coefficient). For example, using the 0.2 reflection coefficient, the SWR is (1 + 0.2)/(1 − 0.2) = 1.2/0.8 = 1.5 SWR. This may be expressed as 1:1.5, 1.5:1, or simply as 1.5, depending on the meter scale. In practical terms, an SWR of 1.5 is poor, since it means that at least 20 percent of the power is being reflected.

SWR can be converted to reflection coefficient by dividing (SWR − 1) by (SWR + 1). For example, using 1.5 SWR, the reflection coefficient is: (1.5 − 1)/(1.5 + 1) = 0.5/2.5 = 0.2 reflection coefficient.

In commercial SWR meters used for RF work, it is not necessary to calculate

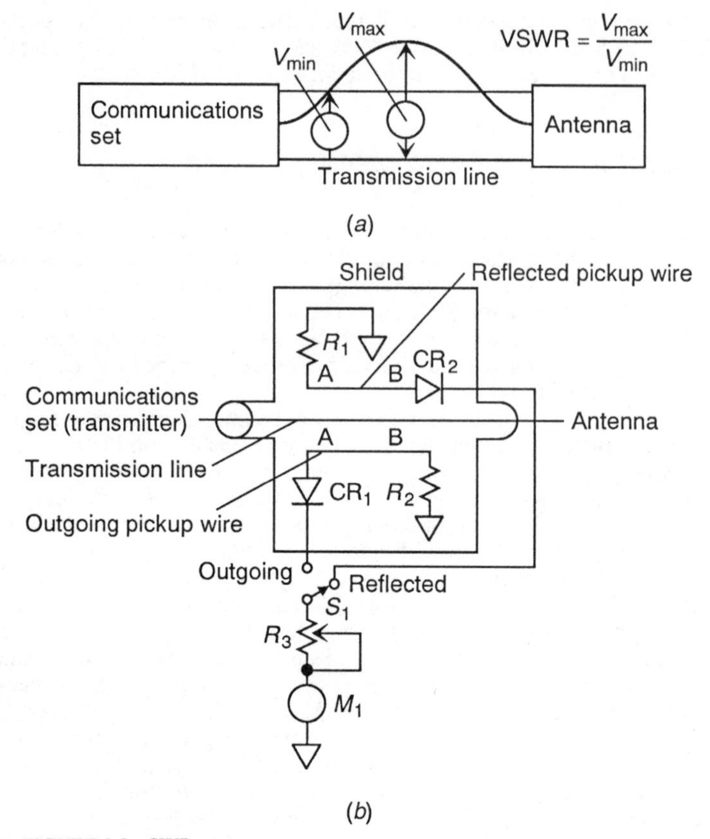

FIGURE 3.8 SWR measurements.

either reflection coefficient or SWR. This is done automatically by the SWR meter. (The meter is actually reading the reflection coefficient, but the scale indicates SWR. If you have a reflection coefficient of 0.2, the SWR readout is 1.5.)

There are a number of SWR meters used in RF work. Some communications sets, such as amateur radio and CB, have built-in SWR meters and circuits. The SWR function is often combined with other measurement functions (field strength, power output, etc.). The special test set described in Sec. 3.13 includes an SWR measurement feature since SWR is so important to proper operation of communications equipment.

Basic SWR meter circuits are quite simple, and it is possible to build them in the shop. However, it is not usually practical. The basic circuit requires that a *directional coupler* be inserted in the transmission line. Even under good conditions, a mismatch and some power loss may result. A poorly designed coupler may result in considerable power loss, as well as inaccurate readings. It is more practical to use the built-in meter (or a commercial meter) for SWR.

The basic SWR meter circuit is shown in Fig. 3.8*b*. Operation of the circuit is as follows. As shown, there are two *pickup wires*, both parallel to the center conductor of the transmission line. Any RF voltage on either of the parallel pickups is rectified and applied to the meter through switch S_1. Each pickup wire is terminated in the impedance of the transmission line by corresponding resistors R_1 and R_2 (typically 50 to 52 Ω).

The outgoing signal (transmitter to antenna) is absorbed by R_1, so there is no outgoing voltage on the reflected pickup wire beyond point A. However, the outgoing voltage remains on the transmission line at the outgoing pickup wire. This signal is rectified by CR_1 and appears as a reading on the meter when S_1 is in the outgoing position.

The opposite occurs for the reflected voltage (antenna to transmitter). There is no reflected voltage on the outgoing pickup wire beyond point B because the reflected voltage is absorbed by R_2. The reflected voltage does appear on the reflected pickup wire beyond this point and is rectified by CR_2. The reflected voltage appears on meter M_1 when S_1 is in the reflected position.

In use, switch S_1 is set to read the outgoing voltage, and resistor R_3 is adjusted until the meter needle is aligned with some "set" or "calibrate" line (near the right-hand end of the meter scale). Switch S_1 is then set to read the reflected voltage, and the meter needle moves to the SWR position.

SWR meters often do not read beyond 1:3 because a reading above 1:3 indicates a poor match. Make certain that you understand the scale used on the SWR meter. For example, a typical SWR meter is rated at 1:3, meaning that the scale reads from 1:1 (perfect) to 1:3 (poor). However, the scale indications are 1, 1.5, 2, and 3. These scale indications mean 1:1, 1:1.5, 1:2, and 1:3, respectively. The scale indications between 1 and 1.5 are the most useful since a good antenna system (antenna and lead-in) typically shows 1.1 or 1.2. Anything between 1.2 and 1.5 is on the borderline.

3.10 DIP METER AND ADAPTER

The dip meter (or grid-dip meter) was a common tool in early radiocommunications service work, particularly in amateur radio. Although the dip meter has many uses, the most useful function is in presetting "cold" resonant RF circuits (no power applied to the circuits). This makes it possible to adjust the resonant RF circuits of badly tuned equipment or equipment where new coils and transformers must be installed as a replacement.

As an example, it is possible that the replacement coil or transformer is tuned to an undesired frequency when shipped from the factory. Using a dip meter, it is possible to install the coil, tune the coil to the correct frequency, then apply power to the circuit and adjust the circuit for "peak" as described in the service literature. (Most service literature assumes that the circuits are not badly tuned or only require peaking.)

The dip meter has all but disappeared as a service instrument. However, a *dip adapter circuit* can be used in place of a dip meter.

3.10.1 Dip Adapter

Figure 3.9 shows the basic dip adapter circuit. The circuit is essentially an RF generator with an external pick-up coil, a frequency counter, a diode, and an external meter. When the coil is held near the RF circuit to be tested, and the gen

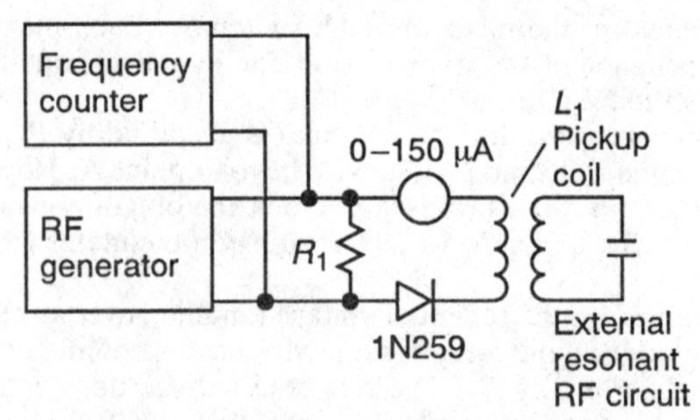

FIGURE 3.9 Dip adapter circuit.

erator is tuned to the resonant frequency of the RF circuit, part of the RF energy is absorbed by the RF circuit, and the meter reading "dips." The procedure can be reversed, where the RF generator is set to a desired frequency, and the RF circuit is tuned to produce a dip indication on the meter.

In the circuit in Fig. 3.9, R_1 should match the impedance of the RF generator (typically 50 Ω). Both diode CR_1 and the meter should match the output of the RF generator. The pick-up coil L_1 consists of a few turns of insulated wire. The accuracy of the dip-adapter circuit depends on counter accuracy (or on the RF-generator dial accuracy if the counter is omitted).

3.10.2 Setting Resonant Frequency with a Dip Adapter

The following procedure applies to both series and parallel resonant RF circuits.

1. Couple the dip adapter to the resonant RF circuit using coil L_1. Usually, the best coupling has a few turns of L_1 passed over the coil of the resonant RF circuit. Make certain that no power is applied to the RF circuit (or to the equipment containing the RF circuit).

2. Set the RF generator to the desired resonant frequency, as indicated on the counter. Adjust the RF generator output-amplitude control for a convenient reading on the adapter meter.

3. Tune the resonant RF circuit for a maximum dip on the adapter meter. As discussed in Chap. 2, the resonant RF circuits are usually tuned by means of adjustable slugs in the coil and/or adjustable capacitors.

4. Most resonant RF circuits are designed so as not to tune across both the fundamental frequency and any harmonics. However, it is possible that the RF circuit will tune to a harmonic and produce a dip. To check this condition, tune the RF circuit for maximum dip and set the RF generator to the first harmonic (twice the desired resonant frequency) and to the first subharmonic (one-half the resonant frequency). Note the amount of dip at both harmonics. The harmonics should produce substantially smaller dips than the fundamental resonant frequency.

5. For maximum accuracy, check the dip frequency from both high and low sides of the resonant RF-circuit tuning. A significant difference in frequency readout from either side indicates overcoupling between the dip-adapter circuit and the resonant RF circuit under test. Move coil L_1 away from the test RF circuit until the dip indication is just visible. This amount of coupling should provide maximum accuracy. (If there is difficulty in finding a dip, overcouple the adapter until a dip is found, then loosen the coupling and make a final check of frequency. Generally, the dip is more pronounced when approached from the direction that causes the meter reading to rise.)

6. If there is doubt as to whether the adapter is measuring the resonant frequency of the RF circuit, or some nearby circuit, ground the RF circuit under test. If there is no change in the adapter dip indication, the resonance of another circuit is being measured.

7. The area surrounding the RF circuit being measured should be free of wiring scraps, solder drippings, and so on, since the resonant RF circuit can be affected (especially at high frequencies), resulting in inaccurate frequency readings. Keep fingers and hands as far away from the adapter coil as possible (to avoid adding body capacity to the RF circuit under test).

8. All other factors being equal, the nature of the dip indication provides an *approximate* indication of the RF-circuit Q. Generally, a sharp dip indicates high Q, whereas a broad dip shows a low Q.

9. The dip adapter may also be used to measure the frequency to which a resonant RF circuit is tuned. The procedure is essentially the same as that for presetting the resonant frequency (steps 1 through 8), except that the RF generator is tuned for maximum dip with the RF circuit still cold (no power applied). The resonant frequency to which the RF circuit is tuned can then be read from the counter. When making this test, watch for harmonics (which also produce dip indications).

3.11 SPECTRUM ANALYZERS AND FM DEVIATION METERS

3.11.1 Spectrum Analyzers

Figure 3.10*a* shows the basic spectrum analyzer circuit. Spectrum analyzers are most often used in RF work where FM is involved, although analyzers can be useful in AM and SSB circuits.

A spectrum analyzer is essentially a narrowband receiver, electrically tuned over a given frequency range, combined with a scope or display tube. As shown in Fig. 3.10*a*, the oscillator is swept over a given range of frequencies by a sweep-generator circuit. Since the IF amplifier passband remains fixed, the input circuits and mixer are swept over a corresponding range of frequencies.

For example, if the intermediate frequency is 10 kHz, and the oscillator sweeps from 100 to 200 kHz, the input is capable of receiving signals in the range 110 to 210 kHz. The output of the IF amplifier is further amplified and supplied to the vertical-deflection plates of the display tube. The horizontal plates receive a signal from the same sweep-generator circuit used to control the oscillator frequency. As a result, the length of the horizontal sweep represents the total sweep-frequency spectrum (if the sweep is from 110 to 210 kHz, the left-hand end of the display-tube horizontal trace represents 110 kHz, and the right-hand end represents 210 kHz). Any point along the horizontal trace represents a corresponding frequency (with the midpoint representing 160 kHz, in this example).

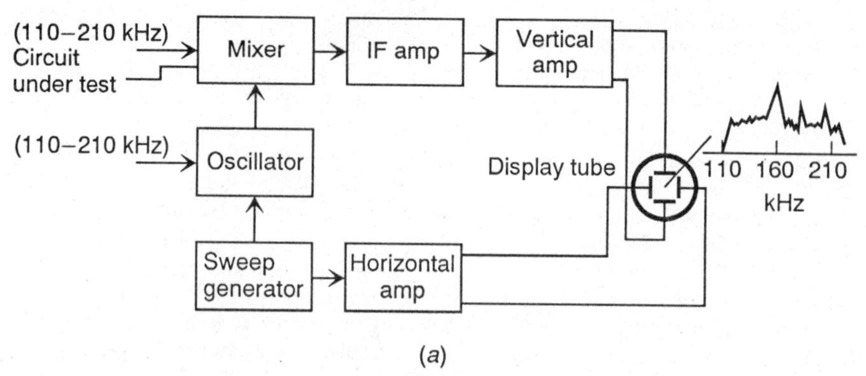

FIGURE 3.10 Spectrum analyzer circuits and displays.

3.11.2 Time Amplitude versus Frequency Amplitude

To gain a better understanding of the usefulness and application of a spectrum analyzer in RF work, it is important to understand what the spectrum display is and how to interpret it. Figure 3.10*b*, 3.10*c*, and 3.10*d* shows the relationship of time-amplitude and frequency-amplitude displays. A conventional scope produces a time-amplitude display. For example, pulse rise time and width are read directly on the horizontal axis of a scope tube. A spectrum analyzer produces a frequency-amplitude display where signals (unmodulated, AM, FM, or pulse) are broken down into individual components and displayed on the horizontal axis.

In Fig. 3.10*b*, both the time-amplitude and frequency-amplitude coordinates are shown together. The example given is that showing the addition of a fundamental frequency and the second harmonic. In Fig. 3.10*c*, only the time-amplitude coordinates are shown. The solid line (which is the composite of fundamental F_1 and $2F_1$ is the only display that appears on a conventional scope. In Fig. 3.10*d*, only the frequency-amplitude coordinates are shown. Note how the components (F_1 and $2F_1$) of the composite signal are clearly seen here.

3.11.3 Practical Spectrum Analysis

Spectrum analyzers are often used in conjunction with Fourier and transform analyses. Both of these techniques are quite complex and beyond the scope of this book (and the author). Instead, we concentrate on the practical aspects of spectrum analysis during RF tests. That is, we discuss what display results from a given input signal and how the display can be interpreted.

Unmodulated Signal Displays. If the spectrum-analyzer oscillator sweeps through an unmodulated or CW signal slowly, the resulting response on the analyzer screen is simply a plot of the analyzer IF-amplifier passband. A pure CW signal has, by definition, energy at only one frequency and should therefore appear as a single spike on the analyzer screen (Fig. 3.10*e*). This occurs provided that the total sweep width (the so-called *spectrum width*) is wide enough compared to the IF-passband of the analyzer. As spectrum width is reduced, the spike response

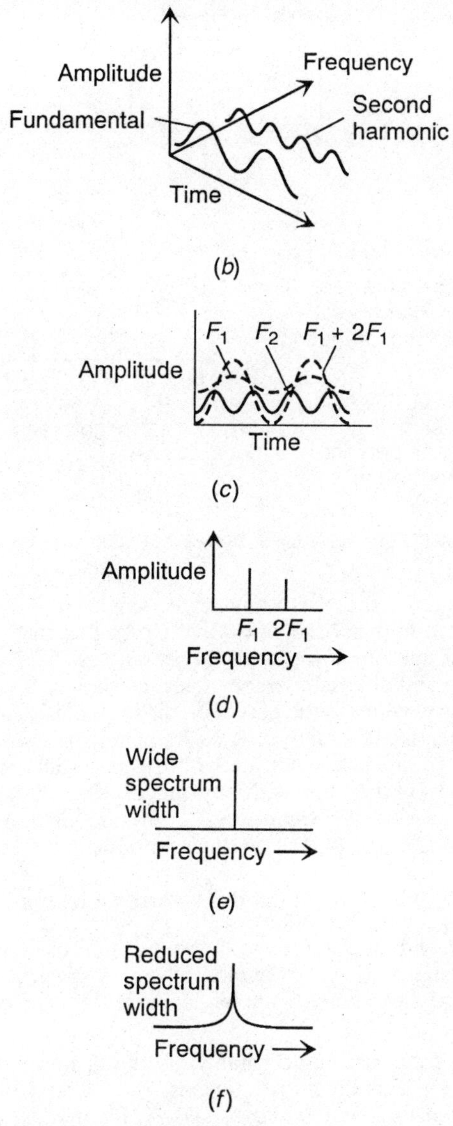

Figure 3.10 *(continued)*

begins to spread out until the IF-passband characteristics begin to appear, as shown in Fig. 3.10*f*.

Amplitude-Modulated Signal Displays. A pure sine wave represents a single frequency. The spectrum of a pure sine wave is shown in Fig. 3.11 and is the same as the unmodulated signal display in Fig. 3.10*e* (a single vertical line). The height of line F_0 represents the power contained in the single frequency.

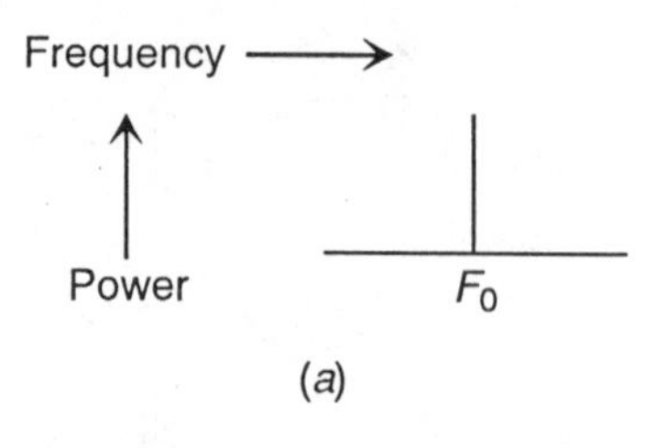

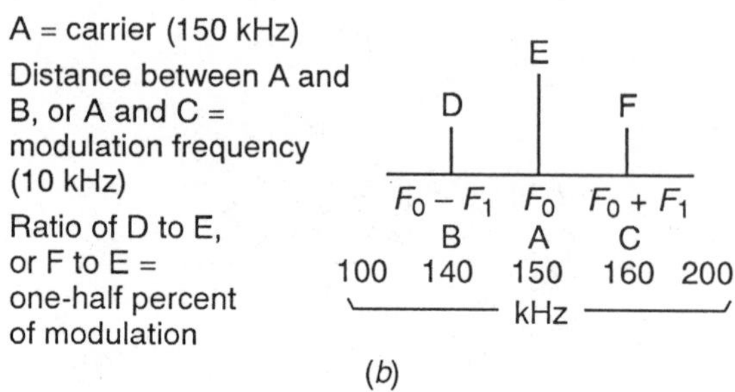

FIGURE 3.11 Spectrum analyzer AM/FM displays.

Figure 3.11*b* shows the spectrum for a single sine-wave frequency F_0, amplitude-modulated by a second sine wave F_1. In this case, two *sidebands* are formed, one higher and one lower than the frequency F_0. These sidebands correspond to the sum and difference frequencies, as shown. If more than one modulating frequency is used (as is the case with most practical amplitude-modulated signals), two sidebands are added for each frequency.

Note that if the frequency, spectrum width, and vertical response of the analyzer are calibrated (as they are with any modern laboratory instrument), it is possible to find (1) the carrier frequency, (2) the modulation frequency, (3) the modulation percentage, and (4) the nonlinear modulation (if any) and incidental FM (if any).

An amplitude-modulated spectrum display can be interpreted as follows:

The carrier frequency is determined by the position of the center vertical line F_0 on the horizontal axis. For example, if the total spectrum is from 100 to 200 kHz, and F_0 is in the center, as shown in Fig. 3.11*b*, the carrier frequency is 150 kHz.

The modulation frequency is determined by the position of the sideband line $F_0 - F_1$, or $F_0 + F_1$, on the horizontal axis. For example, if sideband $F_0 - F_1$ is at 140 kHz, and F_0 is at 150 kHz as shown, the modulating frequency is 10 kHz. Under these conditions, the upper sideband $F_0 + F_1$ should be 160 kHz. The distance between the carrier line F_0 and either sideband is sometimes known as the *frequency dispersion* and is equal to the modulation frequency.

The modulation percentage is determined by the ratio of the sideband amplitude to the carrier amplitude. The amplitude of either sideband with respect to the carrier amplitude is *one-half* of the percentage of modulation. For example, if the carrier amplitude is 100 mV and either sideband is 50 mV, this indicates 100 percent modulation. If the carrier amplitude is 100 mV and either sideband is 33 mV, this indicates 66 percent modulation.

Nonlinear modulation is indicated when the sidebands are of unequal amplitude or are not equally spaced on both sides of the carrier frequency. Unequal amplitude indicates nonlinear modulation that results from a form of undesired frequency modulation combined with amplitude modulation.

Incidental FM is indicated by a shift in the vertical signals along the horizontal axis. For example, any horizontal "jitter" of the signals indicates rapid frequency modulation of the carrier.

The rules for interpreting amplitude-modulated spectrum-analyzer displays are summarized in Fig. 3.11*b*.

In practical tests, carrier signals are often amplitude-modulated at many frequencies simultaneously. This results in many sidebands (two for each modulating frequency) on the display. To resolve this complex spectrum, you must make sure that the analyzer bandwidth is less than the lowest modulating frequency or less than the difference between any two modulating frequencies, whichever is the smaller.

Overmodulation also produces extra sideband frequencies. The spectrum for overmodulation is very similar to multifrequency modulation. However, overmodulation is usually distinguished from multifrequency modulation by the facts that (1) the spacing between overmodulated sidebands is equal, while multifrequency sidebands may be arbitrarily spaced (unless the modulating frequencies are harmonically related); and (2) the amplitude of the overmodulated sidebands decreases progressively out from the carrier, but the amplitude of the multifrequency-modulated signals is determined by the modulation percentage of each frequency and can be arbitrary.

Frequency-Modulated Signal Displays. The mathematical expression for an FM waveform is long and complex, involving a special mathematical operator known as the *Bessel function.* However, the spectrum representation of the FM waveform is relatively simple, as shown in Fig. 3.11.

Figure 3.11*c* shows an unmodulated-carrier spectrum waveform. Figure 3.11*d* shows the relative amplitudes of the same waveform when the carrier is frequency-modulated with a deviation of 1 kHz (modulation index of 1.0). Figure 3.11*e* shows the relative amplitudes of the waveform when the carrier is frequency-modulated with a deviation of 5 kHz (modulation index of 5.0). Note that the modulation index is given by: modulation index = maximum frequency deviation/modulation frequency.

The term *maximum frequency deviation* is theoretical. If a CW signal F_C is frequency-modulated at a rate F_R, an infinite number of sidebands result. These sidebands are located at intervals of $F_C \pm N_F$, where N = 1, 2, 3, and so on.

However, as a practical matter, only the sidebands containing *significant power* are usually considered. For a quick approximation of the bandwidth occupied by the significant sidebands, multiply the sum of the carrier deviation and the modulating frequency by 2: bandwidth = 2(carrier deviation + modulating frequency).

As a guideline, when using a spectrum analyzer to find the maximum deviation of an FM signal, locate the sideband where the amplitude *begins to drop and continues to drop* as the frequency moves from the center. For example, in Fig. 3.11*e,* sidebands 1, 2, 3, and 4 rise and fall, but sideband 5 falls, and all sidebands after 5 continue to fall. Since each sideband is 1 kHz from the center, this indicates a practical or significant deviation of 5 kHz. (It also indicates a modulation index of 5.0, in this case.)

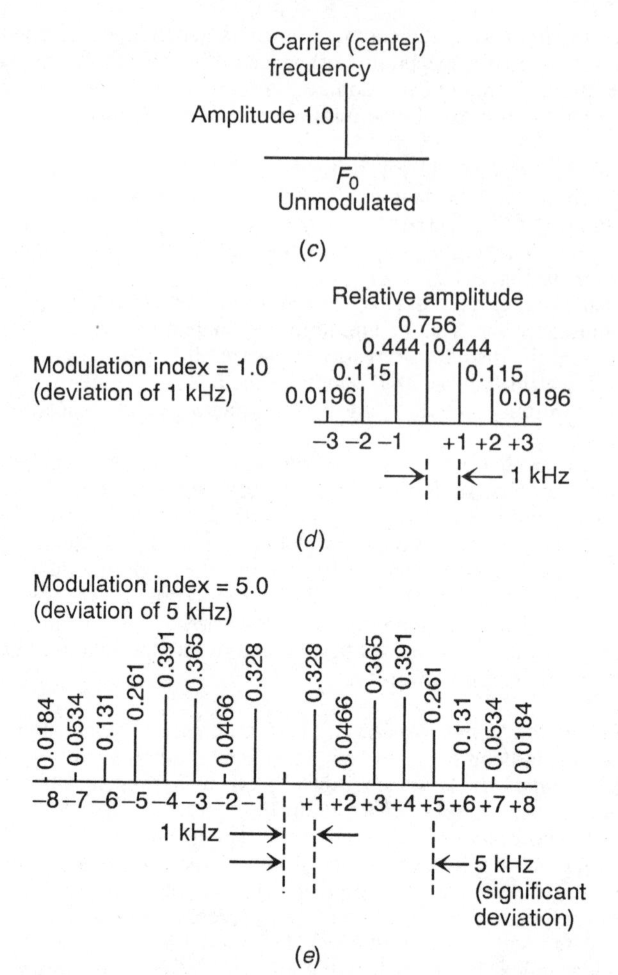

FIGURE 3.11 (*Continued*) Spectrum analyzer AM/FM displays.

As in the case of AM, the center and modulation frequencies for FM can be determined with the spectrum analyzer:

The FM carrier frequency is determined by the position of the center vertical line on the horizontal axis. (The centerline is not always the highest amplitude, as shown in Fig. 3.11.

The FM modulating frequency is determined by the position of the sidebands in relation to the center line or the distance between sidebands (frequency dispersion).

3.11.4 FM-Deviation Meter

Unless a service shop specializes in FM, or complex broadcast work, an FM-deviation meter can be used as a substitute for a spectrum analyzer. The operating controls and procedures for an FM-deviation meter are far less complex than those of a spectrum analyzer. Typically, FM-deviation meter controls include a meter scale marked in terms of frequency, a tuning control, and a zero control.

In use, the FM deviation meter is connected to monitor the output of the RF circuit (typically a communications transmitter). The transmitter is first keyed without modulation so that the meter can be tuned to the exact carrier frequency. Then the transmitter is frequency-modulated with a tone (typically in the range of 1 to 5 kHz), and the exact amount of frequency modulation is indicated on the FM-deviation meter.

3.12 MISCELLANEOUS RF TROUBLESHOOTING EQUIPMENT

There are many items of equipment that make life easier for the RF service technician but are not absolutely essential for all communications troubleshooting. The following are some examples.

3.12.1 Base-Station Set and Antenna

The uses of a known-good communications set and base station (or shop) antenna are obvious. You can check operation of a suspected set on the good shop antenna. If the set performs properly with the shop antenna but not with the set's antenna, the problem is localized. You can reverse the procedure and test the suspected antenna with a known-good set. Also, you can communicate between the shop and another station (either mobile or base station) to check operation before and after service.

Walkie-Talkie CB. A walkie-talkie CB may also be used for communication from the shop to remote locations for field-strength and other tests. Remember that the walkie-talkie must be licensed under Part 95 of the FCC regulations if you communicate with other CB stations (unless you communicate only between unlicensed walkie-talkies). A walkie-talkie may also be used to track down electrical interference.

3.12.2 Power Supply and Isolation Transformer

A well-equipped RF service shop should have at least two power supplies: one ac power supply, variable from about 100 to 130 V and one dc power supply, variable from about 10 to 15 V. There are a number of commercial power supplies that meet these requirements, so we do not discuss the circuits.

Most commercial power supplies include a transformer. Often this is a variable auto-transformer (or variac). The use of a transformer in the power supply eliminates the need for an *isolation transformer* (Chap. 6).

3.12.3 Distortion Meters

Some service shops include distortion meters or distortion analyzers. There are two basic types: the *intermodulation distortion analyzer* and the *harmonic distortion analyzer*. Before you rush to buy either of these instruments, consider the following.

The basic purpose of any distortion meter or analyzer is to *measure the amount of distortion* (usually as a percentage), not to locate the cause of distortion. Generally, if there is sufficient distortion in RF equipment (say in the audio and modulation circuits) to be a problem, you will hear the distortion in both transmission and reception.

3.12.4 Communications Receivers, Spectrum Analyzers, and TV Sets

In addition to checking that an RF circuit produces the correct signal on all frequencies, it is helpful to know that the circuit is not producing any other signals. For example, the final RF amplifier in a transmitter may break into oscillation (if not properly neutralized) and produce signals at undetermined frequencies. These signals may not show up on the communications channel being used but may interfere with other communications or with TV. (TV interference is a very common problem in CB communications.) All such extra signals (generally referred to as *spurious signals* in FCC regulations and service literature) are illegal and certainly undesirable.

The ideal instrument for detecting undesired signals from an RF circuit is the spectrum analyzer described in Sec. 3.11. Although the spectrum analyzer is ideal, it is also very expensive and is generally restricted to laboratory use or commercial broadcast work. You can do essentially the same job with a good communications-type receiver. The receiver should have a BFO, as well as an S-meter (Sec. 3.5.4). In addition to using the communications receiver for signal checks, you can monitor transmissions of sets being serviced (and check WWV signals).

One of the most frequent types of interference caused by communications sets is on TV channels (especially CB communications). A TV set in the shop quickly indicates if a communications set being serviced is causing any interference. The shop TV set may help settle some disputes concerning TV interference problems.

Before you become overconfident with your use of a TV set for interference checks, remember the following. Most interference enters the TV set through the IF amplifiers, and the IF amplifiers on all TV sets do not operate on the same frequency. Some TV sets use the range 22 to 28 MHz, whereas other sets use the range 41 to 47 MHz (and some very old TV sets use other IF ranges). So it is possible for a communications set to produce interference on one TV and not on another, with both TV sets located in the same room, and tuned to the same channel.

3.13 SPECIAL RF TEST SETS

There are a number of test sets designed specifically for RF service, particularly for communications-set service. Some of the sets are for field use, whereas others are for the bench or shop. Still other sets may be used in either the shop or field. The following paragraphs describe one such special test set. Remember that this is not the only test set available, now and in the future, but represents a typical test set that incorporates the most important required functions for RF service work. The information here may be used as a guide in selecting the right test set, or combination of sets, for your particular RF service needs.

3.13.1 Multipurpose RF Tester

Figure 3.12 shows a multipurpose tester suitable for checking communications transceivers in the range of 25 to 50 MHz. The instrument is particularly suited for testing CB equipment. The tester measures power outputs up to 25 W (or 250 W when an external dummy load is used), SWR up to 1:3, percentage of modu-

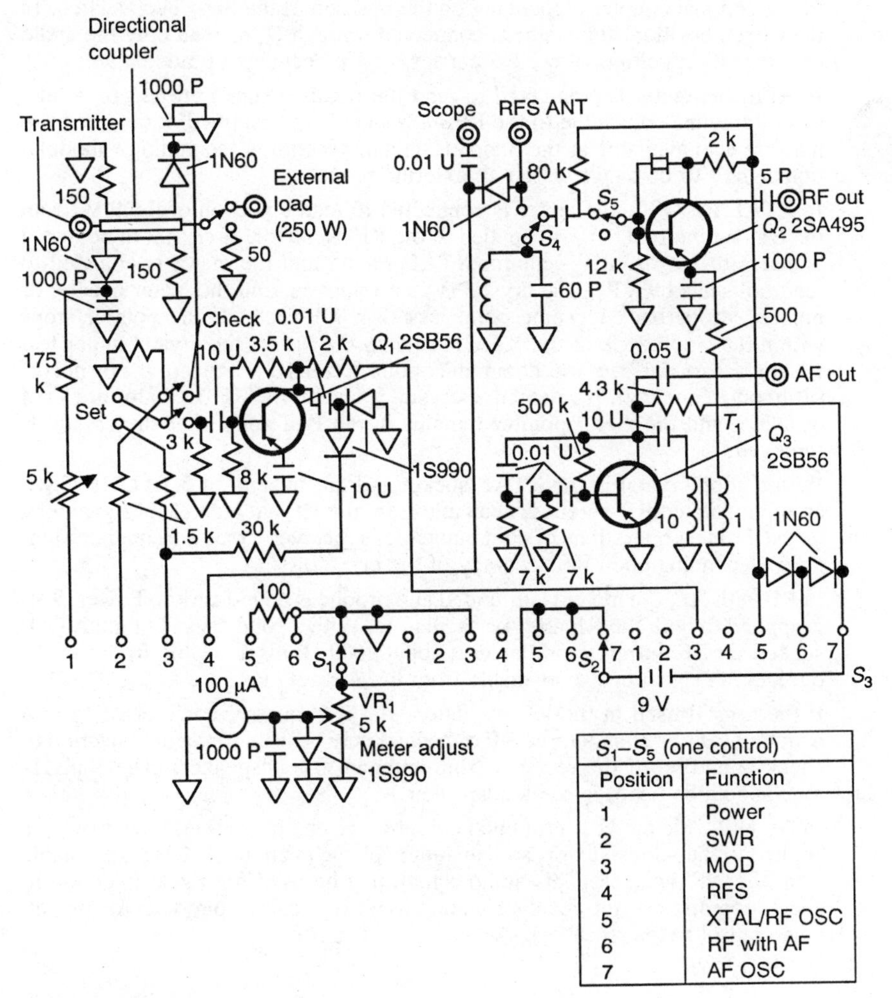

S_1–S_5 (one control)	
Position	Function
1	Power
2	SWR
3	MOD
4	RFS
5	XTAL/RF OSC
6	RF with AF
7	AF OSC

FIGURE 3.12 Multipurpose RF tester.

lation up to 100 percent, relative field strength, and crystal activity (on a good-bad basis). The tester also provides a built-in 25-W, a dummy load, a crystal-controlled RF oscillator at 27 MHz, and an audio oscillator at 1000 Hz (which can also be used to modulate the 27-MHz RF oscillator).

The same meter is used for all functions. Operation of the meter is controlled by the selector switches as follows:

In Power, the meter is connected to read the forward voltage applied to the dummy load (25 W) or to an external load (up to 250 W).

In SWR, the meter is connected to read the forward voltage and reflected voltages in the directional coupler, depending on the position of the Set-Check switch.

In MOD, the meter is connected to read both forward and reflected voltages in the directional coupler, depending on the position of the Set-Check switch. In the Check position, the meter is connected through Q_1 to read only the audio or modulation voltage of the RF carrier, as a percentage of modulation.

In RFS, the meter is connected to read the rectified signal present on a telescopic antenna connected to the RFS ANT jack. The rectified or detected signals are also available at the Scope terminal, permitting the audio or modulation signals to be displayed on an external scope.

In XTAL/RF OSC, the meter is connected to read a portion of the 9 V (from the test-set internal battery) applied to the RF oscillator. A crystal to be tested is inserted into the RF-oscillator XTAL socket and the meter is adjusted to read full-scale by VR_1. The crystal is then removed, and the meter needle, or pointer, drops back to some point less than full scale. If the pointer stops within the Good zone of the XTAL scale on the meter, the crystal under test is satisfactory for use in a communications set or RF circuit (but not necessarily on-frequency). If a defective crystal is tested, the RF oscillator does not oscillate, and the meter pointer remains in the Bad zone after the crystal is removed.

With a good crystal in the XTAL socket and the selector at XTAL/RF OSC, an unmodulated RF output is available from the RF out jack. This signal may be used to test operations of communications receivers or as a frequency standard (depending upon the accuracy of the crystal).

In RF with AF, the meter is grounded and produces no indication. Power (9 V) is applied to both the RF and AF oscillators. With a good crystal in the XTAL socket, an RF output (modulated at about 1000 Hz) is available from the RF out jack. A signal is also available from the AF out jack.

If the crystal used in the RF oscillator is at a frequency corresponding to a communications channel, the RF out signal may be used for signal injection to check operation of the receiver from antenna to loudspeaker. (The 1000-Hz tone should be heard in the loudspeaker.)

In AF OSC, the meter is grounded and produces no indication. Power (9 V) is applied to the AF oscillator, and an audio voltage at about 1000 Hz is available from the AF out jack. This audio signal may be used for signal injection to check operation of the receiver audio circuit (typically from detector or volume control to loudspeaker).

CHAPTER 4
BASIC RF TESTS

This chapter covers basic test procedures for RF equipment. These procedures can be applied to complete equipment (such as a transmitter) or to specific circuits (such as the RF circuits of a receiver). The procedures can also be applied to RF circuits at any time during design or experimentation.

For the experimenter or hobbyist, the tests described in this chapter should be made when the circuit is first completed in experimental form. If the test results are not as desired, the component values should be changed as necessary to get the desired results. Also, RF circuits should always be retested in final form (with all components soldered in place). This shows if there is any change in circuit characteristics because of the physical relocation of components.

Although this procedure may seem unnecessary, it is especially important at higher radio frequencies. Often, there is capacitance or inductance between components, from components to wiring, and between wires. These stray "components" can add to the reactance and impedance of circuit components. When the physical location of parts and wiring is changed, the stray reactances change and alter circuit performance.

4.1 BASIC RF-VOLTAGE MEASUREMENTS

As discussed in Chap. 3, when voltages to be measured are at radio frequencies and are beyond the frequency capabilities of the meter circuits or scope amplifiers, an RF probe is required. Such probes rectify the RF signals into a dc output which is almost equal to the peak RF voltage. The dc output of the probe is then applied to the meter or scope input and is displayed as a voltage readout in the normal manner.

If a probe is available as an accessory for a particular meter, that probe should be used in favor of any experimental or homemade probe. The manufacturer's probe is matched to the meter in calibration, frequency compensation, and so on. If a matching probe is not available for a particular meter or scope, probes can be made up as described in Sec. 3.4 and shown in Fig. 3.5.

4.2 BASIC RESONANT-FREQUENCY MEASUREMENTS

RF equipment is based on the use of resonant circuits consisting of a capacitor and a coil (inductance) connected in series or parallel, as discussed in Chap. 2

and shown in Fig. 2.1. At the resonant frequency, the inductance (L) and capacitive (C) reactances are equal, and the LC circuit acts as a high impedance (if a parallel circuit) or a low impedance (if a series circuit). In either case, any *combination of capacitance and inductance has some resonant frequency.*

A meter can be used in conjunction with an RF signal generator to find the resonant frequency of either series or parallel LC circuits. The generator must be capable of producing a signal at the resonant frequency of the circuit, and the meter must be capable of measuring the frequency. If the resonant frequency is beyond the normal range of the meter, an RF probe must be used. The following steps describe the measurement procedure:

1. Connect the equipment as shown in Fig. 4.1. Use the connections in Fig. 4.1*a* for parallel-resonant LC circuits or the connections in Fig. 4.1*b* for series-resonant LC circuits.

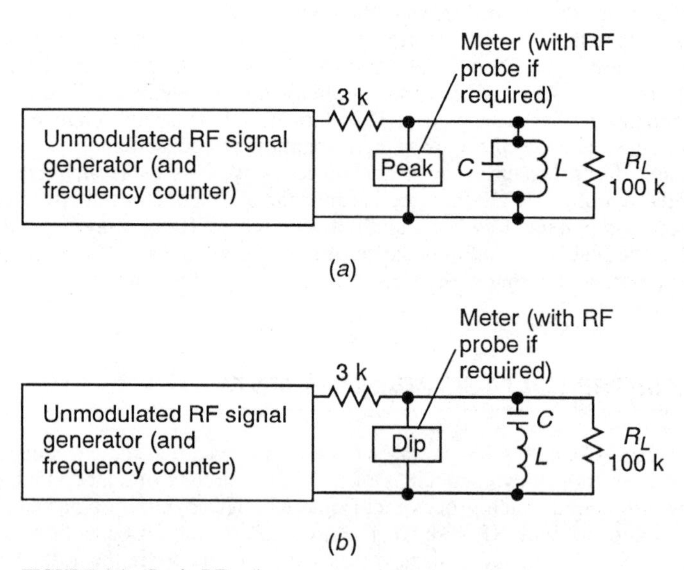

FIGURE 4.1 Basic RF voltage measurement.

2. Adjust the generator output until a convenient midscale indication is obtained on the meter. Use an unmodulated signal output from the generator.
3. Starting at a frequency well below the lowest possible frequency of the circuit under test, slowly increase the generator output frequency. If there is no way to judge the approximate resonant frequency, use the lowest generator frequency.
4. If the circuit being tested is parallel-resonant, watch the meter for a maximum, or peak, indication.
5. If the circuit being tested is series-resonant, watch the meter for a minimum, or dip, indication.

6. The resonant frequency of the circuit under test is the one at which there is a maximum (for parallel) or minimum (for series) indication on the meter.
7. There may be peak or dip indications at harmonics of the resonant frequency. Therefore, the test is most efficient when the approximate resonant frequency is known.
8. The value of load resistor R_L is not critical. The load is shunted across the LC circuit to flatten, or broaden, the resonant response (to lower the circuit Q), causing the voltage maximum or minimum to be approached more slowly. A suitable trial value for R_L is 100 k. A lower value of R_L sharpens the resonant response, and a higher value flattens the curve.

4.3 BASIC COIL INDUCTANCE MEASUREMENTS

A metor can be used in conjunction with an RF signal generator and a fixed capacitor (of known value and accuracy) to find the inductance of a coil. The generator must be capable of producing a signal at the resonant frequency of the test circuit, and the meter must be capable of measuring the frequency. If the resonant frequency is beyond the normal range of the meter, an RF probe must be used. The following steps describe the measurement procedure:

1. Connect the equipment as shown in Fig. 4.2. Use a capacitive value such as 10 μF, 100 pF, or some other even number to simplify the calculation.

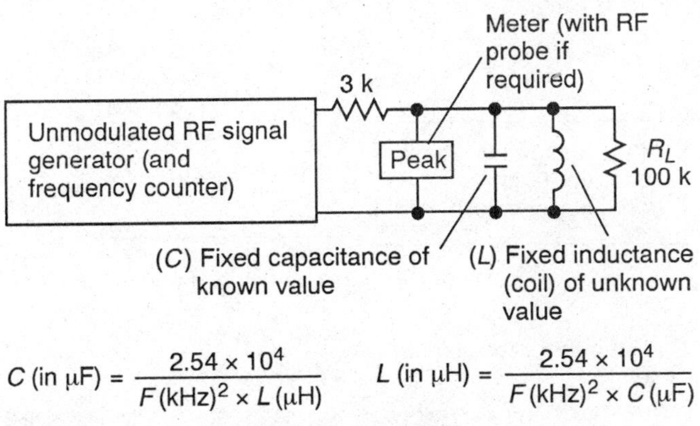

$$C \text{ (in } \mu\text{F)} = \frac{2.54 \times 10^4}{F(\text{kHz})^2 \times L(\mu\text{H})} \qquad L \text{ (in } \mu\text{H)} = \frac{2.54 \times 10^4}{F(\text{kHz})^2 \times C(\mu\text{F})}$$

FIGURE 4.2 Basic coil inductance measurements.

2. Adjust the generator output until a convenient midscale indication is obtained on the meter. Use an unmodulated signal output from the generator.
3. Starting at a frequency well below the lowest possible resonant frequency of the inductance-capacitance combination under test, slowly increase the generator frequency. If there is no way to judge the approximate resonant frequency, use the lowest generator frequency.

4. Watch the meter for a maximum, or peak, indication. Note the frequency at which the peak indication occurs. This is the resonant frequency of the circuit.
5. Using the resonant frequency, and the known capacitance value, calculate the unknown inductance using the equation in Fig. 4.2.
6. Note that the procedure can be reversed to find an unknown capacitance value, when a known inductance value is available.

4.4 BASIC COIL SELF-RESONANCE AND DISTRIBUTED-CAPACITANCE MEASUREMENTS

No matter what design or winding method is used, there is some distributed capacitance in any coil. When the distributed capacitance combines with the coil inductance, a resonant circuit is formed. The resonant frequency is usually quite high in relation to the frequency at which the coil is used. However, since self-resonance may be at or near a harmonic of the frequency to be used, the self-resonant effect may limit the usefulness of the coil in *LC* circuits. Some coils, particularly RF chokes, may have more than one self-resonant frequency.

A meter can be used in conjunction with an RF signal generator to find both the self-resonant frequency and the distributed capacitance of a coil. The generator must be capable of producing a signal at the resonant frequency of the circuit, and the meter must be capable of measuring voltages at that frequency. Use an RF probe if required. The following steps describe the measurement procedure:

1. Connect the equipment as shown in Fig. 4.3.

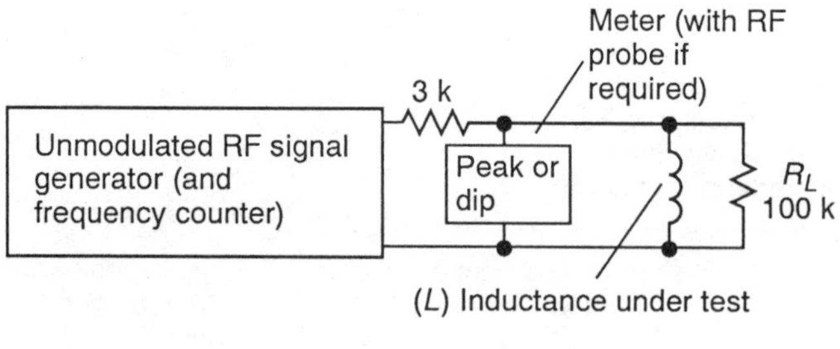

$$C \text{ (distributed capacitance in pF)} = \frac{2.54 \times 10^4}{F(\text{MHz})^2 \times L\,(\mu\text{H})}$$

FIGURE 4.3 Basic coil self-resonance and distributed-capacitance measurements.

2. Adjust the generator output amplitude until a convenient midscale indication is obtained on the meter. Use an unmodulated signal output from the generator.
3. Tune the signal generator over the entire frequency range, starting at the lowest frequency. Watch the meter for either peak or dip indications. Either a peak or a dip indicates that the inductance is at a self-resonant point. The generator output frequency at that point is the self-resonant frequency. Make cer-

tain that peak or dip indications are not the result of changes in generator output level. Even the best laboratory generators may not produce a flat (constant level) output over the entire frequency range.

4. Since there may be more than one self-resonant point, tune through the entire signal-generator range. Try to cover a frequency range up to at least the third harmonic of the highest frequency involved in a resonant-circuit design.
5. Once the resonant frequency is found, calculate the distributed capacitance using the equation in Fig. 4.3. For example, assume that a coil with an inductance of 7 μH is found to be self-resonant at 50 MHz: C (distributed capacitance) = $(2.54 \times 10^4)/(50^2 \times 7) = 1.45$ pF.

4.5 BASIC RESONANT-CIRCUIT Q MEASUREMENTS

As discussed in Chap. 2, a resonant circuit has a Q, or quality, factor. From a practical test standpoint, a resonant circuit with a high Q produces a sharp resonance curve (narrow bandwidth), whereas a low Q produces a broad resonance curve (wide bandwidth), as shown in Fig. 2.2.

The Q of a resonant circuit can be measured using a signal generator and a meter with an RF probe. A high-impedance digital meter generally provides the least loading effect on the circuit and thus provides the most accurate indication.

Figure 4.4*a* shows the test circuit in which the signal generator is connected directly to the input of a complete stage. Figure 4.4*b* shows the indirect method of connecting the signal generator to the input.

When the stage, or circuit, has sufficient gain to provide a good reading on the meter with a nominal output from the generator, the indirect method (with isolating resistor) is preferred. Any signal generator has some output impedance (typically 50 Ω). When this resistance is connected directly to the tuned circuit, the Q is lowered, and the response becomes broader. (In some cases, the generator output seriously detunes the circuit.)

Figure 4.4*c* shows the test circuit for a single component (such as an IF transformer). The value of the isolating resistance is not critical and is typically in the range of 100 *k*. The procedure for determining Q using any of the circuits in Fig. 4.4 is as follows:

1. Connect the equipment as shown in Fig. 4.4. Note that a load is shown in Fig. 4.4*c*. When a circuit is normally used with a load, the most realistic Q measurement is made with the circuit terminated in that load value. A fixed resistance can be used to simulate the load. The Q of a resonant circuit often depends on the load value.
2. Tune the signal generator to the circuit resonant frequency. Operate the generator to produce an unmodulated output.
3. Tune the generator frequency for maximum reading on the meter. Note the generator frequency.
4. Tune the generator below resonance until the meter reading is 0.707 times the maximum reading. Note the generator frequency. To make the calculation more convenient, adjust the generator output level so that the meter reading is some even value, such as 1 or 10 V, after the generator is tuned for maximum. This makes it easy to find the 0.707 mark.

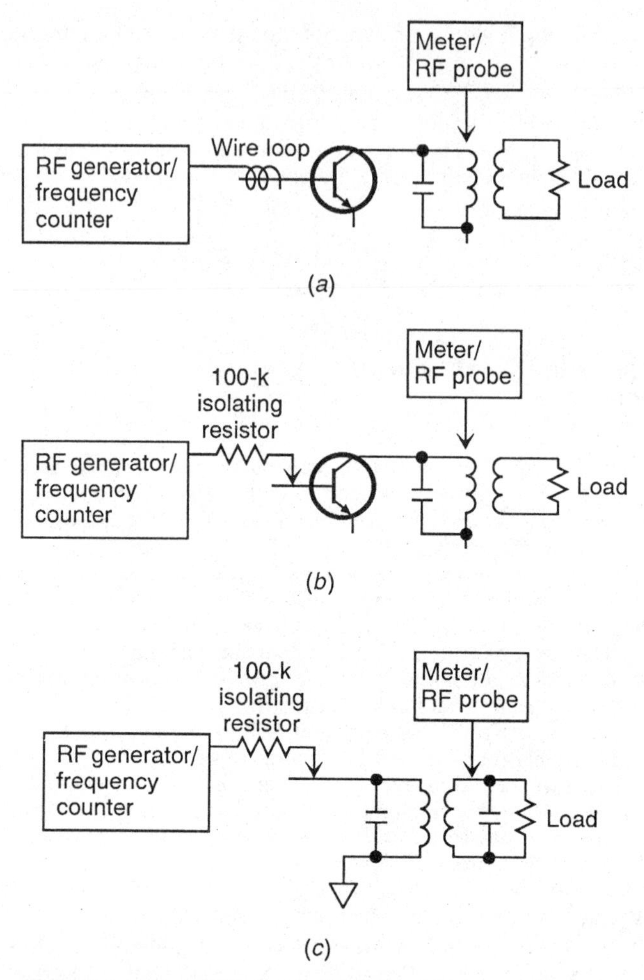

FIGURE 4.4 Basic resonant-circuit Q measurement.

5. Tune the generator above resonance until the meter reading is 0.707 times the maximum reading. Note the generator frequency.
6. Calculate the resonant-circuit Q using the equation in Fig. 2.2. For example, assume that the maximum meter indication occurs at 455 kHz (F_R), the below-resonance indication is 453 kHz (F_2), and the above-resonance indication is 458 kHz (F_1). Then, Q = 455/(458 − 453) = 91.

4.6 BASIC RESONANT-CIRCUIT IMPEDANCE MEASUREMENTS

Any resonant circuit has some impedance at the resonant frequency. The impedance changes as frequency changes. This includes transformers (tuned and untuned), RF tank circuits, and so on. In theory, a series-resonant circuit has zero impedance, while a parallel-resonant circuit has infinite impedance, at the resonant frequency. In practical RF circuits, this is impossible since there is always some resistance in the circuit.

It is often convenient to find the impedance of an experimental resonant circuit at a given frequency. Also, it may be necessary to find the impedance of a component in an experimental circuit so that other circuit values can be designed around the impedance. For example, an IF transformer presents an impedance at both the primary and secondary windings. These values may not be specified on the transformer datasheet.

The impedance of a resonant circuit or component can be measured using a signal generator and a meter with an RF probe, as shown in Fig. 4.5. A high-impedance digital meter provides the least loading effect on the circuit and thus produces the most accurate indication.

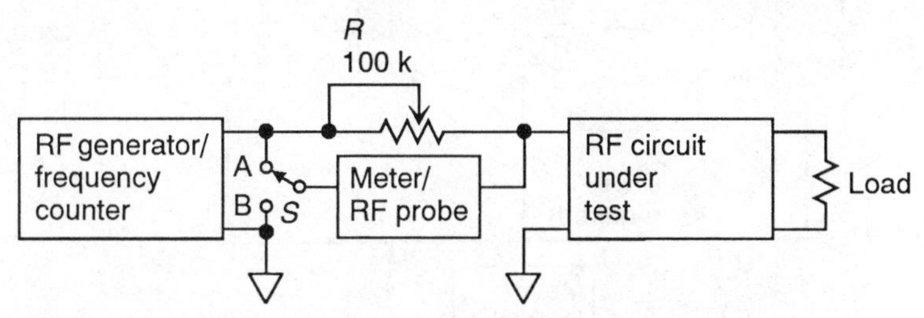

FIGURE 4.5 Basic resonant-circuit impedance measurements.

If the circuit of a component under measurement has both an input and output (such as a transformer), the opposite side or winding must be terminated in the normal load as shown. If the impedance of a tuned circuit is to be measured, tune the circuit to peak or dip, then measure the impedance at resonance. Once the resonant impedance is found, the signal generator can be tuned to other frequencies to find the corresponding impedance (if required).

The RF generator is adjusted to the frequency (or frequencies) at which impedance is to be measured. Switch *S* is moved back and forth between positions A and B, while resistance *R* is adjusted until the voltage reading is the same in both positions of the switch. Resistor *R* is then disconnected from the circuit, and the dc resistance of *R* is measured with an ohmmeter. The dc resistance of *R* is then equal to the impedance at the circuit input.

Accuracy of the impedance measurement depends upon the accuracy with which the dc resistance is measured. A noninductive resistance must be used. The impedance found by this method applies only to the frequency used during the test.

4.7 BASIC TRANSMITTER RF-CIRCUIT TESTING AND ADJUSTMENT

It is possible to test and adjust transmitter RF circuits using a meter and RF probe. If an RF probe is not available (or as an alternative), it is possible to use a circuit such as shown in Fig. 4.6*a*. This circuit is essentially a pickup coil which is placed near the RF-circuit inductance and a rectifier that converts the RF into a dc voltage or measurement on a meter. The basic procedures are as follows:

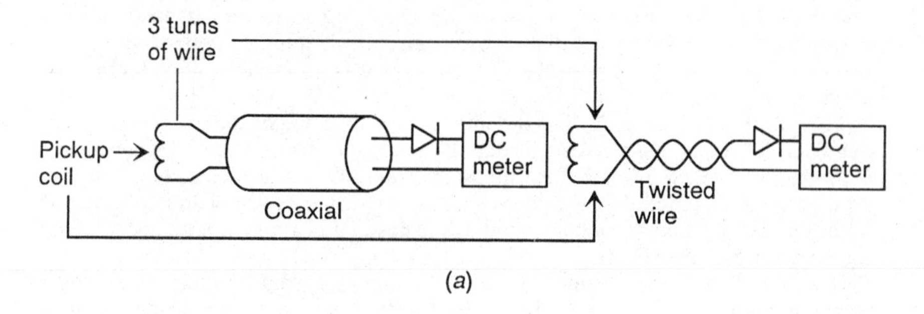

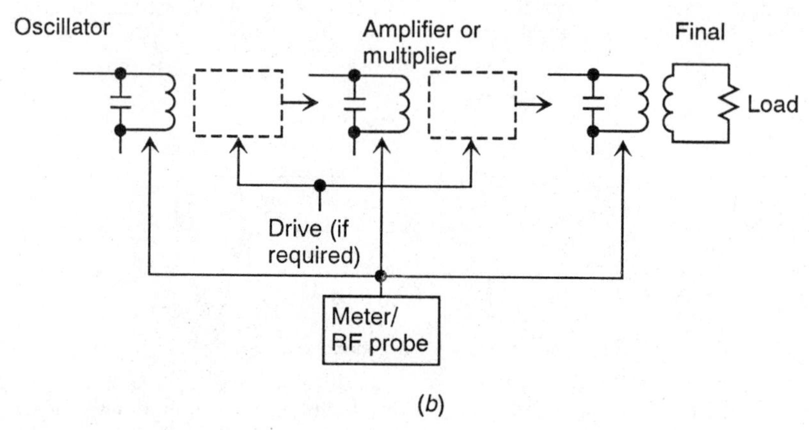

FIGURE 4.6 Basic transmitter RF-circuit testing and adjustments.

1. Connect the equipment as shown in Fig. 4.6*b*. If the circuit being measured is an RF amplifier, without an oscillator, a drive signal must be supplied by means of a signal generator. Use an unmodulated signal at the correct operating frequency.
2. In turn, connect the meter (through an RF probe or the special circuit in Fig. 4.6*a*) to each stage of the RF circuit. Start with the first stage (this is usually the oscillator if the circuit under test is a complete transmitter) and work toward the final (or output) stage.
3. A voltage indication should be obtained at each stage. Usually, the voltage indication increases with each RF-amplifier stage as you proceed from oscillator to the final amplifier. However, some stages may be frequency multipliers and provide no voltage amplification.
4. If a particular stage is to be tuned, adjust the tuning control for a maximum reading on the meter. If the stage is to be operated with a load (such as the final amplifier into an antenna), the load should be connected or a simulated load should be used. A fixed, noninductive resistance provides a good simulated load at frequencies up to about 250 MHz.
5. Note that this tuning method or measurement technique does not guarantee that each stage is at the desired operating frequency. It is possible to get maximum readings on harmonics. However, it is conventional to design RF-

transmitter circuits so that the circuits cannot tune to both the desired operating frequency and a harmonic. Generally, RF-amplifier tank circuits tune on either side of the desired frequency but not to a harmonic (unless the circuit is seriously detuned or the design calculations are hopelessly inaccurate).

4.8 BASIC RECEIVER RF-CIRCUIT TESTING AND ADJUSTMENT

The RF circuits of most present-day TV sets and AM/FM tuners use some form of frequency synthesis, or FS. The same is true of many modern radio receivers such as those used in communications and amateur radio. The procedures for testing and adjustment of frequency-synthesis RF circuits are discussed in Sec. 4.10.

This section describes basic receiver RF-circuit tests using a *meter and signal generator* and applies primarily to older discrete-component RF circuits. Note that these same circuits can also be tested and adjusted (or aligned) using a *sweep generator and scope,* as discussed in Sec. 4.9.

4.8.1 Basic RF-Circuit Alignment Sequence

Both AM and FM receivers require alignment of the RF and IF stages. An FM receiver also requires alignment of the detector stage (discriminator or ratio detector). The normal sequence for alignment in a complete FM receiver is (1) detector, (2) IF amplifier and limiter stages, and (3) RF and local-oscillator (mixer/converter). The alignment sequence for an AM receiver is (1) IF stages and (2) RF and local oscillator. The following procedures can be applied to a complete receiver or to individual stages.

If a complete receiver is being tested, and the receiver includes an AVC-AGC circuit, the AGC must be disabled. This is best done by placing a fixed bias, of opposite polarity to the signal produced by the detector, on the AGC line. The fixed bias should be of sufficient amplitude to overcome the bias signal produced by the detector (usually about 1 or 2 V). When such bias is applied, the stage gain is altered from the normal condition. Once alignment is complete, make certain to remove the bias.

4.8.2 FM-Detector Alignment and Testing

1. Connect the equipment as shown in Fig. 4.7*a* (for a discriminator) or Fig. 4.7*b* (for a ratio detector).
2. Set the meter to measure dc voltage.
3. Adjust the signal generator frequency to the intermediate frequency (usually 10.7 MHz for a broadcast FM receiver). Use an unmodulated output from the signal generator.
4. Adjust the secondary winding (either capacitor or tuning slug) of the discriminator transformer for zero reading on the meter. Adjust the transformer slightly each way and make sure the meter moves smoothly above and below

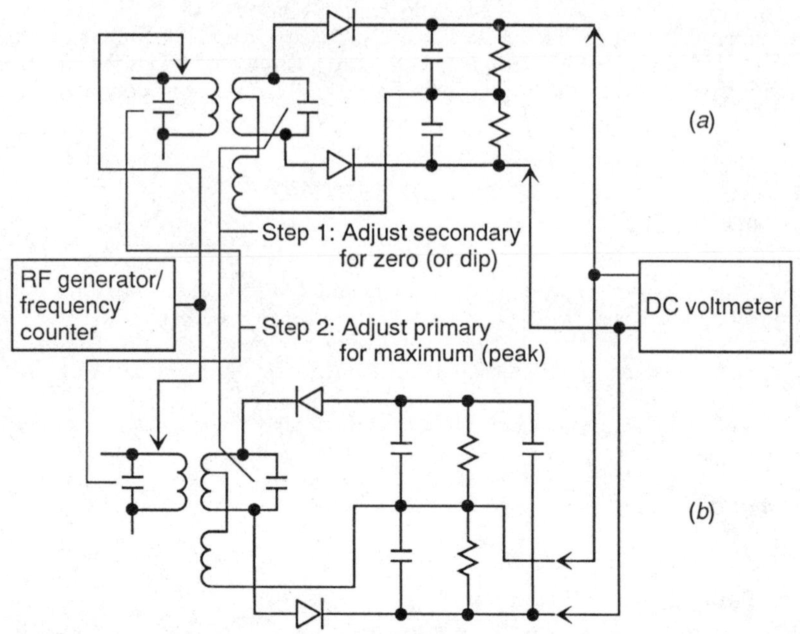

FIGURE 4.7 Basic FM-detector alignment and testing.

the exact zero mark. (A *meter with a zero-center scale* is most helpful when adjusting FM detectors of the type shown in Fig. 4.7.)

5. Adjust the signal generator to some point below the intermediate frequency (to 10.625 MHz for an FM detector with a 10.7-MHz IF). Note the meter reading. If the meter reading goes downscale against the pin, reverse the meter polarity or test leads (the RF probe is not used for FM-detector alignment).
6. Adjust the signal generator to some point above the intermediate frequency, exactly equivalent to the amount set below the IF in step 5. For example, if the generator is set to 0.075 MHz below the IF (10.7 − 0.075 = 10.625), set the generator to 10.775 (10.7 + 0.075 = 10.775).
7. The meter should read approximately the same in both steps 5 and 6, except that the polarity is reversed. For example, if the meter reads seven scale divisions below zero for step 5 and seven scale divisions above zero for step 6, the detector is balanced. If a detector circuit under test cannot be balanced under these conditions, the fault is usually a series mismatch of diodes or other components.
8. Return the generator output to the intermediate frequency (10.7 MHz).
9. Adjust the primary winding of the detector transformer (either capacitor or tuning slug) for a maximum reading on the meter. This sets the primary winding at the correct resonant frequency of the IF amplifier.
10. Repeat steps 4 through 8 to make sure that adjustment of the transformer primary has not disturbed the secondary setting. (The two settings usually interact.)

4.8.3 AM and FM Alignment and Testing

The alignment procedures for the IF stages of an AM receiver are essentially the same as those for an FM receiver. However, the meter must be connected at different points in the corresponding detector, as shown in Fig. 4.8. In either case the meter is set to measure direct current, and the RF probe is not used. In those cases where IF stages are being tested without a detector (such as during design), an RF probe is required. As shown in Fig. 4.8, the RF probe is connected to the secondary of the final IF output transformer.

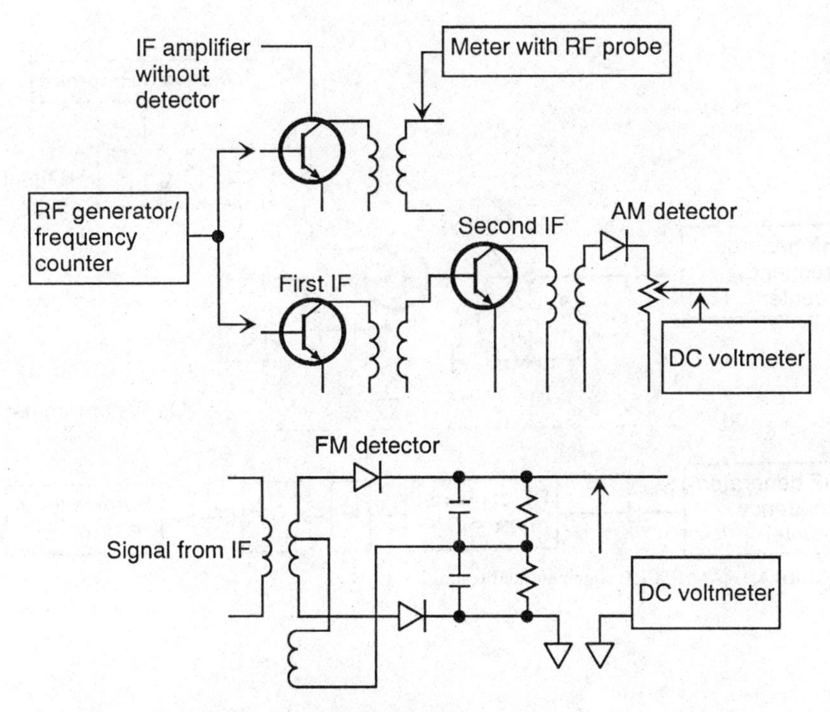

FIGURE 4.8 AM/FM IF alignment and testing.

1. Connect the equipment as shown in Fig. 4.8.
2. Set the meter to measure direct current at the appropriate test point (with or without an RF probe as applicable).
3. Place the signal generator in operation and adjust the generator frequency to the receiver intermediate frequency (typically 10.7 MHz for FM and 455 kHz for AM). Use an unmodulated output from the signal generator.
4. Adjust the windings of the IF transformers (capacitor or tuning slug) in turn, starting with the last stage and working toward the first stage. Adjust each winding for maximum reading.
5. Repeat the procedure to make sure the tuning on one transformer has no effect on the remaining adjustments. Adjust each winding for maximum reading.

4.8.4 AM and FM RF Alignment and Testing

The alignment procedures for the RF stages (RF amplifier, local oscillator, mixer and converter) of an AM receiver are essentially the same as for an FM receiver. Again, it is a matter of connecting the meter to the appropriate test point. The same test points used for IF alignment can be used for aligning the RF stages, as shown in Fig. 4.9. However, if an individual RF stage is to be aligned, the meter must be connected to the secondary winding of the RF-stage output transformer, through an RF probe. The procedure is as follows:

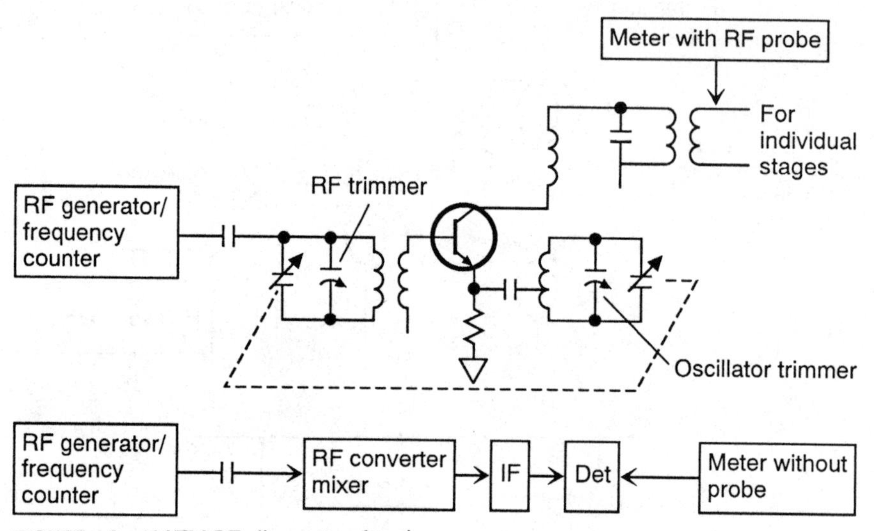

FIGURE 4.9 AM/FM RF alignment and testing.

1. Connect the equipment as shown in Fig. 4.9.
2. Set the meter to measure direct current at the appropriate test point (with or without an RF probe as applicable).
3. Adjust the generator frequency to some point near the high end of the receiver operating frequency (typically 107 MHz for a broadcast-FM receiver and 1400 kHz for a broadcast-AM receiver). Use an unmodulated output from the signal generator.
4. Adjust the RF-stage trimmer for maximum reading on the meter.
5. Adjust the generator frequency to the low end of the receiver operating frequency (typically 90 MHz for FM and 600 kHz for AM).
6. Adjust the oscillator-stage trimmer for maximum reading on the meter.
7. Repeat the procedure to make sure the resonant circuits "track" across the entire tuning range.

4.9 RECEIVER RF-CIRCUIT TESTING AND ADJUSTMENT USING SWEEP TECHNIQUES

The response characteristics of non-FS receiver RF circuits can be checked, or the IF stages aligned, using a sweep-generator and scope combination. The sweep generator must be capable of sweeping over the entire IF range. If maximum accuracy is desired, a marker generator must also be used. Before going into the specific testing and adjustment procedures, let us review the basic sweep-generator and scope test technique, as well as some sweep-generator features.

4.9.1 Sweep and Marker Generators

The main purpose of sweep and marker generators in RF service is *sweep-frequency alignment.* A sweep and marker generator capable of producing signals of the appropriate frequency is used with a scope to display the bandpass characteristics of an RF circuit (RF tuner, IF, video amplifier, etc.).

The sweep portion of the generator is essentially an FM generator. When the sweep generator is set to a given frequency, this is the *center frequency.* The rate at which the frequency modulation takes place is typically 60 Hz. The sweep width, or the amount of variation from the center frequency, is determined by a control, as is the center frequency.

The marker portion is essentially an RF generator with highly accurate dial markings that can be calibrated precisely against internal or external signals. Usually, the internal signals are crystal controlled. The marker signals necessary to pinpoint frequencies when making sweep-frequency alignments are usually produced by a built-in *marker adder.* The basic sweep-frequency alignment procedure is described in Sec. 4.9.2.

In addition to the basic sweep and marker outputs, the generator may have other special features. For example, the generator may provide a *variable bias* to disable the AGC circuits and a *blanking circuit* that permits a zero reference line to be observed on the scope during the retrace period.

4.9.2 Basic Sweep-Frequency Alignment Procedure

The relationship between the sweep and marker generator and the scope during sweep-frequency alignment is shown in Fig. 4.10. If the equipment is connected as shown in Fig. 4.10*a,* the scope sweep is triggered by a sawtooth output from the generator. The scope's internal sweep is switched off, and the scope sweep selector and sync selector are set to external.

Under these conditions, the scope sweep represents the total sweep spectrum, as shown in Fig. 4.10*c,* with any point along the horizontal trace representing a corresponding frequency. For example, the midpoint on the trace represents 15 kHz. If you want a rough approximation of frequency, adjust the horizontal gain control until the trace occupies an exact number of scale divisions on the scope screen (such as 10 cm for the 10- to 20-kHz sweep signal). Each centimeter division then represents 1 kHz.

If the equipment is connected as shown in Fig. 4.10*b,* the scope sweep is triggered by the scope's internal circuits (both the sweep and sync selectors are set

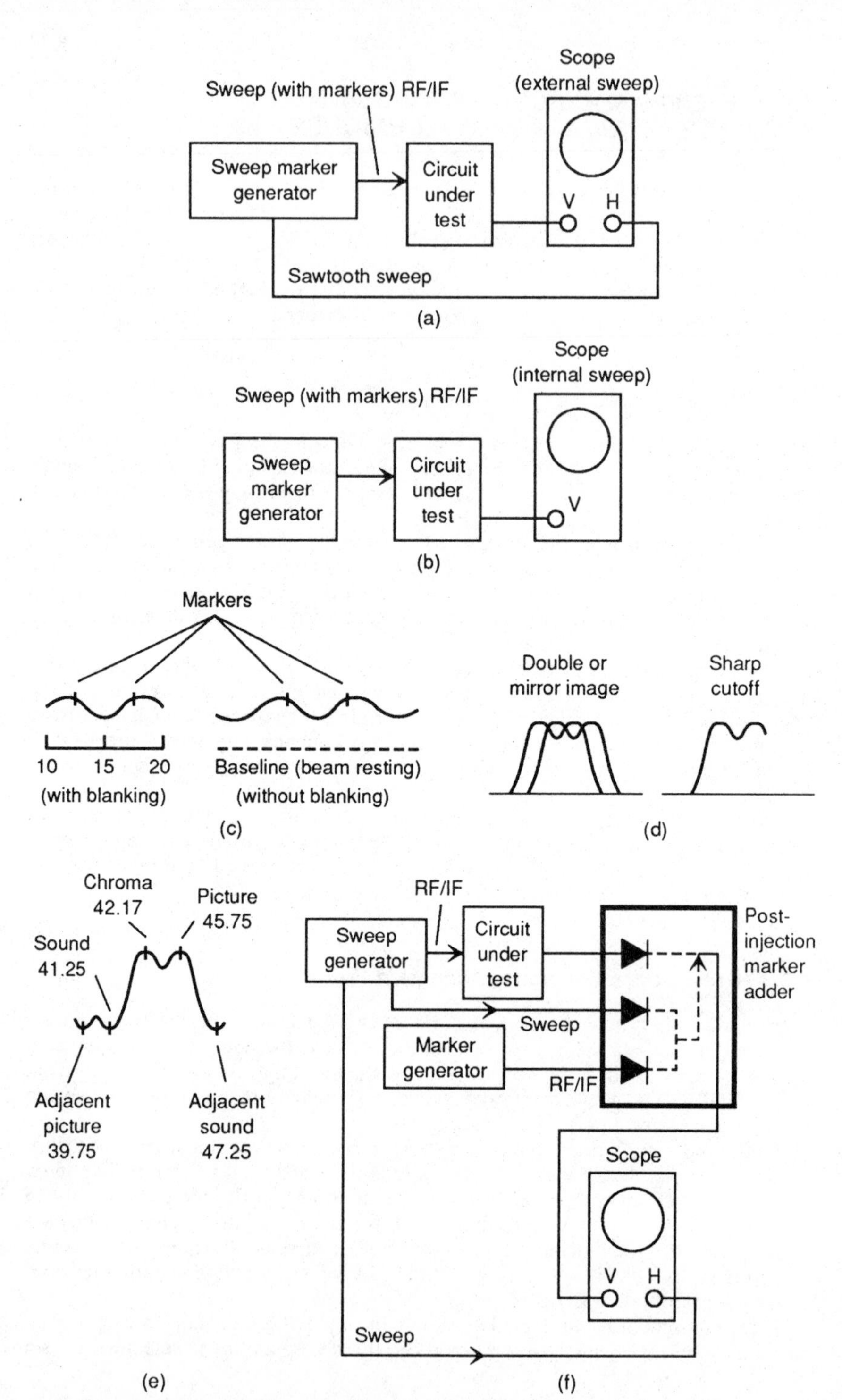

FIGURE 4.10 Basic sweep-frequency alignment procedure.

to internal). Certain conditions must be met to use the test connections shown in Fig. 4.10*b*. If the scope has a triggered sweep, there must be sufficient delay in the vertical input, or part of the response curve may be lost. If the scope is not a triggered sweep, the generator must be swept at the same frequency as the scope (usually at the line frequency of 60 Hz). Also, the scope or generator must have a *phasing control* so that the two sweeps can be synchronized. If the phase adjustment is not properly set, the sweep curve may be prematurely cut off, or the curve may appear as a double or mirror image, as shown in Fig. 4.10*d*.

As shown in Fig. 4.10*c*, the markers provide accurate frequency measurement. Although some older generators have variable markers, present-day generators have a number of markers at precise, crystal-controlled frequencies. Such fixed-frequency markers are illustrated in Fig. 4.10*e*, which shows the bandpass response curve of a typical RF circuit (a VCR tuner and VIF package). The markers can be selected (one or several at a time) as needed.

The response curve (as shown by the scope trace) depends on the RF circuit under test. If the circuit has a wide bandpass characteristic, the generator is set so that the sweep is wider than that of the circuit. (The bandpass of the circuit shown in Fig. 4.10*e* is about 6 MHz.) Under these conditions, the trace starts low at the left, rises toward the middle, and then drops off at the right.

The sweep and marker-scope method of alignment tells at a glance the overall bandpass characteristic of the circuit (sharp response, irregular response at certain frequencies, and so on). The exact frequency limits of the bandpass can be measured with the markers (often called *pips*).

4.9.3 Direct Injection versus Postinjection

There are two basic methods for injection of marker signals. With *direct injection,* the sweep-generator and marker-generator signals are mixed *before* the signals are applied to the circuit. This method is sometimes called *preinjection* and has generally been replaced by postinjection.

With postinjection, as shown in Fig. 4.10*f*, the sweep-generator output is applied to the circuit. A portion of the sweep-generator output is also mixed with the marker-generator output in a mixer-detector circuit known as a *postinjection marker adder.* The mixed and detected output from both generators is then mixed with the detected output from the circuit. The scope vertical input represents the detected values of all three signals (sweep, marker, and circuit output).

Most present-day sweep and marker generators have some form of built-in postinjection marker-adder circuits. The postinjection method (sometimes known as the *bypass injection* technique) for adding markers is usually preferred for consumer-electronics RF service because postinjection minimizes the chance of overloading the circuits and permits use of a narrowband scope (typically a less expensive scope). At one time, postinjection marker-adders were available as separate units, and some are still in use today.

4.9.4 Typical Sweep and Marker Generator Features

The following are features found on typical sweep and marker generators.

There are RF outputs, with equivalents of all IF and chroma (color) markers (Fig. 4.10*e*) available for one or more VHF TV channels (typically 3 and 4). This makes it possible to connect the RF output to the antenna terminals of a TV set or VCR and, without further input reconnections, evaluate alignment conditions of all tuned signal-processing RF circuits (including IF and video).

There are several (typically 10) *crystal-controlled markers,* and postinjection markers can be added. All markers can be used simultaneously or individually. The markers shown in Fig. 4.10*e* are typical.

A video-sweep output permits direct sweep alignment of the video circuits where specified by the manufacturer. With some generators it is necessary to use the IF sweep for signal injection and then monitor the video circuits for response.

The generator may include *pattern polarity reversal* and *sweep reversal* features which permit you to match scope displays shown in service literature (positive or negative, left- or right-hand sweep, etc.).

In some generators, the *markers can be tilted* to horizontal or vertical positions, permitting easy identification. For example, if the sides of a bandpass display are steep (vertical), a horizontal marker is easier to identify. A vertical marker shows up better on the flat top (horizontal) of the same pattern.

The generator may also include a number of features that are primarily for TV service: built-in amplifiers and filters, marker outputs for spot alignment of traps and bandpass circuits, and visual reproductions of idealized alignment curves on the front panel to indicate desired marker positions.

4.9.5 Basic Sweep-Generator and Scope Testing and Adjustment Procedure

The following steps describe the basic procedure for using a sweep generator with a scope.

1. Connect the equipment as shown in Fig. 4.10*f.*
2. Place the scope and sweep generation in operation. Allow at least a 15-min warmup before proceeding with any test or alignment.
3. Switch off the scope internal sweep, and set the scope sweep selector and sync selector to external so that the horizontal sweep is taken from the generator sweep and the length of the sweep represents total sweep-frequency spectrum.
4. Use markers (built in or from an adder) for the most accurate frequency measurement.
5. Switch the sweep-generator blanking control (if any) on or off as desired. Some sweep generators do not have a blanking function. With the blanking function in effect, there is a zero-reference line on the trace (Fig. 4.10*c*). With the blanking function off, the horizontal baseline does not appear. The sweep-generator blanking function is not to be confused with the scope blanking (which is bypassed when the sweep signal is applied directly to the scope horizontal oscillator).

4.9.6 Basic Sweep-Generator and Scope IF Alignment

1. Connect the equipment as shown in Fig. 4.11.
2. Place the scope in operation. Switch off the internal sweep. Set the scope sweep selector and sync selector to external.

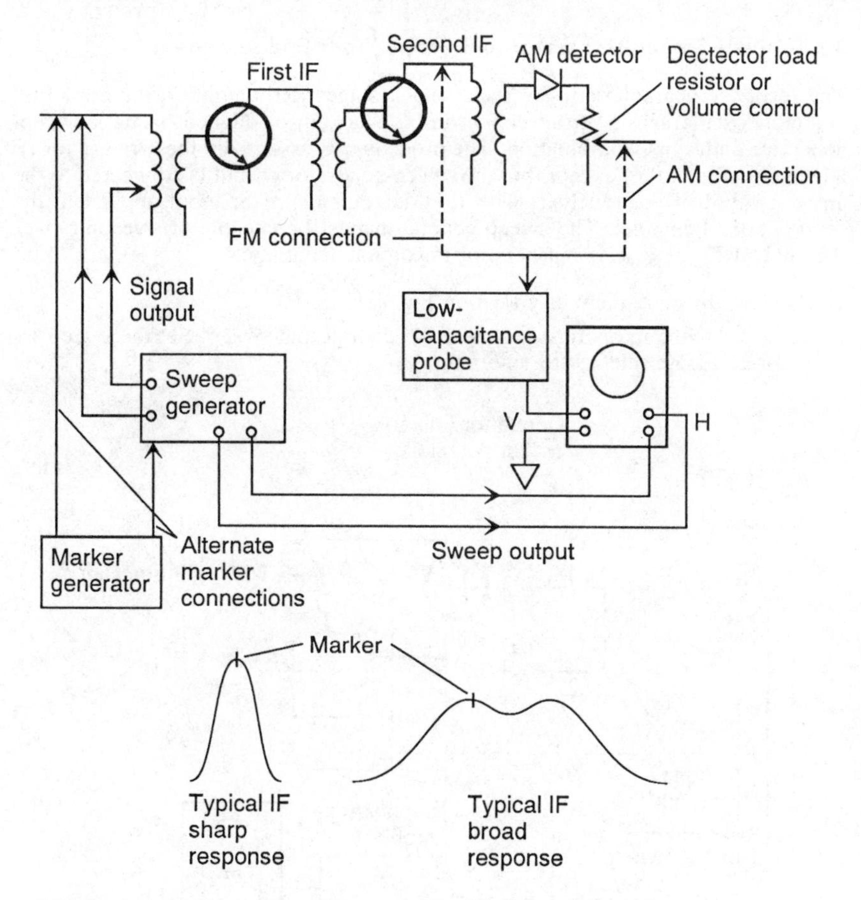

FIGURE 4.11 Basic sweep-generator and scope IF alignment.

3. Place the sweep generator in operation. Switch the sweep-generator blanking control on or off as desired. Adjust the sweep generator to cover the complete IF range. Usually, the AM IF center frequency is 455 kHz and requires a sweep about 30 kHz wide. The FM IF center frequency is 10.7-MHz and requires a sweep of about 300 kHz.
4. Check the IF response curve appearing on the scope against those of Fig. 4.11 or (even better) against the receiver specifications.
5. Use the markers to determine the frequencies at which the IF response occurs.
6. The amplitude of any point on the response curve can be measured directly on the scope (assuming that the vertical display is voltage-calibrated (as it is on most scopes).
7. Adjust the IF alignment controls to produce the desired response curve, as specified in the receiver service data and as discussed throughout the remainder of this section.

4.9.7 Basic Sweep-Generator and Scope Front-End Alignment

The response characteristics of receiver RF stages (RF amplifier, mixer or first detector, oscillator) or "front end" can be checked, or aligned, using a sweep-generator and scope combination. The procedure is essentially the same as for IF alignment (Sec. 4.9.6) except that the sweep-generator output is connected to the antenna input of the receiver, with first-detector or mixer input applied to the scope vertical channel. The sweep generator must be capable of sweeping over the entire RF range. Use markers for maximum accuracy.

1. Connect the equipment as shown in Fig. 4.12.
2. Place the scope in operation. Switch off the internal sweep. Set the sweep selector and sync selector to external.

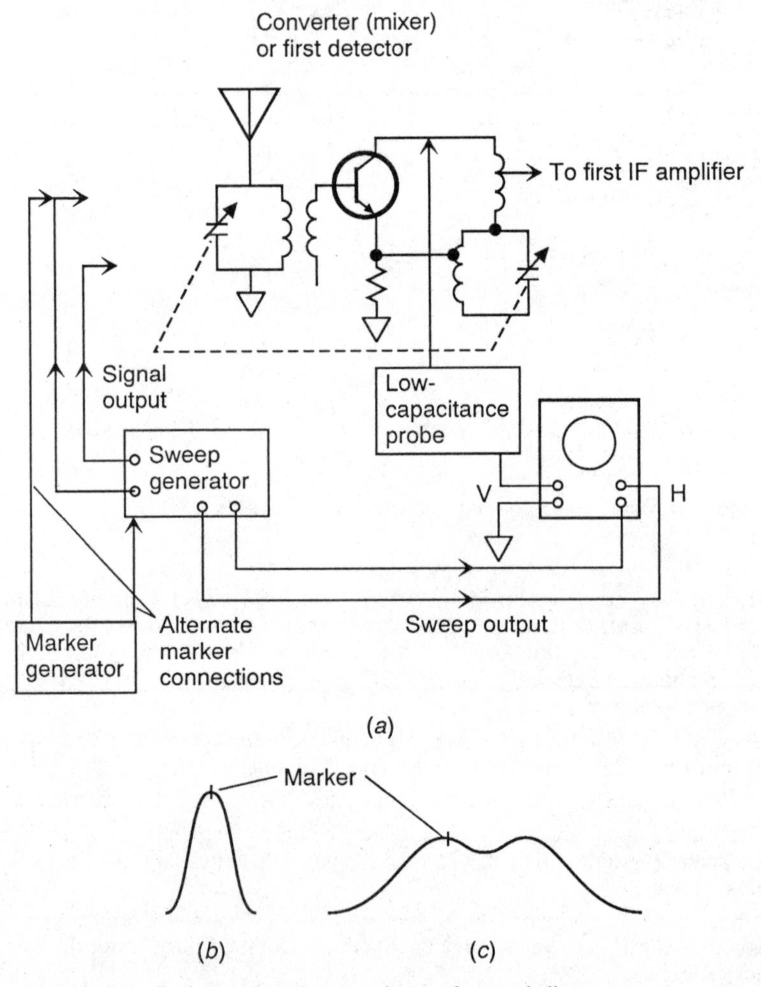

FIGURE 4.12 Basic sweep-generator and scope front-end alignment.

3. Place the sweep generator in operation. Switch the sweep-generator blanking control on or off as desired. Adjust the sweep generator to cover the complete RF range. The center frequency depends on the receiver. Usually, an AM receiver needs a 30-kHz sweep width, and an FM receiver needs about 300 kHz.
4. Check the RF response curve appearing on the scope against those in Fig. 4.12 or against the receiver specifications.
5. Use the markers to determine the frequencies at which the RF response occurs.
6. The amplitude of any point on the response curve can be measured directly on the scope (assuming that the vertical display is voltage-calibrated).
7. Adjust the RF alignment controls to produce the desired response curve, as specified in the receiver service data and as discussed throughout the remainder of this section. Usually, the RF response of an AM receiver is similar to that in Fig. 4.12*b,* whereas an FM receiver has a broad response similar to that in Fig. 4.12*c*.

4.9.8 Basic Sweep-Generator and Scope FM-Detector Alignment

The detector of an FM receiver (either discriminator or ratio detector) can be aligned using the sweep-generator and scope combination. The test connections are similar to front-end and IF alignment. The sweep-generator output is connected to the last IF stage input, whereas the scope vertical channel is connected across the FM-detector load resistor. The sweep generator must be capable of sweeping over the entire IF range. Use markers for maximum accuracy.

1. Connect the equipment as shown in Fig. 4.13*a*.
2. Place the scope in operation. Switch off the internal sweep. Set the scope sweep selector and sync selector to external.
3. Place the sweep generator in operation. Switch the sweep-generator blanking control on or off as desired. Set the sweep-generator frequency to the receiver IF (usually 10.7 MHz). Adjust the sweep width to about 300 kHz.
4. Check the detector response curve appearing on the scope against that in Fig. 4.13*b* or against the receiver specifications.
5. Adjust the last IF-stage and detector alignment controls so that peaks 2 and 4 of the response curve are equal in amplitude above and below the zero line. Also, points 1, 3, and 5 of the response curve should be on the zero reference line.
6. Use the markers to determine the frequencies at which the detector response occurs.
7. The amplitude of any point on the response curve can be measured directly on the scope (assuming that the vertical display is voltage-calibrated).

4.9.9 Basic TV RF-Circuit Alignment

There are four separate steps for overall alignment of black and white sets (and the black and white circuits of color sets): (1) tuner or RF alignment, (2) picture (video) IF alignment, (3) trap alignment, and (4) sound IF and FM-detector alignment. These procedures are described in the following sections.

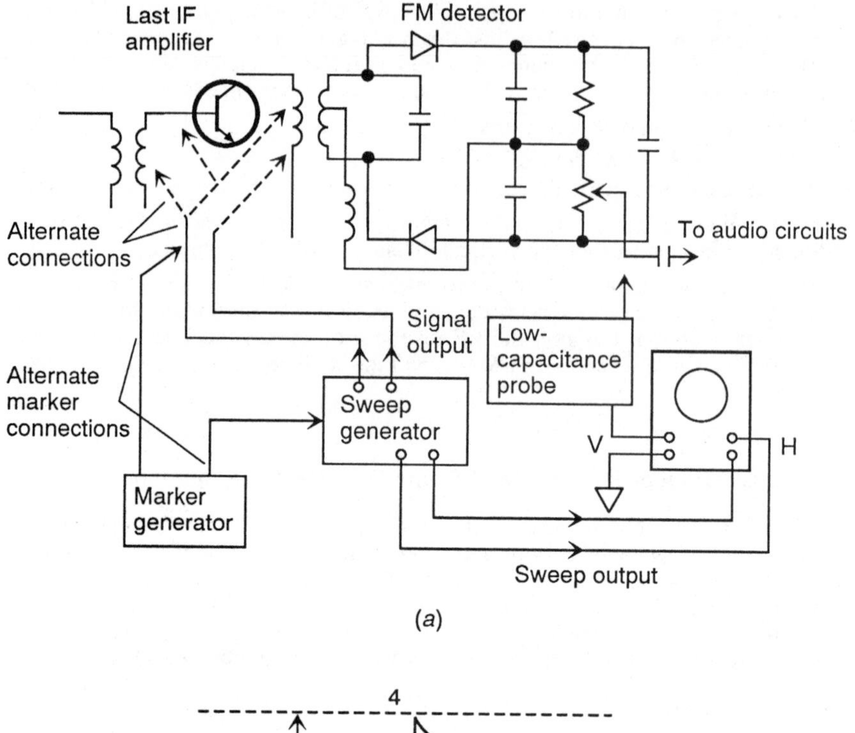

FIGURE 4.13 Basic sweep-generator and scope FM-detector alignment.

Remember that the following descriptions are general in nature and that the steps apply primarily to sets with discrete-component circuits. However, present-day sets with replaceable circuit boards still require alignment. The following basic procedures can be used as a guide in performing alignment for any set. However, always follow the specific procedures recommended in the service literature.

4.9.10 RF-Tuner Alignment

Figure 4.14 shows the basic test connections and typical waveshape for postinjection alignment of an RF tuner found in (non-FS) TV sets. Figure 4.15 shows the direct-injection alignment connections. The primary purpose of alignment is to obtain a response curve of proper shape, frequency coverage, and

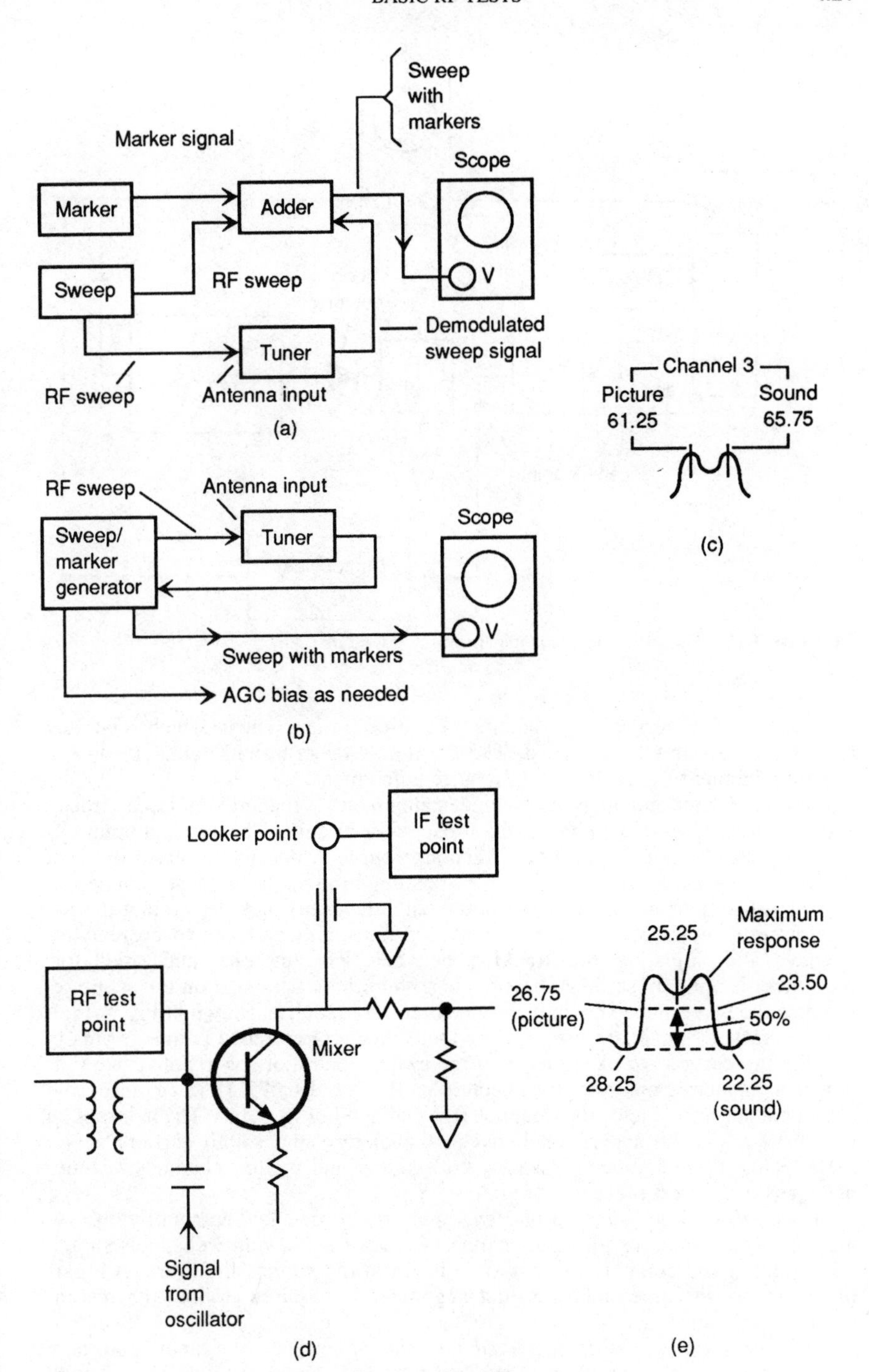

FIGURE 4.14 Basic postinjection alignment of an RF tuner.

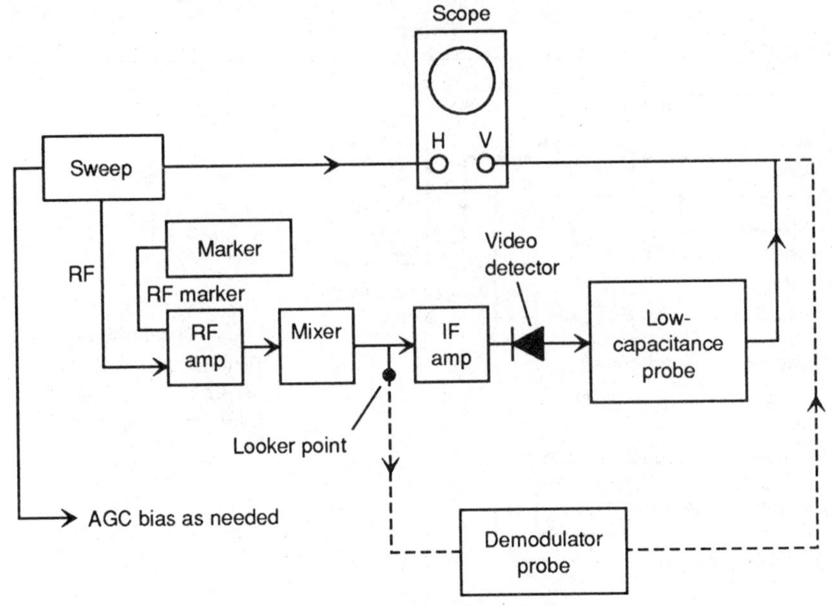

FIGURE 4.15 Basic direct-injection alignment of an RF tuner.

gain. Most RF tuners merely require "touch-up" alignment in which relatively few of the adjustments are used. The FS tuners described in Sec. 4.10 do not require alignment (as such) but do require adjustment.

Complete front-end alignment includes alignment of the antenna-input circuits and adjustment of the amplifier and RF-oscillator circuits. The antenna input circuits are usually aligned to give a response curve which has a sharp drop-off slightly below Channel 2 and which is flat up through the highest channel involved. Alignment of the amplifier and oscillator stages includes setting the oscillator frequencies for all channels, setting one or more traps to correct frequency, and adjusting the tracking between RF amplifier and oscillator. Alignment is done by setting adjustments so that the waveshape on the scope resembles the waveshape shown in service literature (such as shown in Fig. 4.14*c*).

The marker signals are used to provide frequency-reference points as aids in shaping the curve. For example, with the sweep generator set to deliver an output on Channel 3, markers are injected at 61.25 and 65.75 MHz (picture- and sound-carrier frequencies for Channel 3), as shown in Fig. 4.14*c*. The markers on the curve show the separation between the picture and sound carriers of 4.5 MHz. Since the RF tuner must pass both sound and picture channels, a tuner bandpass of about 6 MHz is required.

If you are working with a supposedly good tuner, check alignment by observing the response curves for each channel. Curves for individual channels should be examined and compared with those shown in the service literature. If a particular response curve indicates that alignment is required, follow the recommended procedure.

If you are working with a suspected defective tuner, leave the tuner set to only one channel until the trouble is located and cleared. Then other channels can be compared with the initial channel for sensitivity, switching noise, and general performance.

RF-Tuner Alignment Notes. The following notes describe typical alignment and test connections for an RF tuner:

1. Do not start alignment until the set *and all test equipment* have warmed up (even with solid-state equipment). Leave the tuner oscillator in operation during alignment. It is possible to make some tuner adjustments with the oscillator disabled. However, the lack of oscillator injection voltage at the mixer alters the mixer bias, resulting in an increase in amplitude of the response curve and distortion of the waveshape.

If there has been extensive work on the RF tuner (particularly by untrained "technicians"), it may be necessary to check the oscillator frequency on one or more channels with a counter. However, this is generally not required.

If you find serious misalignment of the tuner when making the test setup or if you find considerable difficulty or failure in alignment, this usually indicates a defect in the tuner. Likewise, if alignment fails to produce correct tuner curves, you should submit the tuner to full troubleshooting (or, more realistically, replace the tuner at outrageous expense to the customer).

2. Connect the equipment as shown in Fig. 4.14 for postinjection alignment. The connections in Fig. 4.14*a* assume that the sweep generator, marker generator, and marker adder are three separate units, while Fig. 4.14*b* assumes that a single sweep and marker generator (with adder) is used (typical for present-day test equipment). With either arrangement, it is necessary to connect the output to the scope vertical input.

The output test point may be across the load resistor of the video detector (IF output) or at the RF-tuner output (at the "looker" point). If you connect at the video detector, you get an "overall" response curve (RF, IF, detector). By connecting directly to the looker point, you can check the shape of the tuner curve independently of other circuits (which is usually the better method if you suspect problems in the tuner).

Figure 4.14*d* shows two typical output test points for an RF tuner. Note that the signals at the base test point are at the RF frequency, whereas signals at the collector test point are IF. With either test point, the signals are demodulated (in the marker-adder or generator) for presentation on the scope vertical input.

3. Connect the equipment as shown in Fig. 4.15 for direct-injection alignment. With this arrangement, both the sweep and marker-generator outputs are connected directly to the RF tuner input (at the antenna terminals). The output to the scope can be taken from the video detector (for an overall response curve) or at the tuner looker point.

Note that if the video-detector output is chosen, a low-capacitance probe can be used with the scope (since the video detector demodulates both the sweep and marker signals). However, if you monitor the output at the tuner looker point (to get a separate tuner response curve), you must use a demodulator probe because the tuner output is usually at a frequency beyond the bandpass of the scope.

4. With either alignment method, it is necessary to disable the AGC line. Figures 4.14 and 4.15 assume that bias voltages to disable the AGC circuits are available from the sweep and marker generator (which is usually the case with present-day generators). Always use the recommended bias voltage (amplitude, polarity, etc.) and apply the bias at the recommended test point. In some sets, there is a common AGC line for both tuner and IF amplifier. In other sets, the tuner bias is taken from a separate AGC line.

5. With the test connections made as shown in Figs. 4.14 and 4.15, adjust the sweep and marker generator to the correct channel. Set the sweep width to maximum or as recommended in the service literature. Typically, the sweep width is 10 to 15 MHz for most TV tuners. Adjust the scope controls for a trace similar to that shown in Fig. 2.14*c* (or as shown in literature).

6. The test setup is now complete and ready for test and/or alignment as described in the service literature. Typical non-FS TV alignment procedures are described in Secs. 4.9.14 through 4.9.20.

4.9.11 IF Amplifier and Trap Alignment

As in the case of the tuner, the primary purpose of IF-amplifier alignment is to get a response curve of proper shape, frequency coverage, and gain. The purpose of trap alignment is to remove undesired signals from the IF circuits. Again, most IF amplifiers require only touch-up alignment. Of course this may not be true if there has been extensive work on the IF circuits.

If a TV set is to give wideband amplification to the television signal, the picture (video) IF system of the set must pass a frequency band of about 3.5 to 4 MHz. This is necessary to ensure that all video information is fed through to the picture tube and that the resultant picture has full definition. (The bandpass of color sets must be essentially flat to beyond 4 MHz to ensure that the color information contained in the color sidebands is not lost. Likewise, the bandwidth of S-VHS TV sets must be even greater.)

The desired bandpass is obtained by proper alignment of the IF adjustment controls (typically coil tuning slugs but possibly trimmer capacitors), with sweep and marker signals fed to the IF amplifiers so that a response curve (with markers) is produced on the scope screen. The basic sweep and marker techniques described in Sec. 4.9.10 are used. However, input to the IF amplifiers is at the RF-tuner looker point or at a separate IF input cable (in some sets). Alignment is done by setting adjustments so that the waveshape on the scope resembles the waveshape shown in service literature. Typical alignment procedures are described in Secs. 4.9.14 through 4.9.20.

IF Alignment Notes. The following notes describe typical alignment problems and techniques. They should be compared with any notes found in literature and with example procedures described in the following paragraphs of this section.

Figure 4.14*e* is a typical IF amplifier waveshape obtained with a sweep and marker setup. Note that the frequency relationship of the sound carrier to the picture carrier is sometimes reversed in the IF amplifiers because the tuner oscillator usually operates at a frequency higher than that of the transmitted carrier.

For example, assume that (for Channel 9) a picture carrier is transmitted at 211.25 MHz and that the corresponding sound carrier is transmitted at 215.75 MHz. Further assume that the IF frequency is 26.75 MHz (typical for older black and white sets). If the tuner oscillator is at 238 MHz, this signal combines with the picture carrier of 211.25 MHz to produce a difference-IF of 26.75 MHz. However, the sound carrier of 215.75 MHz also combines with the 238-MHz oscillator signal to produce a difference-IF signal of 22.25 MHz, which is 4.5 MHz *below* the picture IF.

Some sweep and marker generators compensate for this condition by sweeping from high to low frequencies. However, always consult the service literature for proper response curves and marker frequencies, and check the generator literature for the method of producing the display.

No matter how the display is obtained, the following two characteristics of the IF-response curve (Fig. 4.14*e*) should be noted: (1) the amplitude of the picture carrier is set at about 50 percent of the maximum response, and (2) the sound-carrier amplitude must be at 1 percent (or less) of the maximum response.

The sound carrier is kept at this low level to prevent interference with the video signal. The *skirt selectivity* (or selectivity at the high and low extremes of the approximate 6-MHz bandpass) of the IF response is made sharp enough to reject the sound part of the composite signal. In some sets, an absorption circuit (consisting of a trap tuned to the sound IF) is used to get the necessary selectivity. In other sets, additional traps are tuned to frequencies of *adjacent* sound and picture channels. All of these traps have a pronounced effect on the shape of the response curve.

The picture carrier is placed at about 50 percent of maximum because of the nature of the TV broadcast transmission. If the IF circuits are adjusted to put the carrier too high on the response curve, the effect is a general decrease in picture quality caused by the resulting low-frequency attenuation. If the picture carrier is placed too low on the curve, there is a loss of low-frequency video response (that usually shows up as *poor definition* in the picture). *Loss of blanking* and *poor sync* can also occur when the picture carrier is too low on the response curve.

Note that in addition to the picture and sound frequencies, two additional marker frequencies are shown on the curve in Fig. 4.14*e* (23.5 and 28.25 MHz). These marker frequencies are used to align the IF traps. Most IF amplifiers have one or more traps, depending on the type of set. The setup for alignment of the IF traps is essentially the same as for alignment of the IF stages. However, certain problems may occur.

Because the IF-amplifier response is very low at the trap frequencies (when properly adjusted), the marker may be difficult to see on the response curve. In some cases, traps are set with a meter rather than a scope. The meter is connected to measure the signal at the video-detector load resistor (IF output), the marker generator is set for the trap frequency, and the trap is tuned for a minimum voltage reading on the meter.

Typically, the traps are set first, then the IF amplifier circuits are aligned. Since any adjustment of the circuits usually detunes the traps (slightly), the traps may require touch-up. Always follow the exact procedures given in service literature.

4.9.12 Alignment of 4.5-MHz Sound Traps

The video-amplifier circuits of all TV sets (black and white, as well as color) have a 4.5-MHz sound trap. This trap is tuned to reject any sound signals (at 4.5 MHz) that may pass through the video amplifier (so as to prevent the signals from appearing on the picture tube). Adjustment of the sound trap is usually not critical for black and white sets. However, the problem of *sound in picture* for a color set may be quite critical.

For a typical black and white set, the sound trap can be adjusted simply, without test equipment, once the RF and IF stages are aligned. To adjust the trap, tune in a strong TV broadcast signal, and set the contrast control at maximum. Adjust the fine-tuning control until a beat pattern (herringbone) is visible on the picture-tube screen. Then adjust the sound trap (usually a coil tuning slug) for *minimum beat interference.* Reset the fine-tuning control, making sure that the beat interference is completely gone when the fine tuning is set back to normal (for best picture and sound, usually at midrange). Do not try this on a color set or any set with FS tuning.

4.9.13 Checking Response of Individual Stages

Although it is not necessary to check individual RF or IF stages in present-day IC equipment, the need may arise in certain circuits (particularly in discrete-component circuits). The response of individual stages, or of two or more stages together, may be checked using a sweep and marker combination. Basically, the sweep signal is fed into the stage *immediately preceding* the stage being checked. The response curve is then checked on a scope connected across the output.

Figure 4.16 shows how the sweep and marker technique can be used to check the response of individual stages. In this case, the stages are IF and the output is monitored at the video-detector load resistor. As in the case of other IF-circuit tests, the AGC line must be displayed for accurate results.

1. Connect the equipment as shown in Fig. 4.16. Disable the AGC line using the bias recommended in service literature.
2. Connect the marker generator (or the marker output of a combined sweep and marker generator) to the RF-tuner looker point.
3. Connect the sweep generator to the input of the IF stages *ahead* of the stage being checked, which isolates it from the test equipment. Note that the sweep generator output should not be connected to the input of the stage being checked because even a slight loading of the input circuit may cause a change in circuit impedance and result in distortion of the normal response characteristics. The test connections in Fig. 4.16 (sweep generator at test point B) are to check response of the third IF amplifier.
4. Connect a resistor (typically 500 to 1000 Ω) across the primary of the transformer ahead of the stage being checked (primary of T_3). The resistor acts to "swamp" the primary winding and prevents inductive reactance of the winding from affecting the bandpass characteristics of the stage being checked (third IF). With the connections in Fig. 4.16, the response curve is determined primarily by the third-IF bandpass characteristics but is also affected by the video-detector response.
5. To check the bandpass characteristics of the video detector only, move the generator from point B to point C, and place the swamping resistor across the primary of T_4.
6. To check response of the second IF, third IF, and video detector stages together, move the generator to point A, and connect the swamping resistor across the primary of T_2.

4.9.14 Color-TV RF Tuned Circuits

To process the signals properly, the RF, IF, and chroma sections of color TV sets must have certain bandwidth characteristics (usually superior to those of black and

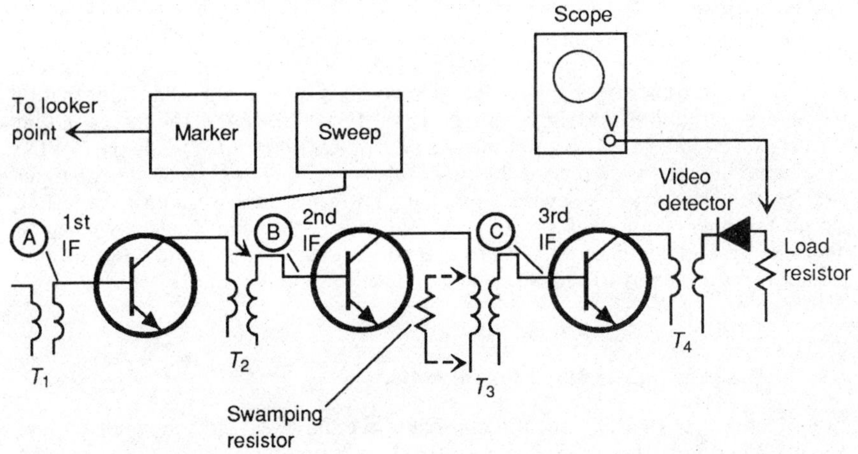

FIGURE 4.16 Sweep and marker technique used to check response of individual stages.

white TV). The IF bandpass is obviously narrower than the tuner bandpass and thus contributes most to bandpass shaping. In sets with circuits similar to those in Fig. 1.19 (discrete-component and/or tuned IC), the bandwidths are determined by the number of amplifiers and associated circuits. In present-day IC sets such as shown in Fig. 1.20, the bandwidths are set by IC and filter characteristics.

4.9.15 The Need for Sweep Alignment

The IF circuits in Fig. 1.20 do not require sweep alignment. Such circuits are often found in sets with FS tuners which also do not need sweep alignment (but do require adjustments, as described in Sec. 4.10).

The circuits in Fig. 1.19 (or any *tuned* circuit, IF or RF) must be properly aligned. If such circuits drift or are misaligned or if the gain of one or more states changes, the signals are affected in many ways. Signal levels may be too low, the bandwidth may be too narrow, the signals may begin to interfere with each other, or if traps are misaligned, the overall performance may be degraded by interference from undesired signals, such as adjacent-channel sound or picture-carrier signals. Most of these problems can be overcome by proper alignment.

4.9.16 Recommended Sweep-Alignment Techniques

The service literature of most manufacturers recommends sweep alignment for all tuned circuits similar to that in Fig. 1.19. However, the exact method and the order of steps to be performed are not standard. The following notes summarize typical recommendations.

Generally, alignment starts by injecting the sweep signal at the RF input (antenna terminals). You then monitor the IF and chroma outputs for comparison against service-literature waveform patterns. If the IF is good but the chroma is not, the problem is between the video-detector output and the bandpass-amplifier output (input to the color demodulators). If both IF and chroma outputs are abnormal, it is most likely that the IF requires a touch-up, particularly if the response is poor on the *slope affecting chroma response* (Sec. 4.9.17).

You seldom find an alignment problem in the RF portion of the tuner (unless there has been tinkering) because the bandwidth is so much greater than that of the IF section (about 6 MHz compared to 4 MHz). However, the mixer output circuit, which is located on the tuner, may require attention. This is part of the tuned matching network (called the *tuner link,* or simply *link*) between the tuner and the first IF stage. A separate prealignment procedure is given for the link circuit by some manufacturers. (In present-day sets, the tuner is usually a sealed package, and no alignment is possible. A possible exception is the output transformer from the tuner to the IF package, sometimes called the IFT.)

4.9.17 Diagnosing the Need for Alignment

You must realize that alignment alone is not the universal cure for poor picture quality. Before attempting to diagnose the need for alignment, you must be sure that the convergence, purity, and focus (and even high-voltage regulation) are good and that the set is properly degaussed. You should also eliminate the possibility of interference from other test equipment (caps left off RF-generator outputs, etc.).

With these problems out of the way, the quickest method to determine the overall condition of the set is to use sweep alignment (with markers), as described in Secs. 4.9.9 through 4.9.13. As the sweep signal is processed through the tuned circuit, the signal is shaped by the gain and bandpass characteristics of the various sections (RF, IF, chroma, etc.).

Figure 4.17*b* shows the sweep signal with basic response curves (waveforms) of the RF tuner, IF amplifiers, and chroma-bandpass circuits. The bandwidths shown are approximately to scale. These outlines are similar to the curves that you get if you monitor corresponding points in the set during sweep alignment.

Figure 4.17*b* includes some reference frequencies (for Channel 4) to show the importance of proper alignment. For example, note that the chroma frequencies are on the slope of the IF-response curve. This area is the most critical because improper IF alignment on the slope affects the amplitude and shape of the chroma-response curve. In turn, this affects color-picture quality.

4.9.18 Analyzing the Response Curve

Using the procedures in Secs. 4.9.9 through 4.9.13 and/or the service-literature instructions, you must now determine if the curves obtained are satisfactory or if the set must be realigned. If alignment is required, to what extent? Is a touch-up required or a complete realignment?

Figure 4.18 shows typical IF and chroma bandpass-response curves as the curves might appear in the service literature of a non-FS color TV set. Note that the reference-marker locations are shown with given tolerances. This means that the response curves obtained may vary within these limits and still give satisfactory performance.

Figure 4.18*c* and 4.18*d* show some allowable variations based on the limits in Fig. 4.18*a* and Fig. 4.18*b*. You must evaluate the response curves obtained with the allowable tolerances in mind. The areas to examine (aside from tilt across the

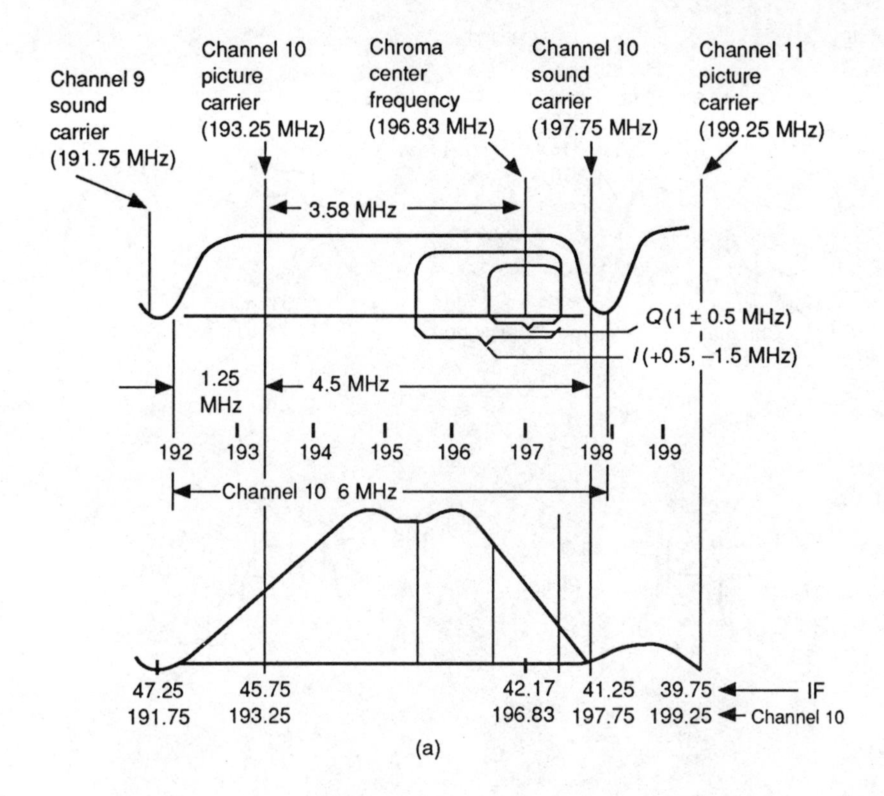

(a)

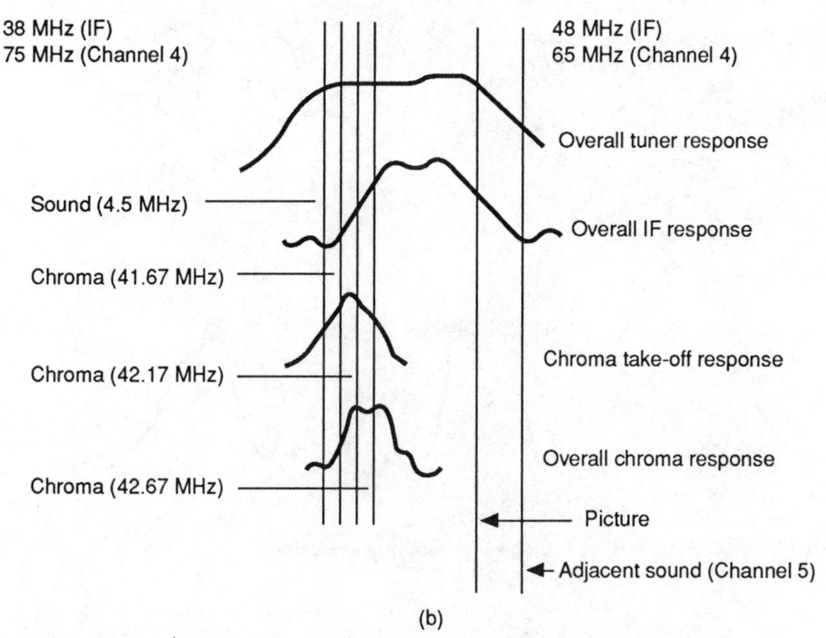

(b)

FIGURE 4.17 Simplified TV frequency spectrum of Channels 4 and 10.

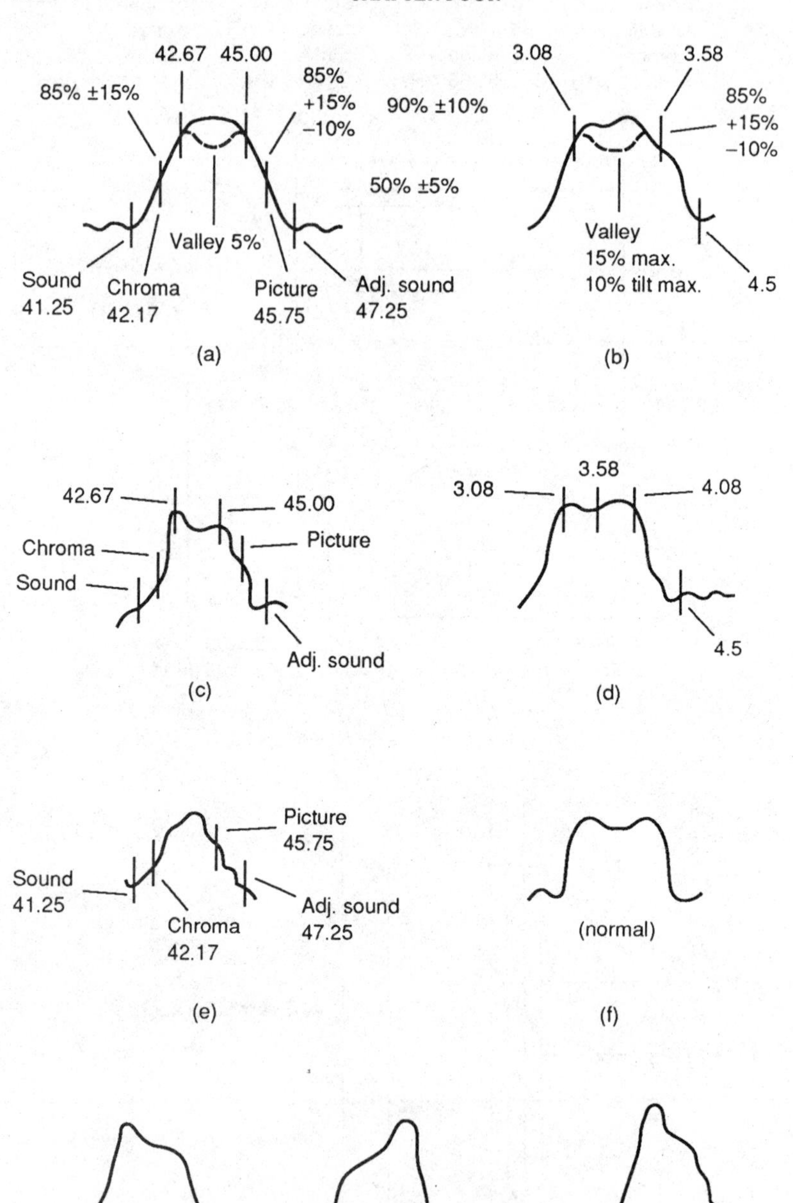

FIGURE 4.18 Typical IF and chroma bandpass-response curves.

curve top) are the areas of the trap frequencies, such as sound and adjacent sound and the picture and chroma markers at the 50-percent reference points.

If a trap has been detuned toward the center of the response curve, the overall response is pulled down. For example, if the sound trap is tuned near the chroma frequency, the curve response at the chroma frequency is as shown in Fig. 4.18*e*.

For stagger-tuned circuits, the curve is tilted if the circuit (tuned to the approximate center of the response curve) is misaligned. This is shown in Fig. 4.18*g* and 4.18*h*. Also, if one of the outside tuned circuits is tuned toward the center of the response curve, the curve peaks toward the center with reduced bandwidth and excessive gain, as shown in Fig. 4.18*i*.

By realizing the effect of mistuning traps and other circuits, you can usually locate the mistuned circuit if the approximate alignment frequency of each circuit is given in the service literature. One way to get this information is to check the alignment procedures thoroughly for *prealignment* instructions. Then make a cross-reference between marker frequencies and the tuned circuits to be adjusted. If prealignment of traps and transformers is specified, most or all of the information related to tuned-circuit frequencies is included in that section of the service instructions.

4.9.19 Alignment Touch-Up Procedures

Touch-up procedures are used when your analysis of the curve indicates that the curves are recognizable but marginal on response limits (such as excessive tilt, abnormal peaking, etc.). If the curves fall within the limits indicated, no alignment is required. If you decide to align, the extent of alignment must be determined. If the response curves are marginal at several or all points but are still recognizable, a touch-up can be performed to correct excessive tilts and to restore response levels at various points on the response curves.

4.9.20 Complete IF and Chroma Alignment

When complete IF and chroma alignment is required, use the basic procedures in Secs. 4.9.9 to 4.9.13, and observe the following notes (which supplement the service information, if any).

Trap Alignment. The standard trap frequencies are the adjacent-picture carrier (39.75 MHz), the sound carrier (41.25 MHz), and the adjacent-sound carrier (47.25 MHz), as shown in Fig. 1.19. However, several manufacturers have additional trap frequencies (such as 35.25 and 38.75 MHz found on older sets). All traps except the sound trap (41.25 MHz) are located at the input to the first IF. The sound trap is usually located in the last IF or in the output circuit of the last IF.

Prealignment. Some alignment procedures specify a pretuning of all IF bandpass circuits, as well as the traps. Each circuit is tuned for a maximum output as indicated on the scope using modulated markers. That is, you inject the required marker signal using the generator function and then tune the specified control for a maximum trace (typically a 400-Hz trace). If prealignment is recommended, remember to use only one marker at a time and turn off the marker when alignment of a particular circuit is complete.

Marker Height. In some service instructions, the phrase "set the marker height" is used extensively. This may be misleading since actual marker height is not adjustable. (Marker height is not to be confused with marker amplitude, which is adjustable on some generators.) The phrase means that you are adjusting the IF response so that at a particular marker frequency the response is at *some percentage down from maximum.* For example, to set the height of the 42.17-MHz marker at 50 percent amplitude, you adjust the IF to alter the curve so that at 42.17 MHz the response is 50 percent of maximum.

Chroma Markers. When aligning the chroma circuits, the 3.58-MHz oscillator within the TV set may produce a "marker" on the scope display. In some cases, a low-frequency beat may be produced between the set and the generator markers.

4.10 FREQUENCY-SYNTHESIS RF-CIRCUIT TESTING AND ADJUSTMENT

This section describes the testing and adjustment procedures for the RF circuits of a typical AM/FM tuner with frequency synthesis, using the tuner discussed in Chap. 1 (Fig. 1.13) as an example. We begin with a description of an FM stereo generator and the FM stereo broadcast system.

4.10.1 FM Stereo Generator

An FM stereo generator is essential for troubleshooting the FM portion of an AM/FM tuner. This is because an FM generator simulates the very complex modulation system used by FM stereo broadcast stations. Without an FM stereo generator, you are totally dependent on the constantly changing signals from such stations, making it impossible to adjust the FM portion of the tuner or to measure frequency response after adjustment.

FM Stereo Modulation System. Before we describe the characteristics of an FM stereo generator, let us review the basic FM stereo modulation system (which is quite complex when compared to that of the AM broadcast system). It is essential that you understand the FM system to troubleshoot RF circuits in the FM portion of any AM/FM tuner.

Figure 4.19*a* shows the composite audio-modulating signal used for FM stereo. This FM system permits stereo tuners and receivers to separate audio into left and right channels and permits mono FM tuners to combine left- and right-channel audio into a single output.

Figure 4.19*b* shows the block diagram of an FM stereo modulator and transmitter. Left- and right-channel audio signals are applied through preemphasis networks to a summing network that adds the two signals. A low-pass filter limits this signal to the 0- to 15-kHz audio band, which is the maximum authorized for FM broadcast service. This (L + R) signal contains both left- and right-channel audio in a 0- to 15-kHz baseband.

The left-channel audio is applied to another summing network, along with the right-channel audio, which is inverted. The summing network effectively subtracts the two signals. The 0- to 15-kHz (L − R) signal is fed to a balanced mod-

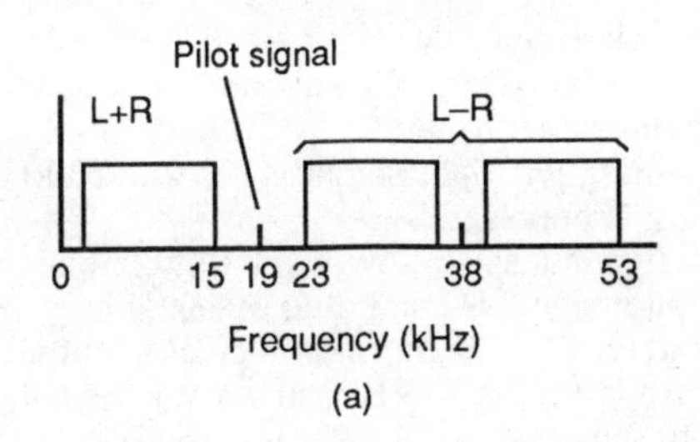

(a)

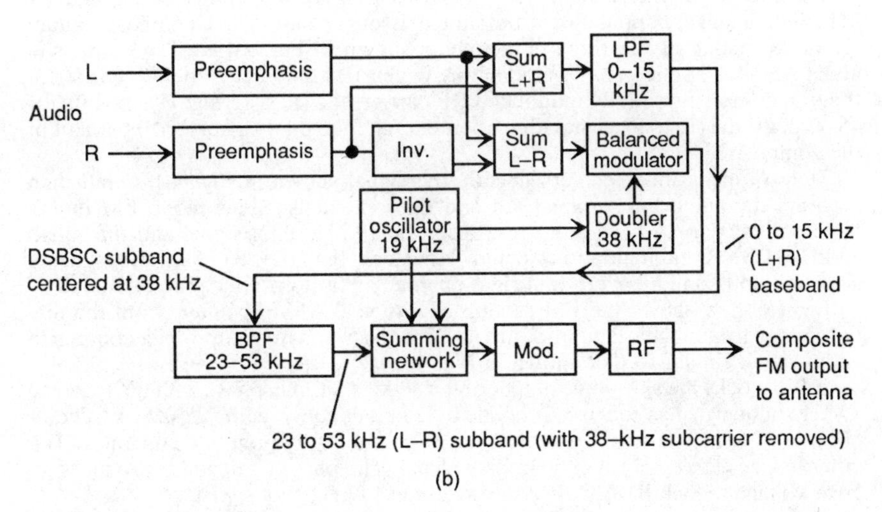

(b)

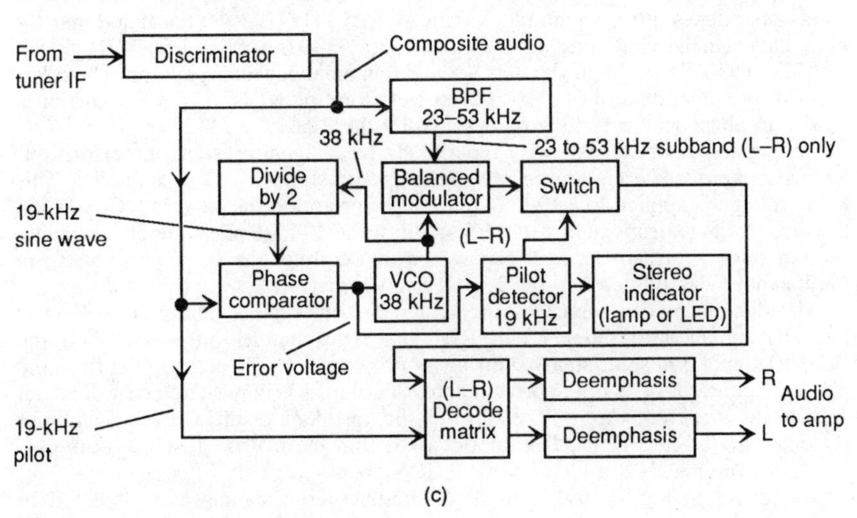

(c)

FIGURE 4.19 FM stereo modulation system.

ulator along with a 38-kHz sine wave. The balanced modulator produces a double sideband suppressed carrier (DSBSC) subband, centered around 38 kHz. A bandpass filter limits the signal to the 23- to 53-kHz range (± 15 kHz of the 38-kHz carrier). The resulting left-right signal is a 23- to 53-kHz subband, with the 38-kHz subcarrier fully suppressed.

When FM stereo is broadcast, a low-level 19-kHz pilot signal is transmitted simultaneously. The pilot signal is generated by a stable, crystal-controlled oscillator operating at 19 kHz. The 19-Hz pilot oscillator output is also applied to a frequency doubler, providing the 39-kHz carrier for the balanced modulator.

The 0- to 15-kHz baseband (L + R), 23- to 53-kHz subband (L − R), and 19-kHz pilot signal are applied to a summing network, resulting in a composite audio signal consisting of the three components shown in Fig. 4.19*a*. The composite audio signal is applied to the modulator, which FM-modulates the RF carrier of the transmitter. For a fully modulated RF carrier, the (L + R) signal accounts for 45 percent, the (L − R) signal for 45 percent, and the pilot signal for 10 percent of the composite signal.

Stereo tuners and receivers decode the signal shown in Fig. 4.19*a* and then separate the audio into original left and right channels. Since mono FM tuners have only a 0- to 15-kHz audio response, the 19-kHz pilot signal and the 23- to 53-kHz (L − R) subband are rejected. However, the (L + R) baseband signal is accepted and combines left- and right-channel audio to produce mono audio.

Figure 4.19*c* shows the FM section of a typical AM/FM tuner, from the discriminator to the audio-amplifier inputs. The discriminator output is a composite audio signal similar to that shown in Fig. 4.19*a*.

A PLL locks the sine-wave output of the 38-kHz VCO in phase with the received 19-kHz pilot signal as follows. A divide-by-2 circuit converts the 38-kHz VCO output to a 19-kHz sine wave, which is one of the inputs to a phase comparator. The other input is the received 19-kHz pilot signal. The phase comparator produces an error voltage to lock the VCI in phase with the 19-kHz pilot signal.

The composite audio signal from the discriminator is applied through a 23- to 53-kHz bandpass filter, which blocks the 0- to 15-kHz (L + R) baseband and 19-kHz pilot signals, while passing only the 23- to 53-kHz subband (L − R) signal. The 23- to 53-kHz subband signal is applied to a balanced demodulator. The other input to the demodulator is a 38-kHz carrier from the VCO. The VCO carrier is locked in phase to the 19-kHz pilot signal by the PLL.

During stereo broadcast, with the 19-kHz pilot signal present, the error voltage from the phase comparator approaches zero as phase lock is achieved. This error voltage is applied to a 19-kHz pilot detector and to the 38-kHz VCO. When the error voltage drops below the threshold of the 19-kHz pilot detector, the stereo indicator is turned on, as is the switch that couples the (L − R) signal from the balanced demodulator to the L − R decode matrix.

The decode matrix separates the L and R components of the (L + R) and (L − R) signals into independent left- and right-channel outputs. When the (L − R) signal is absent, such as during mono reception, the error signal from the phase comparator does not approach zero, and the 19-kHz pilot detector does not actuate the switch or turn on the stereo indicator. As a result, the (L + R) input is not separated into L and R components within the matrix. Instead, both outputs from the matrix are identical (L + R) signals.

As shown in Fig. 4.19*b*, both the L and R audio signals are subjected to preemphasis before reaching the FM-modulator circuits. Likewise, as shown in Fig. 4.19*c*, the L and R outputs from the decode matrix are applied to audio amplifiers through deemphasis networks. The use of preemphasis and deemphasis is mostly to improve the signal-to-noise (S/N) ratio.

Preemphasis increases the highs, while deemphasis increases the lows. FM is usually a form of phase modulation, or PM. The noise-modulation characteristic of PM is not flat but increases with noise frequency. So, with deemphasis in the tuner, the high-frequency noise is reduced to the same level as the low-frequency noise, thus improving the S/N ratio.

Remember that noise modulation is essentially an internal modulation of constant level. By preemphasizing the external audio of the transmitter, an overall flat audio response is possible. This compensates for the deemphasis characteristics of the receiver but does not affect noise modulation.

The time constants of the coupling components in a preemphasis or deemphasis network determine the center frequency and are specified in microseconds (μs); 75 μs is standard for FM broadcasts in the United States, and 50 μs is used in some other countries. The standard rolloff rate for both preemphasis and deemphasis is 6 dB per octave.

FM Stereo Generator Features. A typical FM stereo generator produces a stereo-multiples FM-modulated RF carrier that conforms to FCC regulations and duplicates the type of signal radiated by an FM stereo broadcast transmitter. External or internal modulation audio is converted to a composite audio signal containing an (L + R) 0- to 15-kHz baseband and (L − R) 23- to 53-kHz subband. Generally, the modulating signal (composite audio) is also available for injection directly into audio and stereo decoder circuits.

On most generators, the composite audio signal is continuously adjustable, and a modulation meter is calibrated to measure the rms value of composite audio. The RF output may be internally or externally modulated. In either case, modulation is continuously adjustable up to 75 kHz. A calibrated modulation meter reads FM deviation in kilohertz.

For external modulation, independent left and right input jacks permit stereo modulation through 75- or 50-μs preemphasis networks, or with no preemphasis, on most generators. The 50-Hz to 15-kHz audio input bandwidths equal that of an FM broadcast transmitter, thus permitting full audio-range frequency-response test of tuners.

The typical internal-modulation frequency is 1 kHz, which can be selected in one of five combinations: left channel only, right channel only, (L + R) baseband, (L − R) subband, and left and right (line) with 1-kHz signal applied to left channel and 50- to 60-Hz line voltage applied to the right channel. These internal-modulation combinations permit complete testing of stereo decoder circuits, including channel-balance and channel-separation characteristics (if required).

During either internal or external modulation, a highly stable 19-kHz pilot signal is generated and combined with the composite audio. The pilot signal may be switched off when desired, such as for the testing operation of a pilot detector circuit (if any).

Some FM-stereo generators have a detachable telescoping antenna to simulate an FM-stereo broadcast transmitter. On most generators, the RF output may be turned off whenever desired, thus generating the composite audio only.

4.10.2 FM IF Adjustments (Preliminary)

Figure 4.20*a* shows the FM IF test and adjustment points. The purpose of this procedure is to set the IF circuits of the FM section of an AM/FM tuner (of the frequency-synthesis type). The procedure can be used at any time but is of most value when performed before any extensive troubleshooting. (The procedure just

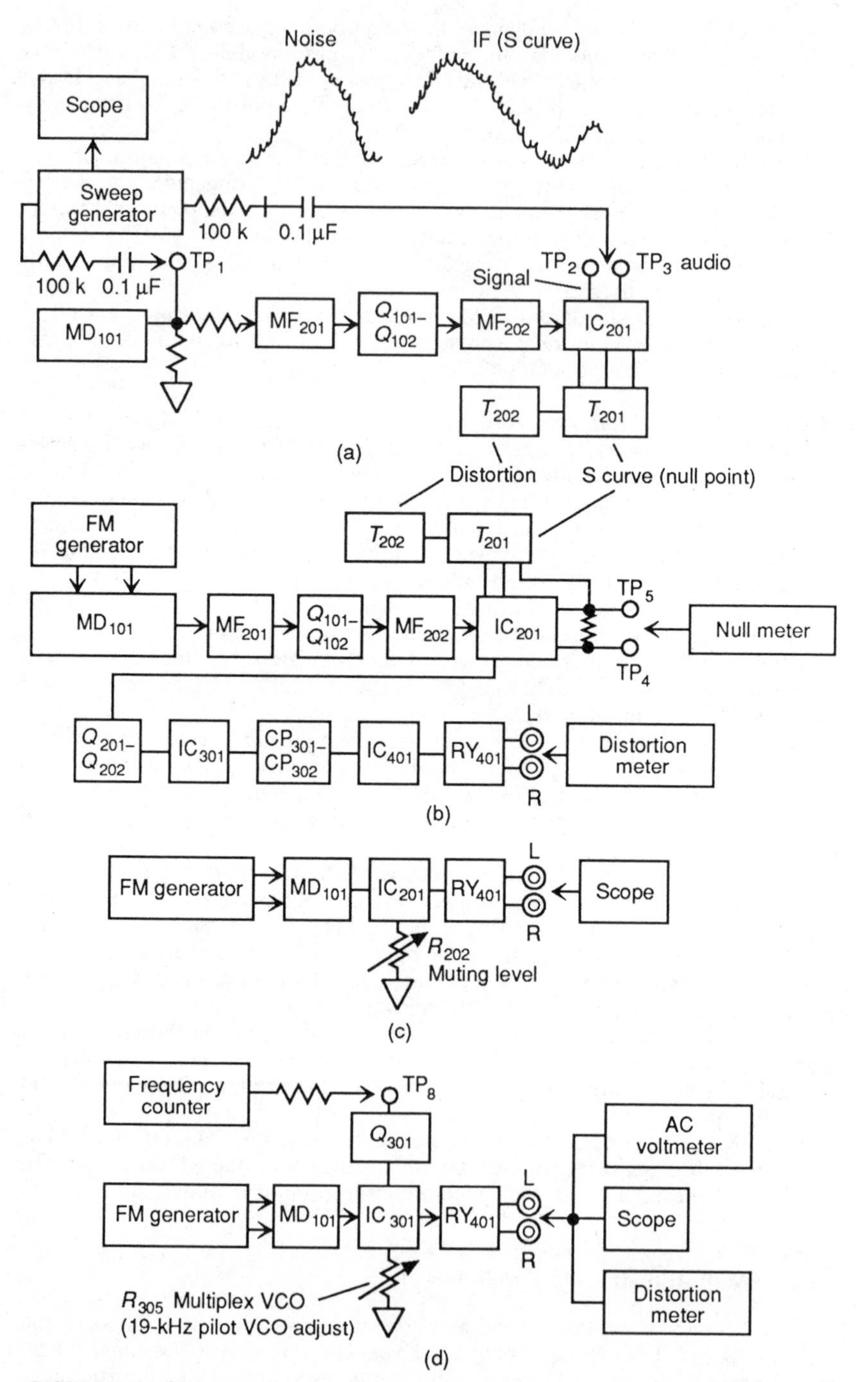

FIGURE 4.20 FS tuner FM test and adjustment points.

might cure a number of problems.) A preferred procedure is described in Sec. 4.10.3.

With the sweep generator connected to TP_2, adjust the sweep generator for 10.7 MHz with a 200-kHz sweep width. Adjust the generator output (amplitude) for a weak signal that produces noise patterns as shown on the waveforms in Fig. 4.10*a*.

Using a nonmetallic alignment tool, adjust the IF tuning (IFT) of the MD_{101} FM tuner so the IF waveform (S curve) in Fig. 4.10*a* is maximum.

Move the sweep generator to TP_3 but leave the frequency and output as previously set. If necessary, reduce scope gain. Alternately, adjust T_{201} for a symmetrical S curve and T_{202} for linearity between the positive and negative peaks of the IF waveform.

4.10.3 FM IF Adjustments (Preferred)

Figure 4.20*b* shows the test and adjustment points. First perform the preliminary adjustment as described in Sec. 4.10.2. Adjust the FM generator for 97.9 MHz, modulated by 1 kHz with 75-kHz deviation (100 percent modulation, mono). Adjust the generator output (amplitude) for 65.2 dBf. The term *dBf* (found in some FM-tuner specifications) is the power level measured in decibels referenced to 1 *femtowatt* (10^{-15}). If you are fortunate, the service literature will spell out a voltage reading (usually in the microvolt range).

Adjust the tuner to 97.9 MHz (as indicated on the front-panel display), and adjust T_{201} for 0 V ± 50 mV on the null meter (connected between TP_4 and TP_5).

Adjust T_{202} for *minimum distortion* on the distortion meter (connected to the left-channel audio-output jack).

Work between T_{201} and T_{202} until you get minimum distortion and (it is hoped) 0 V ± 50 mV on the null meter.

4.10.4 FM Muting Adjustment

Figure 4.10*c* shows the FM muting test and adjustment points. This adjustment sets the signal threshold for audio muting in the FM auto mode (where the audio should be muted in between stations and on weak stations).

Adjust the FM generator for 97.9 MHz, modulated by 1 kHz and with a 75-kHz deviation. Adjust the generator output for 33 dBf (or the equivalent output voltage).

Adjust the tuner to 97.9 MHz (as indicated on the front-panel display). Place the tuner in the FM auto mode (press auto mode and check that the auto mode indicator turns on).

Adjust R_{202} until the audio is muted. Then slowly readjust R_{202} until the audio is *just unmuted.*

4.10.5 FM Multiplex VCO Adjustment and Distortion Test

Figure 4.10*d* shows the FM multiplex VCO test and adjustment points. This adjustment sets the 19-kHz pilot VCO in IC_{301} and checks the resultant distortion.

Adjust the FM generator for 97.9 MHz with no modulation. Adjust the generator output amplitude for 65.2 dBf (or the equivalent output voltage).

Adjust the tuner to 97.9 MHz (as indicated on the front-panel display). Place the tuner in FM auto mode (press auto mode and check that the auto mode indicator turns on).

Adjust R_{305} for 19 kHz ± 50 Hz as indicated on the frequency counter connected to TP_8.

Leave the tuner and FM generator set at 97.9 MHz. Apply 1-kHz modulation to the FM generator left channel (with a pilot-carrier deviation of 6 kHz and a total deviation of 75 kHz).

Adjust the IFT of MD_{101} (using a nonmetallic alignment tool) for minimum distortion (as indicated by the distortion meter connected to the left-channel audio output jack). Do not adjust the IFT of MD_{101} more than one-quarter turn from the setting established in Sec. 4.10.2.

Remove the left-channel modulation from the FM generator, and apply the same modulation to the right channel. Monitor the right-channel audio output with the distortion meter, and check that the distortion is about the same for both channels.

4.10.6 AM IF Adjustment

Figure 4.21*a* shows the AM IF test and adjustment points. The purpose of this test is to set the IF circuit of the AM section. Adjust the sweep-generator frequency to 450 kHz. Increase the generator output until a waveform appears on the scope. Do not overdrive the IF section. Adjust T_{151} until the waveform is maximum at 450 kHz.

4.10.7 AM Local-Oscillator Confirmation

Figure 4.21*b* shows the AM local-oscillator confirmation test and adjustment points. This test is to check that the tuning voltage (error voltage from PLL IC_{503}) applied to the AM section is correct. Normally, it is not necessary to adjust the local oscillator, but the tuning voltage should be checked after any service in the AM section of the tuner.

Adjust the tuner to 1630 kHz (as indicated by the front-panel display). Check that the DVM (connected to TP_9) reads less than 23 V.

Adjust the tuner to 530 kHz, and confirm that the TP_9 voltage is 1.8 V ± 0.3 V. If the reading at 530 kHz is out of tolerance, adjust L_{152} for an error voltage of 1.8 V ± 0.1 V. Then check that the error voltage is less than 23 V with the tuner at 1630 kHz.

4.10.8 AM Tracking Adjustment

Figure 4.21*c* shows the test and adjustment points. This procedure adjusts the RF input circuit of the AM section for proper tracking across the AM broadcast band.

Adjust the generator for 600 kHz, modulated with 400 Hz at 30 percent. Set the generator output level as necessary for a reading on the ac voltmeter (connected to the left- or right-channel audio output). Use the minimum output from the generator that produces a satisfactory reading.

Adjust the tuner to 600 kHz (as indicated on the front-panel display). Adjust L_{151} for maximum output on the ac voltmeter. Adjust the tuner to 1400 kHz, and adjust CT_{151} for maximum output. If necessary, work between L_{151} and CT_{151} for maximum output at both 600 and 1400 kHz. These two adjustments usually interact.

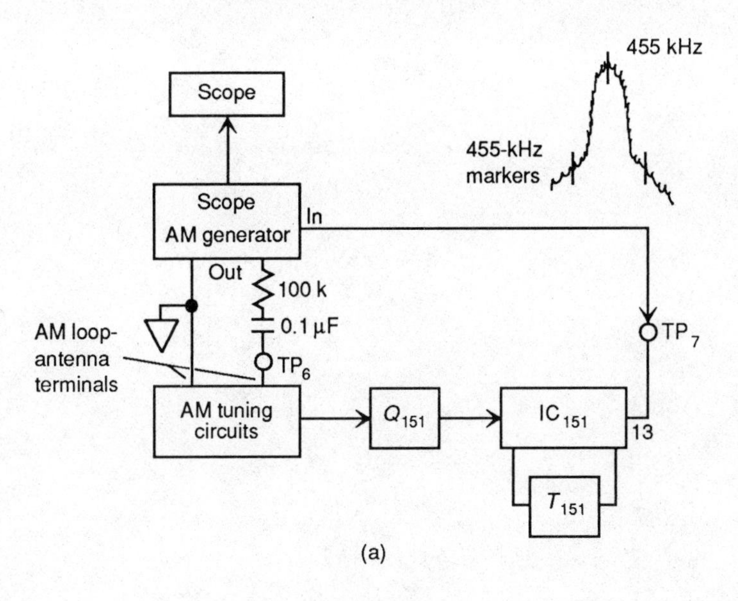

(a)

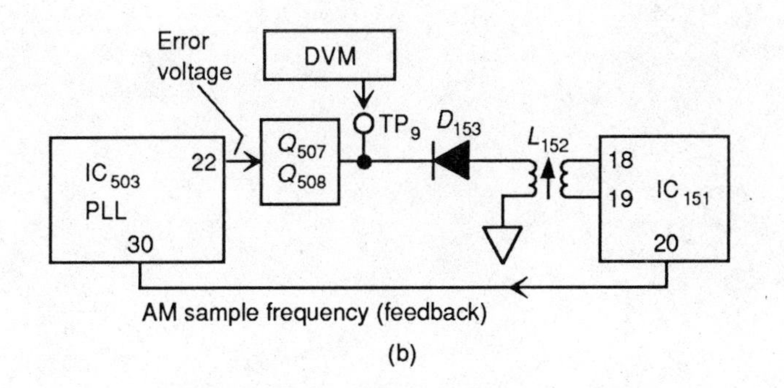

(b)

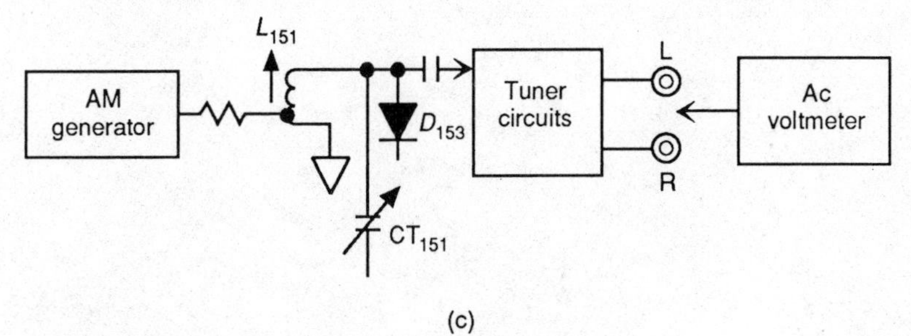

(c)

FIGURE 4.21 FS tuner AM IF test and adjustment points.

CHAPTER 5
ADVANCED RF TESTS

This chapter covers advanced RF-test procedures. Here, the emphasis is on detailed checks for complete RF equipment, such as a *communications set* or *transceiver* (transmitter, receiver, antenna, etc.) rather than on RF-circuit tests. The checks described here establish a basis for the troubleshooting procedures in Chaps. 6 and 7.

It is recommended that you perform the test procedures in the order described here. The sequence is arranged to provide a complete test of the communications set or transceiver with a minimum of steps.

5.1 TRANSMITTER RF-POWER CHECK

This is normally the first check performed on the transmitter of a communications set or transceiver. The test measures transmitter RF output to determine if the power level is normal. The check can be used for both AM and FM transmitters operated in the unmodulated condition. A similar test for SSB transmitters is described in Sec. 5.13.

Note: FCC regulations require that all checks, adjustments, and repairs that affect transmitter power and frequency be performed only by or under the immediate supervision of persons holding a valid First or Second Class Radiotelephone License.

1. Connect the equipment as shown in Fig. 5.1. Make certain that the RF wattmeter (Sec. 3.7) is capable of measuring the maximum output power of the transmitter. If an RF wattmeter is not available, use a dummy load and meter to measure power, as described in Sec. 3.6. If the transmitter is capable of both AM and FM, check the AM output power first (without modulation).
2. Set the transmitter to the first channel (or lowest frequency) to be checked. Operate the transmitter controls to produce the maximum output power.
3. Key the transmitter with the microphone push-to-talk (PTT) switch (or whatever operating control is used). If necessary, cover the microphone so that the transmitter is not modulated.
4. Read transmitter RF power on the RF wattmeter (Sec. 3.7) or on the dummy load and meter (Sec. 3.6). As an example, most AM CB transmitters operate at the maximum allowable 5 W of input power, which results in an RF-output power of 2.5 to 3.5 W. Refer to the manufacturer's specifications for normal RF-output power of other transmitters.

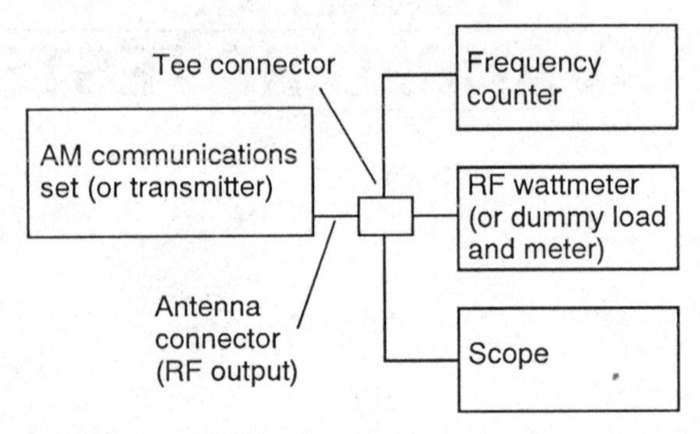

FIGURE 5.1 AM transmitter test connections.

5. Repeat steps 3 and 4 for each transmitter channel (or across the entire operating frequency range). Typically, RF power should be the same on all channels (or at all operating frequencies).
6. Repeat steps 3 and 4 while applying some modulation (speak into the microphone). There should be substantially no change (theoretically) in RF power with modulation. (You will probably notice RF-output peaks when you speak loudly.) This condition is normal for most AM transmitters but is highly undesirable for FM transmitters. As an example, FCC rules limit transmitter output power to under 4 W under any condition of modulation when a CB transmitter is operated in the AM mode. Thus, a CB set that produces 3.5 W with no modulation can produce up to 4 W fully modulated and still be considered as operating properly.

5.2 TRANSMITTER FREQUENCY CHECK

This check measures the accuracy of the transmitter operating frequency. The check should be performed simultaneously with the transmitter RF-power check (Sec. 5.1). Immediately after reading the RF power from the wattmeter, read the transmitter frequency from the frequency counter. The check can be used for both AM and FM transmitters operated in the unmodulated condition. A similar test for SSB transmitters is described in Sec. 5.15.

1. Leave the equipment connected, as shown in Fig. 5.1. Repeat steps 1 through 5 of the transmitter RF-power checks (Sec. 5.1). Make certain that the transmitter is not modulated. If necessary, cover the microphone. If the microphone gain is adjustable, set it for the lowest gain.
2. Read the transmitter frequency from the frequency counter (Sec. 3.5). Check all channels (or all operating frequencies). Note that if the transmitter shows no RF output, or the RF output is very low, you will probably not get an accurate frequency reading on the counter.
3. For an AM transmitter, repeat steps 1 and 2 while applying some modulation (speak into the microphone). In theory, there should be no change in frequency with modulation. Any substantial variation in frequency with modula-

tion indicates problems since the upper sidebands (USBs) should cancel the lower sidebands.

5.3 TRANSMITTER MODULATION CHECKS

These checks show whether or not transmitter modulation is normal. The check for AM transmitters (Sec. 5.3.1) shows modulation by displaying the modulation envelope on a scope, as described in Sec. 3.2. FM transmitters are checked by measuring output frequency deviation with fixed modulating input frequencies using an FM deviation meter (or possibly a spectrum analyzer, Sec. 3.11). The checks should be performed after the transmitter power output and frequency have been checked. A similar test for SSB transmitters is described in Sec. 5.14.

5.3.1 AM Transmitter Modulation Check

1. Connect the equipment as described in Sec. 3.2. Use the modulation measurement method best suited to the available test equipment, following the notes and technique discussed in Sec. 3.2.
2. There are two alternatives for connecting the audio-modulation signal source (audio generator) to the transmitter. You can connect the modulation source directly to the transmitter at the microphone input (or to a modulation input jack if the transmitter is so equipped). This provides the most stable and uniform modulation source. However, such direct connection does not provide for a check of the microphone. Also, on some transmitters, the set must be keyed by a PTT switch on the microphone. The modulation source can be connected to a loudspeaker. In turn, the microphone is placed directly over the loudspeaker. Use whichever method is most convenient.
3. If the transmitter is equipped with adjustable microphone gain, set the gain to midposition.
4. Secure the microphone over the speaker so that the speaker output drives the microphone with a constant tone. Adjust the modulation audio generator to a frequency of 1 kHz, unless another frequency is specified by the transmitter service instructions.
5. Key the transmitter with the microphone PTT switch. Adjust the scope for a stable display of the modulation envelope as described in Sec. 3.2.
6. Vary the gain control of the audio generator as necessary to get a scope display that shows modulation from 0 to 100 percent.
7. Return the gain control of the audio generator to midposition. If the transmitter is equipped with adjustable microphone gain, vary the microphone gain from minimum to maximum setting. The observed modulation percentage should vary as the microphone gain control is adjusted.
8. Normally, the modulation check is necessary on only one channel. However, if a complete check is desired, leave the equipment set up as described in step 6 and adjust for 50 percent modulation. Select each channel in turn, and observe the scope display for any change. Unkey the transmitter while changing channels.

5.3.2 FM Transmitter Modulation Check

1. Connect the equipment as shown in Fig. 5.2. Make certain that the RF wattmeter or dummy load is capable of handling the maximum output power of the transmitter.

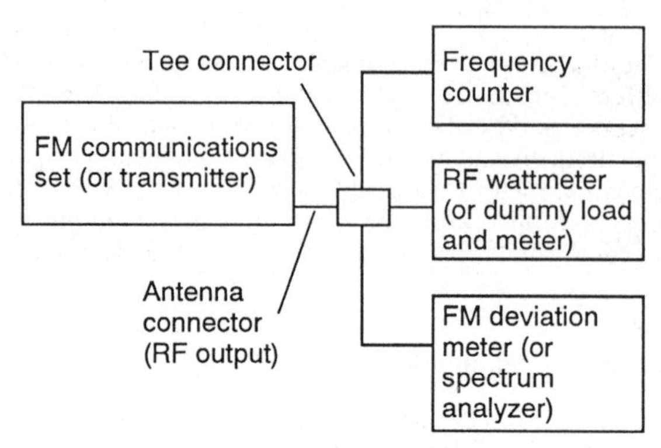

FIGURE 5.2 FM transmitter test connections.

2. Set the transmitter to the first channel (or lowest frequency) to be checked. Operate the transmitter controls to produce the maximum output power (unmodulated).
3. Connect the audio modulation source to the transmitter using the most convenient alternative described in step 2 of Sec. 5.3.1.
4. Key the transmitter, without modulation, so that the FM deviation meter (or spectrum analyzer) can be tuned to the exact carrier frequency.
5. If the transmitter is equipped with adjustable microphone gain, set the gain to midposition. Secure the microphone over the speaker so that the speaker output drives the microphone with a constant tone. Adjust the modulation audio generator to a frequency of 1 kHz, unless another frequency is specified.
6. Key the transmitter and read the amount of deviation on the FM deviation meter or spectrum analyzer (Sec. 3.11). The indicated deviation should be equal to the frequency of the audio generator (1 kHz).
7. If desired, repeat the procedure using other audio generator (modulation) frequencies. A typical FM deviation meter reads up to about 7 or 8 kHz. Typical FM modulation test frequencies are 1 and 5 kHz.
8. If desired, repeat the tests on all channels.

5.4 RECEIVER AUDIO-POWER CHECK

This is normally the first receiver check performed. The check quickly determines if the receiver is totally dead. If it is not, the check determines whether the audio section of the receiver can deliver adequate audio power. The check also accomplishes all the preliminary setup steps necessary for a receiver sensitivity

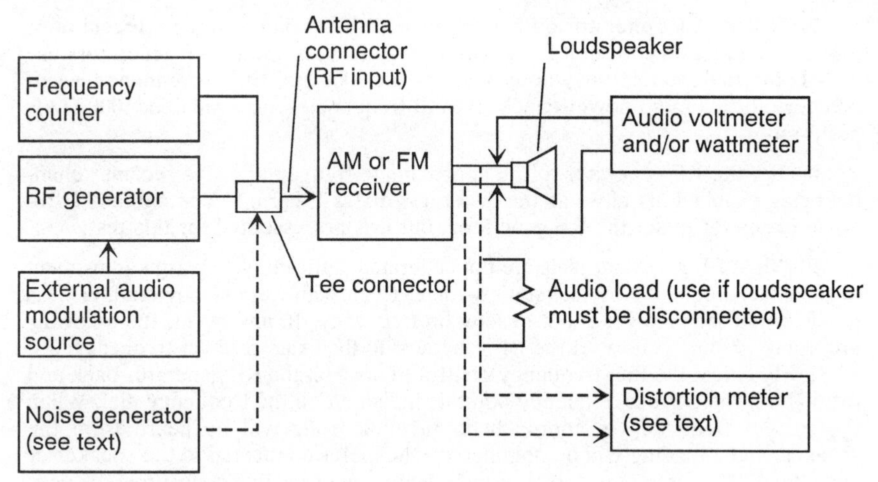

FIGURE 5.3 AM/FM receiver test connections.

check (Sec. 5.5), which should be performed next if the results of this test are satisfactory. The test is designed primarily for AM receivers but can also be used for FM communications (provided that an FM signal generator is available). The last part of the test also provides a quick check of the audio response by comparing the output power at 400 Hz and 2.5 kHz to the 1-kHz reference. This is a good overall check of the receiver's ability to pass all audio signals in the voice-communications range. A similar test for SSB receivers is given in Sec. 5.16.

1. Connect the equipment as shown in Fig. 5.3. If the receiver loudspeaker must be disconnected, make certain to connect an audio load in place of the speaker. Operation of the receiver without a loudspeaker or load will probably result in damage to the final audio transistors or IC. The audio load should be a noninductive (not wire-wound) composition resistor of the same value as the speaker impedance (typically 4, 8, or 16 Ω), capable of dissipating the full audio output power (typically 10 or 15 W but possibly higher).

2. Select the desired channel on the receiver being checked. The check can be performed on any channel and normally needs checking on only one channel.

3. Set the receiver volume control to maximum and the receiver squelch controls to the fully unsquelched position (fully counterclockwise on most communications receivers).

4. If the receiver is equipped with adjustable RF gain, set the gain to maximum.

5. If the receiver is equipped with accessory circuits such as a noise limiter or ignition-noise blanker, switch all such accessory circuits to off. This precaution may exclude some circuits as a possible source of trouble. Accessory circuit operation should be *tested after all basic checks are performed.*

6. With the receiver unsquelched and any noise-limiter circuits off, check that there is receiver background noise on the speaker. The total absence of background noise with the volume control at maximum indicates a totally dead receiver.

7. Set the RF generator output level to 1000 μV. Turn on the RF generator internal modulation and set it for 30 percent modulation. Some RF generators use 400-Hz internal modulation whereas others use 1000 Hz. Either frequency is satisfactory for an audio power check. If both frequencies are available, 1000 Hz is preferred.

8. Set the RF generator to the approximate frequency of the receiver channel being checked, as close as the generator dial will permit. You can use a frequency counter to set the RF generator, but it is not essential for this test.

9. Most RF generator dials are not calibrated sufficiently to permit adjustment to the exact frequency by dial setting alone. Even with a frequency counter, it is possible that the receiver is not exactly on frequency. In any event, the following procedure permits you to set the RF generator to the exact receiver frequency.

Slowly adjust the fine-frequency control (if any) on the RF generator back and forth about the correct frequency point as indicated on the frequency dial. When the correct frequency is approached, an audio tone will be heard from the speaker, and a reading will be obtained on the meter connected to the speaker or audio load. The tuning range over which the tone and reading occurs may be very narrow, and the frequency control of the RF generator may need very careful adjustment. Carefully adjust the frequency control for peak meter reading and peak audio output on the speaker.

If the receiver is completely dead, no tone or meter reading will be obtained. Repeat the test on another channel but with the RF generator output level set at maximum.

10. With the RF generator tuned for peak, at 1000 μV and 30 percent modulation, measure the power output on the meter connected to the speaker or audio load. An audio wattmeter provides a reading directly in watts. If an audio wattmeter is not available, use a voltmeter to calculate the power as follows: power = voltage^2/loudspeaker impedance.

For example, if you measure 4 V across a 4-Ω loudspeaker (or 4-Ω audio load), the power is 4 W. If you get the same 4-V reading across an 8-Ω speaker, the power is 2 W.

11. As a reference, a typical CB or mobile receiver delivers about 2 or 3 W of audio power with the volume control set at maximum and 1000 μV of RF at the input. Of course, some base-station communications receivers deliver much more power. Always check the receiver specifications. Also, receiver audio power specifications usually include a maximum distortion figure (such as 2 W and less than 10 percent distortion, etc.). Distortion checks are described in Sec. 5.10.

12. Once it is established that the receiver is delivering adequate audio power, the following steps provide a quick check of audio response.

13. Reduce the receive volume setting to about one-half the rated maximum power.

14. Note the reading on the meter connected to the speaker or load. If the meter has a decibel scale, set the volume control so that the reading is at some exact decibel reading near the center scale of the meter.

15. Leave the RF generator at the same RF frequency but change the modulating frequency to 400 Hz, with 30 percent modulation. Note the reading on the meter in decibels as compared to step 14.

16. Leave the RF generator at the same RF frequency, but change the modulating frequency to 2.5 kHz, with 30 percent modulation. Note the reading on the meter in decibels as compared to step 14.

17. The readings in steps 15 and 16 should be within the receiver frequency-response specifications, typically within +3 or − 6 dB of step 14.

18. A more precise frequency-response measurement may be made, if desired, by connecting a tunable audio generator to the external-modulation input (if any) of the RF generator, selecting external modulation, and adjusting for 30 percent modulation with a 1000-Hz signal. Then tune the audio generator from about 300 to 3000 Hz (or similar upper and lower limits specified by the service literature), keeping the modulation at 30 percent, and noting the reading on the meter. Ideally, the frequency response should show no substantial dips or peaks, with a smooth response (within specifications) across the audio range (at least across the voice-communications range).

5.5 RECEIVER SENSITIVITY CHECK

This check measures the weakest usable signal level at which the receiver operates. The test is the best overall check of receiver performance that can be made and should be performed immediately after the receiver audio-power check (Sec. 5.4) since the equipment is connected and set up as required for the sensitivity check.

Receiver sensitivity is generally expressed as the signal level required to produce a 10-dB signal-to-noise ratio (or to be more technically accurate, signal-plus-noise-to-noise ratio). For example, a typical receiver-sensitivity specification is 1 μV for 10 dB (S + N)/N. This means that a 1-μV modulated signal into the receiver antenna input should produce an audio output at least 10 dB above the receiver noise level obtained with a 1-μV input signal without modulation. Many receiver specifications include the condition that the 10-dB (S + N)/N sensitivity be obtained at some minimum audio power. For example, a typical specification for overall receiver sensitivity is 1 μV for 10 dB (S + N)/N at 0.5 W audio output.

1. After performing the receiver audio-power check, leave all connections and control settings as they are at the conclusion of that check. The RF generator should already be set to the receiver frequency with 1000-μV output and internal modulation of 30 percent.

2. Turn the receiver volume control to maximum and set the receiver squelch control fully unsquelched (fully counterclockwise on most sets). Squelch circuits and controls are discussed further in Secs. 5.7, 5.8, and 5.18.

3. Reduce the RF-generator output to a convenient low level such as 5 μV. If you think the receiver sensitivity may be normal, set the level to the receiver specifications (typically 1 μV or less). However, if receiver sensitivity is poor, it may be necessary to start with a higher value (such as 5 μV to get a reading in the following steps.

4. Reset the RF generator precisely on frequency. This is very important. If the RF generator is slightly off frequency, the test results make it appear that the receiver has poor sensitivity. Rock the fine-tuning dial of the RF generator back and forth very slowly as the output level is reduced, and carefully adjust for peak reading on the audio-output meter. Peak volume should also be heard from the speaker (but always trust the meter reading rather than your ear). For most RF generators, the frequency must be repeaked after each change in RF output-level control setting. The level control (usually an attenuator) tends to have some pulling effect on frequency.

If no meter reading is obtained with a 5-μV signal, receiver sensitivity is definitely poor (for modern communications receivers). Troubleshooting is required. One area of the receiver circuits to check first is the frequency synthesizer or PLL. Many communications receivers use a frequency synthesizer or PLL to produce the local-oscillator signal. If the oscillator signal is even slightly off frequency, sensitivity appears to be very poor.

5. Note the audio meter reading on a convenient decibel scale.

6. Switch the RF generator from internal modulation to no modulation (unmodulated carrier, continuous wave, CW, or whatever term is used for the RF-generator modulation-control setting).

7. The meter reading will drop. If the receiver has normal sensitivity and a 5-μV signal is used, the meter reading should drop more than 10 dB from the reading in step 5. If you are not familiar with the decibel scales of your particular meter, consider the following notes.

The decibel scale of a meter is used to provide a convenient comparison between two readings, such as the 10-dB S/N ratio measurement. If both readings are taken on the *same meter range,* readings can be taken directly from the scale. For example, if the reading in step 5 is +2 dB and the reading in step 7 is −8 dB, with both readings taken on the same range, the difference is 10 dB. However, for low meter readings (below about −5 dB with the modulation on) a more sensitive meter range should be used to get more accurate readings.

One convenient way to check for a 10-dB difference between steps 5 and 7 is to adjust the meter reading in step 5 to 0 dB using the receiver volume control, regardless of the value in watts. The desired reading in step 7 is then −10 dB (when the modulation is removed). However, this is not always possible since some receiver-sensitivity specifications require that the test be made with a given audio output (0.5 W, 0.75 W, etc.) or with a given volume setting.

8. Return the RF generator to internal modulation. Repeat steps 3 through 7 at progressively lower RF-generator output levels until there is a 10-dB difference in meter readings between steps 5 and 7.

9. Note the setting of the attenuator (or output control) on the RF generator. This setting, in microvolts, is the receiver sensitivity for 10 dB (S + N)/N and should be equal to or lower than the receiver specification. For example, the RF-generator output should be 1 μV (or less) for a specification of 1 μV for 10 dB (S + N)/N at 0.5 W of audio.

10. When you finally get the 10-dB sensitivity reading, note the audio output in watts (which should equal or exceed receiver specifications).

11. The full receiver-sensitivity test need not be performed on all channels, unless you are having a particular problem with one or more channels. However, proper operation on all channels can be checked quickly as follows.

Leave the RF generator set at the 10-dB sensitivity level of step 9. With 30 percent internal modulation, note the audio meter reading. Select each receiver channel in turn. Tune the RF generator to each channel frequency and fine-tune for a peak audio meter reading. You should get the same reading for all channels (theoretically). If you find one or more channels that show a substantially different reading (particularly a low reading), you have a good starting point for troubleshooting.

12. In most cases, it is only necessary to know if the receiver meets or exceeds the specifications for sensitivity. This can be done quickly as follows. Set the RF generator output at the specified sensitivity level with 30 percent modulation (for example, 1 μV with 30 percent modulation). Set the receiver volume control to a convenient level as observed on the meter (0.5 W, 0 dB, etc.). Remove the generator modulation and observe the meter reading. If the reading drops 10 dB or more, the receiver sensitivity is equal to, or better than, the 1-μV specification.

5.6 ADJACENT-CHANNEL REJECTION CHECK

Rejection of adjacent-channel signals is very important to prevent strong signals on adjacent channels from causing interference. This check is comparable to a *receiver selectivity test* and measures the ability of the receiver to reject adjacent-channel signals. The check can be used for both AM and FM receivers. A similar test for SSB receivers is described in Sec. 5.17.

Typically, adjacent-channel rejection is the same for all channels and needs to be checked for only one channel. However, certain component failures can cause low adjacent-channel rejection only on specific channels. The check should be repeated for each channel that shows adjacent-channel interference. Typically, communications receiver channels are separated by 10 kHz or possibly 20 kHz.

1. Perform the receiver sensitivity check (Sec. 5.5), and leave connections and controls as at the conclusion of that check.
2. Set the receiver to the desired channel.
3. Leave the RF generator set for 30 percent internal modulation.
4. Tune the RF generator to the receiver frequency.
5. Set the RF generator output level to the 10-dB (S + N)/N level, which is 1 μV or less. Use the lowest possible signal level. Note the level for reference.
6. Adjust the receiver volume control for a convenient reference level on the audio meter, such as 0.5 W. Use a relatively low volume with respect to maximum-rated audio.
7. Switch the receiver to the adjacent higher channel, but leave the RF generator tuned to the reference channel selected in step 2.
8. Increase the RF generator output level until the audio meter reads the same as the reference level selected in step 6. To be sure that the RF generator remains precisely on the reference frequency, temporarily switch the receiver back to the reference channel (step 2) after the RF output level is readjusted, and retune the RF generator (if necessary). After tuning the RF generator, switch back to the adjacent higher channel. Do not change the receiver volume or other controls.
9. Read the RF generator output level and compare the reading with step 5. The difference between the readings, in decibels, is the adjacent-channel rejection figure. This figure should be at least 30 dB for any receiver. A high-quality communications receiver may measure 60 dB or more. This figure should ex-

ceed the *receiver selectivity specification,* which is usually stated for a 20-kHz bandwidth. If the adjacent-channel rejection measures in the vicinity of 100 dB, the test results are suspect. The receiver probably has poor sensitivity. Use the lowest possible reference level (step 5) to reduce the probability of receiver desensitization.

10. Switch the receiver to the adjacent lower channel.
11. Normally, the audio meter should read the same as in step 8, which indicates that the lower adjacent-channel rejection equals the higher adjacent-channel rejection. If the audio meter reading is different from step 8, readjust the RF generator output level until the audio meter reading equals step 8. Read the RF generator level and compare it with the step 5 level. The difference between the readings, in decibels, is the lower adjacent-channel rejection figure.

5.7 SQUELCH THRESHOLD SENSITIVITY CHECK

Most communications receivers have some form of squelch circuit that restricts background noise until a signal is received. This is done by setting the receiver input signal level at which any audio will pass. For example, the receiver can be set so that no audio is heard until the RF input signal is 1000 μV or more. The squelch threshold is usually set by a front-panel squelch control. Maximum squelch requires a large signal to open the threshold.

Figure 5.4 shows typical squelch circuits. Figure 5.4*a* shows the squelch circuit for an AM communications receiver, while Fig. 5.4*b* shows AM/SSB-receiver squelch circuits.

As shown in Fig. 5.4*a,* the signal level for squelch is sensed at the emitter of Q_1. With no signal, and squelch control R_{87} on full (clockwise), there is a high voltage across R_4 and R_{87}. This allows Q_{18} and Q_{19} to go on. When Q_{19} is on, a high voltage is applied to Q_{12} through CR_9, and Q_{12} becomes reverse biased, thus allowing no audio signals through Q_{12}.

When an incoming signal is detected, Q_{18} and Q_{19} both switch off. The Q_{12} emitter is grounded through R_{57} in the normal manner, and audio passes through Q_{12} to the speaker. With squelch R_{87} off (counterclockwise), Q_{18} and Q_{19} are also off, and audio passes. Potentiometer R_4 is an internal adjustment that sets the squelch level for a given position of R_{87}. Typically, R_{87} is set full clockwise (full on), and R_4 is adjusted so that the squelch does not open with a 1000 μV signal at the input of Q_1.

As shown in Fig. 5.4*b,* squelch control R_{519} sets the operating bias of squelch amplifiers Q_{27} and Q_{28}. Squelch gate Q_{29} controls the flow of audio from the detector (AM or SSB) to the audio amplifier and speaker.

When no signal is being received, R_{519} is set to forward bias Q_{27}, which turns on Q_{28}. This places a positive voltage on the emitter of Q_{29}, which gates off the audio. When a strong signal is received, the negative AGC voltage (Sec. 5.9) overrides the voltage set on Q_{27} by R_{519}, cutting off Q_{27}. In turn, this cuts off Q_{28}, removing the positive voltage from the emitter of Q_{29}, and the audio is gated on. Thus, the voltage set by R_{519} determines the squelch point of the receiver. R_{520} adjusts the range of R_{519}.

A test of squelch threshold sensitivity requires an extremely stable RF generator, preferably one that is crystal-controlled. The check measures the weakest signal required to unsquelch the receiver when the squelch control is set at

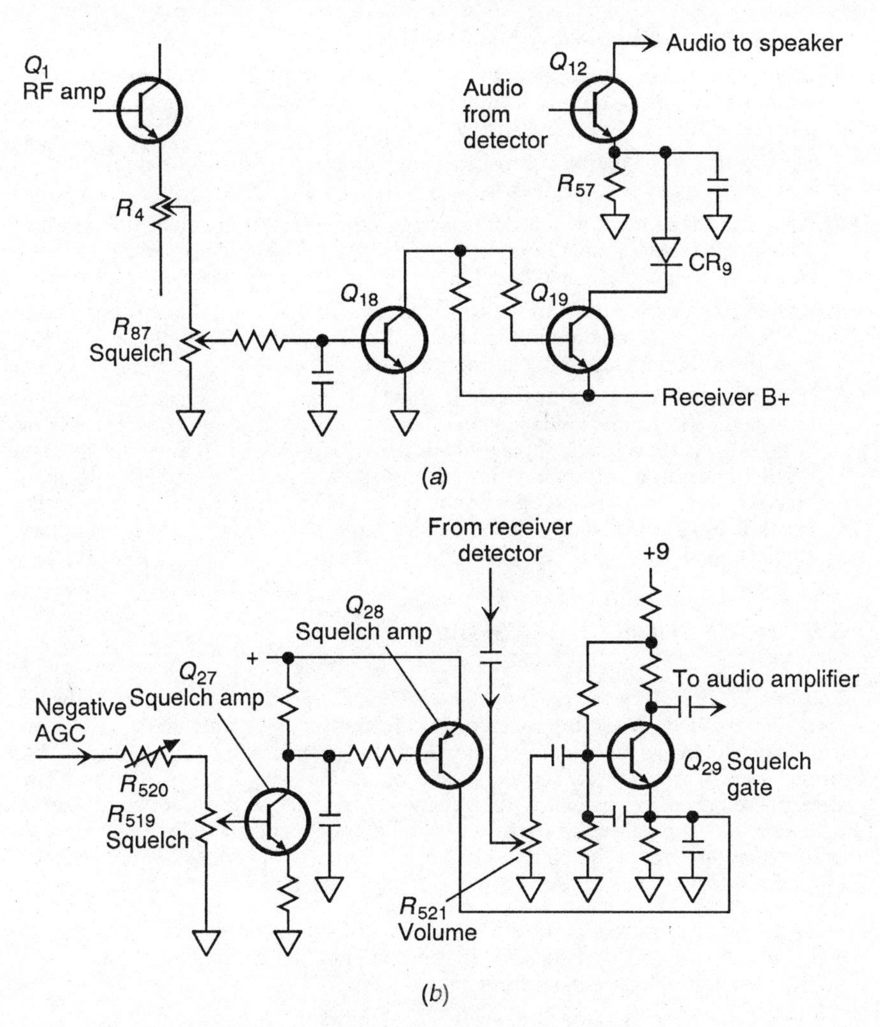

FIGURE 5.4 Typical receiver squelch circuits.

threshold. Perform this check immediately after the adjacent-channel rejection check (Sec. 5.6). Generally, the test needs to be performed only on one channel or on not more than three channels (low, middle, and high channels).

Squelch threshold sensitivity is measured in microvolts. The measured reading in microvolts should be equal to or less than the receiver specifications. Generally, squelch threshold sensitivity should be less than the specified 10-dB (S + N)/N sensitivity of the receiver (which is typically less than 1 μV).

1. Set the receiver squelch control to minimum.
2. Set the RF generator exactly on frequency with 30 percent modulation at 1000 Hz.

3. Set the RF generator output to minimum.
4. If audio output corresponding to the generator modulation is observed at minimum RF-generator output, switch the receiver channel selector to an adjacent channel so that only receiver noise is observed.
5. Set the receiver volume control to a convenient output level as observed on the meter or at the loudspeaker.
6. Adjust the receiver squelch control from minimum to threshold (the point at which the receiver noise output just disappears). Typically, receiver noise should be reduced at least 20 dB when the receiver squelches.
7. Switch the receiver back to the test channel. If audio output corresponding to the RF-generator modulation is observed, the squelch threshold level is less than the minimum generator output level.
8. If the receiver does not unsquelch, slowly increase the RF-generator output level until the receiver unsquelches, and read the RF output level in microvolts. This is the squelch threshold sensitivity. Remember that any changes in the RF generator output level may affect the frequency on many RF generators. Be sure that the RF generator is precisely on frequency. Repeat the check if there is any doubt that the most sensitive reading has been obtained.

5.8 TIGHT SQUELCH SENSITIVITY CHECK

This test requires an extremely stable RF generator, preferably one that is crystal-controlled. When the receiver is adjusted for tight squelch (squelch control fully clockwise on most receivers), weak signals should be blocked, but strong, locally transmitted signals should pass. This check measures the signal strength required to unsquelch the receiver when the squelch control is set at tight squelch. Typically, the sensitivity should not exceed 1000 μV, but it may be as low as 30 μV for some communications receivers. This check should be performed immediately after the squelch threshold sensitivity check (Sec. 5.7).

1. Leave all controls as at the conclusion of the check in Sec. 5.7. The RF generator should already be set for internal modulation at 30 percent and should be exactly on the test-channel frequency.

2. Set the RF-generator output level to minimum.

3. Set the receiver squelch control to tight squelch (fully clockwise). The audio output meter reading should drop to zero.

4. Slowly increase the RF-generator output level until the receiver unsquelches, at which time there should be audio output from the speaker and on the audio meter.

5. In step 4, the receiver may unsquelch at an unacceptably high signal level because the RF generator is pulled slightly off frequency during the measurement. To make sure that the most sensitive reading is obtained, use the following technique.

Reduce the RF generator output level until the receiver squelches. Reduce the squelch control setting so that the receiver unsquelches. Set the RF generator precisely on frequency (get a peak reading on the audio meter). Return the squelch control to tight squelch and slowly increase the generator output until receiver audio output is observed. Read the tight-squelch sensitivity in microvolts from the RF-generator output-level indicator.

5.9 AGC CHECK

Virtually all receivers have an AGC circuit. (This is sometimes known as an AVC circuit.) No matter what term is used, the circuit functions to keep the audio signal level, or constant, in the presence of varying RF signals. The audio at the speaker remains constant when RF input signals at the antenna change because of differences in transmission strength, atmospheric conditions, and so on.

Figure 5.5 shows some typical AGC circuits. With the discrete-component circuit in Fig. 5.5*a,* the output of the receiver detector contains the rectified audio and a dc component proportional to the RF carrier (at the receiver antenna input). The dc component is applied to the base of AGC amplifier Q_7 through the filter network R_{23}, R_7, and C_{40}. This positive voltage turns Q_7 on, causing the collector to go toward ground. Q_1, Q_2, Q_3, and Q_5 receive base bias from the collector of Q_7. The negative-going collector voltage reduces the bias and, consequently, the gain of these stages. At maximum signal level, the negative bias from Q_7 turns Q_1, Q_2, Q_3, and Q_5 completely off.

In most present-day receivers, the AGC voltage is applied to an RF-tuner package and to the IF-amplifier IC package rather than to the individual amplifier stages shown in Fig. 5.5*a*. However, the effect of the AGC voltage is the same.

Note that AGC for the second mixer Q_3 is obtained from the emitter of Q_2. As Q_2 emitter voltage is reduced by the negative-going AGC signal from Q_7, Q_3 is cut off before Q_1, Q_2, or Q_5. This reduces mixer noise more quickly with small increases in signal.

SSB receivers usually require special AGC circuits since there is no carrier present on the SSB signal. As shown in Fig. 5.5*b,* two AGC-detector circuits are used: attack and release.

The circuit for weak signals is called an *attack AGC.* IF signals are applied to the attack-AGC circuit through C_{330}. These signals are rectified by CR_{312} and CR_{313} and are applied to AGC amplifier Q_{20}.

The other circuit receives IF signals through C_{321} and is called the *release AGC.* The IF signals from C_{321} are rectified by CR_{309} and CR_{310} and applied to Q_{20} through CR_{311}.

The AGC feedback voltage is taken from the source of Q_{20}. The level of this feedback is set by R_{361} and R_{362}. During adjustment, R_{362} is set at maximum and R_{361} is adjusted to provide the desired AGC voltage. Then R_{362}, which is a front-panel control, is used to vary the AGC bias and thus set the desired amount of gain.

When a strong signal is received, attack diodes CR_{312} and CR_{313} work quickly and charge C_{335} and C_{336}. At the same time, release AGC diodes CR_{309} and CR_{310} charge C_{334}. Note that C_{334} has no effect on the circuit during AM operation, since C_{334} is returned to ground only when USB or lower side band (LSB) operation is selected.

The charge on C_{334} reverse-biases CR_{311}, which remains in this condition until C_{334} has fully discharged through R_{337}. This delay period is necessary to allow sufficient time for C_{335} and C_{336} to discharge, thus maintaining a constant AGC bias in the presence of very strong or very weak signals (within limits, of course).

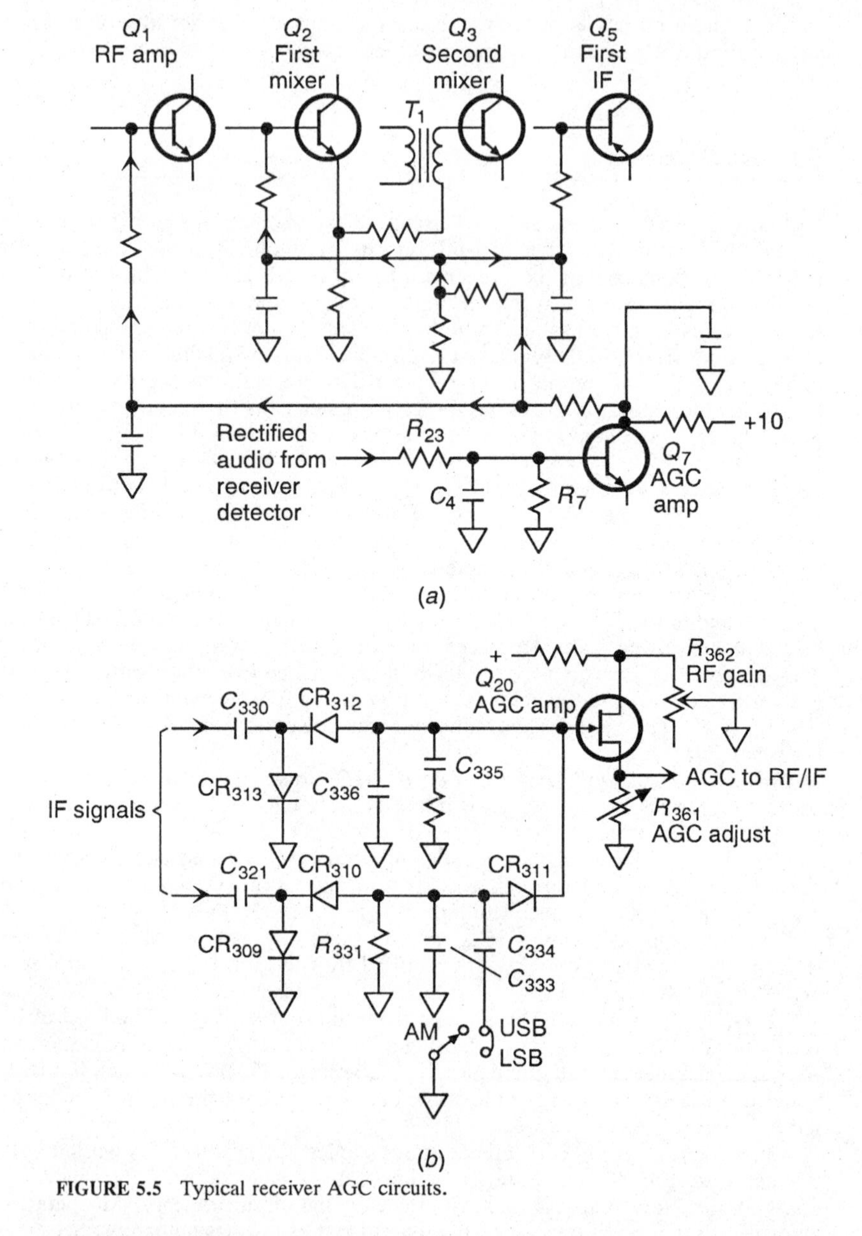

FIGURE 5.5 Typical receiver AGC circuits.

5.10 DISTORTION CHECK

This check measures the percentage of audio distortion of a 1000-Hz test signal. The distortion specification for most communications receivers is rated as a percentage at a given audio output level (for example, less than 10 percent at 2 W). Distortion can be accurately measured only if the modulating signal is undistorted. Thus, if the RF-generator modulation (internal or external) is distorted, it may appear that the receiver has more distortion than is actually the case.

Before performing this distortion check, measure the RF-generator modulation distortion using a distortion meter (Sec. 3.12). If the distortion is severe (something close to the desired maximum distortion figure for the receiver), use another modulating source with low distortion. Subtract any distortion in the modulating source from the measured receiver distortion. This check can be performed after the squelch checks and needs to be performed only on one channel.

1. Leave the equipment connected as for the receiver sensitivity check (Sec. 5.5). Connect a distortion meter in parallel with the audio meter.
2. Set the RF-generator controls for 1000 Hz, 30 percent modulation.
3. Set the RF-generator output level to 1000 μV.
4. Adjust the receiver volume control for the rated audio power output as read on the audio meter.
5. Set up the distortion meter to measure 1000-Hz distortion, and read the percentage of distortion. The distortion should be less than the specified value, with the audio output at the rated wattage.
6. Readjust the receiver volume control to get the maximum-rated distortion (for the receiver being tested), as read on the distortion meter.
7. Read the audio output power on the audio meter. The audio power output (in watts) should equal or exceed the receiver specification at the maximum-rated distortion.

5.11 RECEIVER ANL EFFECTIVENESS CHECK

Most communications receivers have an automatic noise limiter (ANL) circuit that is used to reduce excessive electrical interference, ignition noise, etc. Often, the ANL circuit is controlled by a front-panel ANL on-off switch.

Figure 5.6 shows a typical ANL circuit together with the corresponding waveforms. With this ANL circuit, the input to detector diode D_5 is a 455-kHz amplitude-modulated IF carrier. Diode D_5 detects the negative envelope of the IF carrier, both audio and noise signals. The detected signal is applied to ANL diode D_6.

Under normal operating conditions, D_6 is forward-biased by the voltage divider consisting of R_{76} and R_{77}, and the bias is sufficient to allow audio signals to pass. If excessive noise is present with the audio signal, the detected noise voltage drives the anode of D_6 negative, and D_6 no longer conducts, clipping the audio signal. The ANL switch shorts out D_6 when set to on (switch closed), eliminating the ANL function.

The effectiveness of ANL circuits can be checked only in the presence of noise. Therefore, the RF generator used for an ANL check must also include a noise generator or signal. The effectiveness of an ANL circuit is measured by making a 10-dB (S + N)/N sensitivity measurement without noise (as described in Sec. 5.5) and then again with noise. (Sensitivity is degraded in the presence of noise.) The effectiveness of the ANL circuit determines the degree to which sensitivity is degraded (typically less than about 10 dB for a good-quality communications receiver).

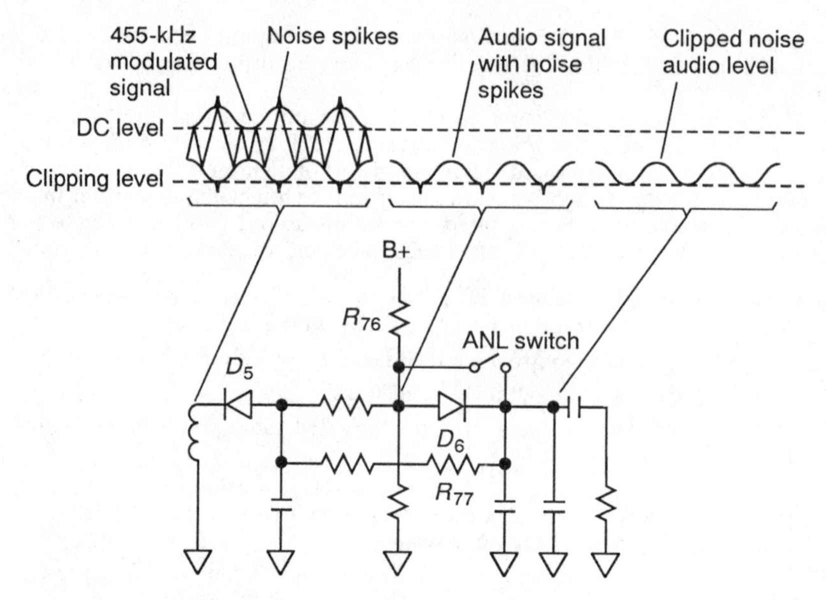

FIGURE 5.6 Typical receiver ANL circuits.

1. Leave the equipment connected as for the receiver sensitivity check (Sec. 5.5).
2. Set the RF-generator controls for 1000 Hz, 30 percent modulation.
3. Set the RF-generator output to the 10-dB (S + N)/N sensitivity level, as determined by the procedures in Sec. 5.5 (typically about 1 μV).
4. Turn on the noise generator. The audio-meter reading should drop.
5. If the receiver is provided with an ANL switch, turn the ANL circuits on. The audio-meter reading should increase when the ANL circuits are on.
6. Readjust the RF-generator output level as necessary to get the 10-dB (S + N)/N reading (the same audio-meter reading as obtained in step 3). Generally, the RF-generator must be increased to get the same audio-meter reading in the presence of noise.
7. Compare the RF-generator output settings in steps 3 and 6. The sensitivity in the presence of noise should be within 10 dB of the sensitivity without noise. For example, if 1 μV is required in step 3 (no noise) to get the desired audio-meter reading, no more than about 3 μV should be required in step 6 (noise) to get the same audio-meter reading (since 10 dB represents a voltage ratio of about 3).

5.12 RECEIVER NOISE-BLANKER EFFECTIVENESS CHECK

Many communications receivers have a noise-blanker, or noise-eliminator, circuit to eliminate most of the noise caused by electrical interference, such as poorly suppressed vehicle ignitions, electric motors, neon signals, and so on.

Figure 5.7 shows a typical noise-blanker circuit. Under normal operating conditions, the noise blanker has no effect on the receiver. When larger noise spikes are received, the spikes are applied to the gate of noise amplifier Q_{12}. The amplified output from Q_{12} is applied to the base of noise blanker Q_{13}. This turns on Q_{13} (which is normally biased off), and the collector voltage of Q_{13} drops.

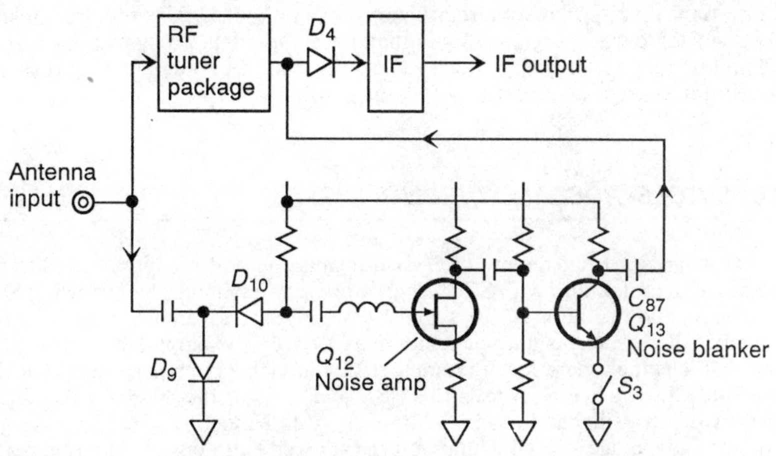

FIGURE 5.7 Typical receiver noise-blanker circuits.

The negative output pulse is coupled through C_{87} to the anode of noise-blanking diode D_4. The negative pulse reverse biases D_4, turning off D_4 for the duration of the pulse. Since all received signals must pass through D_4, the front end of the receiver (RF tuner package in present-day sets) is effectively muted during this time, and the noise signals, or spikes, do not pass.

Switch S_3 is the noise-blanker on-off switch. When S_3 is off (open), the emitter of Q_{13} has no path to ground and is disabled. Diode D_9 is a positive path, or shunt, to ground, so only negative-going noise spikes trigger the noise-blanker circuit.

The effectiveness of noise-blanker circuits can be checked only in the presence of noise. Therefore, the RF generator used for a noise-blanker check must also include a noise generator or source. The effectiveness of a noise blanker is measured by making a 10-dB (S + N)/N sensitivity measurement without noise (as discussed in Sec. 5.5) and again in the presence of noise. (Sensitivity is degraded in the presence of noise.) The effectiveness of the noise-blanker circuit determines the degree to which sensitivity is degraded (typically less than about 3 dB for a quality communications receiver).

1. Leave the equipment connected as for the receiver sensitivity check (Sec. 5.5).
2. Set the RF-generator controls for 1000-Hz, 30 percent modulation.
3. Set the RF-generator output to the 10-dB (S + N)/N sensitivity level, as described by the procedures in Sec. 5.5 (typically about 1 μV).
4. Turn on the noise generator. The audio-meter reading should drop.
5. If the receiver is provided with a noise-blanker switch, turn the noise-blanker circuits on. The audio-meter reading should increase when the noise-blanker circuits are on.

6. Readjust the RF-generator output level as necessary to get the 10-dB (S + N)/N reading (the same audio-meter reading as obtained in step 3). Generally the RF-generator output level must be increased to get the same audio-meter reading in the presence of noise.
7. Compare the RF-generator output settings in steps 3 and 6. The sensitivity in the presence of noise should be within 3 dB of the sensitivity without noise (for a typical noise-blanker circuit). As an example, if 1 μV is required in step 3 (no noise) to get the desired audio-meter reading, no more than about 1.5 μV should be required in step 6 (noise) to get the same audio-meter reading (since 3 dB represents a voltage ratio of about 1.5).

5.13 SINGLE SIDEBAND CIRCUIT TESTS

The following is a brief review of SSB transmission and reception circuits for those readers not familiar with SSB. There are many advantages when using SSB transmission and reception of RF signals. One obvious advantage of SSB is that the number of channels is automatically increased. For example, in a CB set capable of 40-channel operation, the number of channels is increased to 120 if the set includes SSB, since SSB uses the USB and LSB of the 40 channels. Thus, you get two extra sideband channels for each AM channel.

Another advantage of SSB is that you are allowed extra power output in many cases. For example, in CB work you are allowed 12 W PEP for SSB but only 4 W for AM transmission. This, combined with the use of a sideband only (sideband power is essentially lost in AM transmission) you generally get better RF communications with SSB.

5.13.1 Basic SSB Theory

Figure 5.8 is a block diagram of the basic SSB RF transmission system. The microphone signals are amplified and fed to one input of a *balanced modulator* circuit. The other input to the balanced modulator is a fixed-frequency signal from an oscillator. These two signals are mixed, producing upper and lower sidebands. Double sideband (DSB) signals are present at the output of the balanced modulator, but the carrier signal is balanced out.

The balanced modulator is essentially a bridge circuit, and the RF signal from the oscillator is always present. Audio is present only when the microphone is used. When no modulating signal is present, no current flows since R_3 is adjusted to divide or balance the RF input across the bridge. Consequently, there is no RF across the primary of T_1 when audio is absent.

When audio is present, a positive-going audio signal causes CR_2 and CR_4 to conduct, whereas CR_1 and CR_3 are reverse biased. Negative-going audio produces the opposite results (CR_1 and CR_3 on, CR_2 and CR_4 off). Both positive and negative audio signals thus unbalance the bridge and amplitude-modulate the oscillator RF signals to produce upper and lower sidebands but no carrier.

Both sideband signals are applied through T_1 to the filter, which removes one sideband. The lower sideband is removed when USB operation is desired, and vice versa. The SSB signal from the filter is mixed with a fixed-frequency signal from a second oscillator to produce a SSB signal that is above or below the chan-

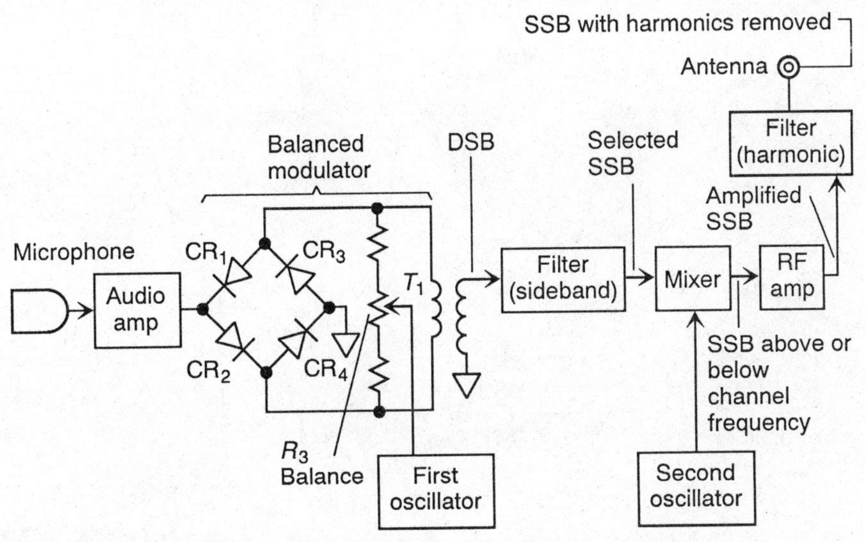

FIGURE 5.8 Basic SSB RF transmission system.

nel frequency. This SSB signal is fed through the usual RF amplifiers and output harmonic filters to the antenna.

As an example, assume that CB channel 1 is used, and the audio is in the range of 300 to 3000 Hz (typical voice-frequency range). The CB channel 26.965-MHz carrier is absent with SSB. When USB is selected, the radiated RF-sideband signals are in the range of 26.965300 to 26.968000 MHz. When LSB is selected, the output is in the range of 26.962000 to 26.964700 MHz.

The SSB receiver is essentially the same as an AM receiver, with two major exceptions. First, the carrier removed by the balanced modulator in the transmitter must be restored. Generally, this is done by mixing the received signal with a fixed-frequency signal at some point after the IF and before the AM detector.

When the fixed-oscillator signal is mixed with the sideband passing through the IF, only the audio remains. The audio is processed in the same manner as the AM audio signals.

An SSB receiver also includes a *clarifier control,* which is comparable to the *Delta-tune* control of AM CB sets. (The Delta-tune control sets the AM receiver to some frequency on either side of the channel to bring in transmitter signals that are slightly off frequency.) The SSB clarifier shifts the local oscillator as necessary so that the audio signal sounds natural. By shifting the oscillator frequency, the missing carrier frequency and the frequencies within the sidebands are matched to produce a normal sound (rather than something like a voice from outer space).

5.13.2 SSB Transmitter

Figure 5.9 is a block diagram of the LSB circuits of a typical SSB transmitter (a CB set in this case). Note that the circuits in Fig. 5.9 involve the use of *frequency synthesis* where the signals of two or more oscillators are combined to produce

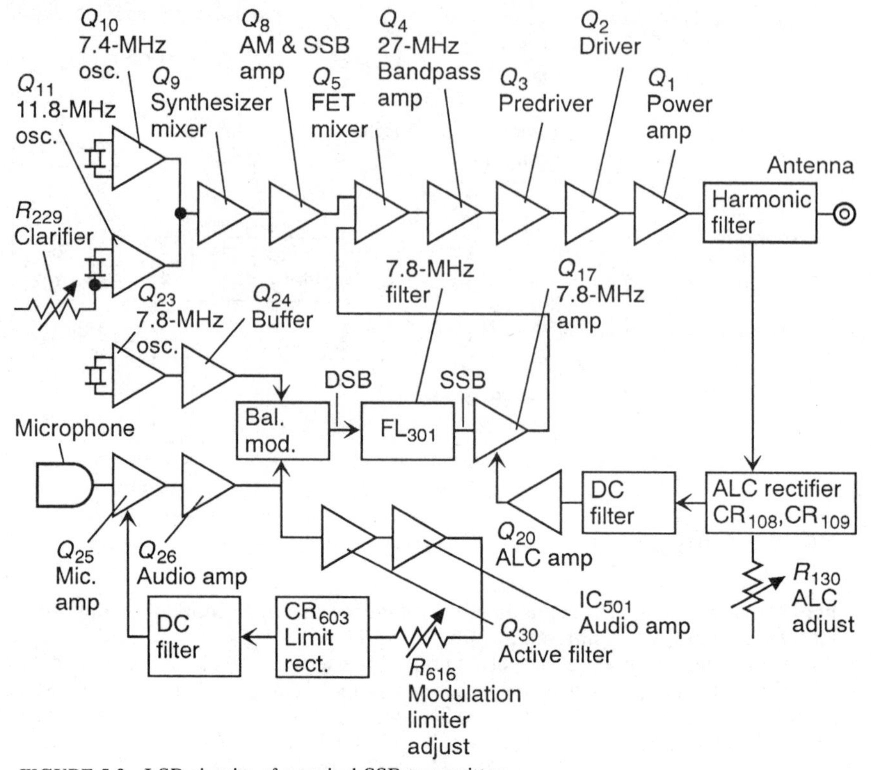

FIGURE 5.9 LSB circuits of a typical SSB transmitter.

various RF carrier signals. Frequency synthesis is found mostly in older communications equipment and has generally been replaced by PLL circuits (Secs. 1.6 and 7.1) in present-day equipment.

As shown in Fig. 5.9, audio signals from the microphone are amplified in Q_{25} and Q_{26} and coupled to the balanced modulator. This audio signal modulates the 7.8025-MHz carrier frequency to produce a suppressed-carrier DSB signal, as discussed in Sec. 5.13.1. One of the two sidebands is cut off by crystal filter FL_{301}, and the other sideband appears as a suppressed-carrier SSB signal.

The SSB signal is amplified by Q_{17} and coupled to one gate of transmitter mixer Q_5. The other gate of Q_5 receives the 19-MHz synthesized signal from Q_8 and Q_9. An LSB-modulated 27-MHz signal appears at the output of Q_5, where the signal is amplified by Q_3 and Q_4. The modulated output is then fed to the antenna through Q_1, Q_2, and the harmonic filter. Note that Q_1 through Q_4 are *wideband linear amplifiers* capable of passing the modulated RF signals.

During CB SSB operations, RF power output is limited to 12 W PEP by *automatic-level control* (ALC) circuits. Part of the RF power output at the harmonic filter is fed back and rectified by CR_{108} and CR_{109}. The rectified voltage is then filtered and fed to the base of Q_{17} through Q_{20} as an adjustable bias. (Q_{20} operates as an AGC circuit during receive.) The bias on Q_{17} is adjusted by the front-panel ALC adjust control R_{130} to set Q_{17} gain (and thus the amplitude of the RF output) so that output does not exceed 12 W PEP.

5.13.3 SSB Receiver

Figure 5.10 is a block diagram of the CB SSB circuits used during AM, LSB, and USB receive. Except for the synthesizer frequencies used, the only major difference between AM and SSB during receive is the detector system. A conventional AM detector and noise limiter is used for AM receive. A *product detector* Q_{21} is used for SSB receive to mix the reintroduced carrier from Q_{23} with the USB or LSB output from the IF amplifiers.

RF signals from the antenna pass through an antenna-switching circuit (Sec. 5.19) and are coupled to RF amplifier Q_{12}. After amplification, the RF signals are fed to one gate of RF mixer Q_{13}, where the signals are mixed with local-oscillator signals from the synthesizer Q_9. Note that the clarifier (Fig. 5.9) is used only during SSB.

IF signals from the mixer are fed into the crystal filter FL_{301}, which passes only the desired signals and cuts out the undesired signals. In AM receive, the carrier plus one sideband are passed through FL_{301}. In SSB, only the sideband passes FL_{301}.

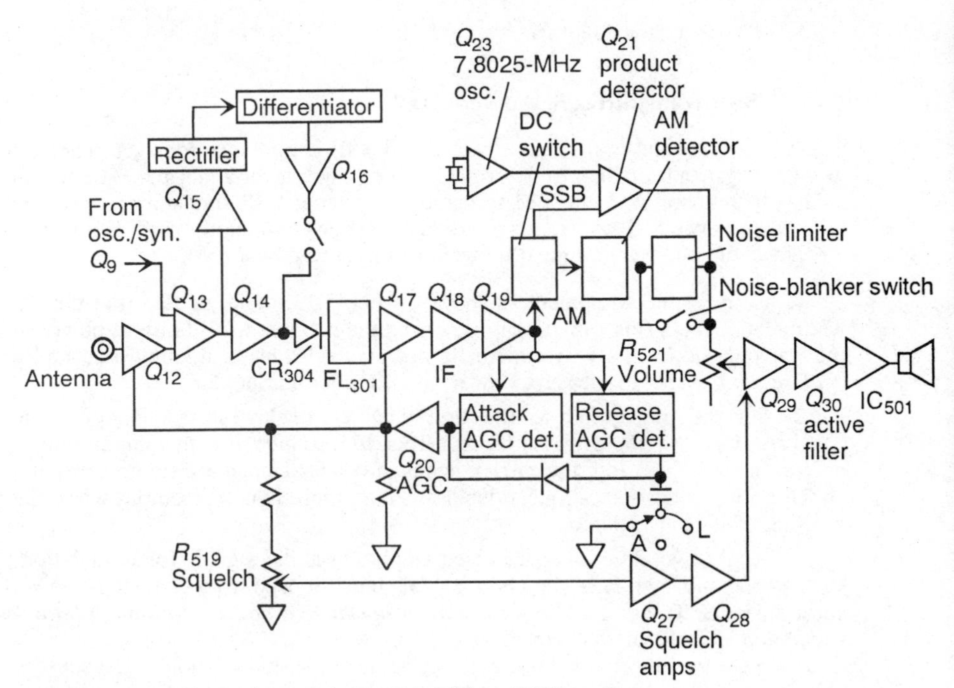

FIGURE 5.10 SSB circuits used during AM, LSB, and USB receive.

The output of FL_{301} is amplified by Q_{17}, Q_{18}, and Q_{19} and is then applied to the detector. For AM receive, the IF output is applied through a switching network to the AM detector and noise limiter. For SSB receive, the IF is mixed with the 7.8025-MHz signal from Q_{23} in SSB product detector Q_{21}.

As shown in Fig. 5.11, the IF signal from Q_{19} is applied to the base of Q_{21}. The emitter of Q_{21} receives a 7.8025-MHz signal from the carrier oscillator. Detected audio from the collector of Q_{21} is fed through volume control R_{521} and Q_{29}. Product detector Q_{21} is active only in SSB mode when the SSB voltage (9 V) is applied.

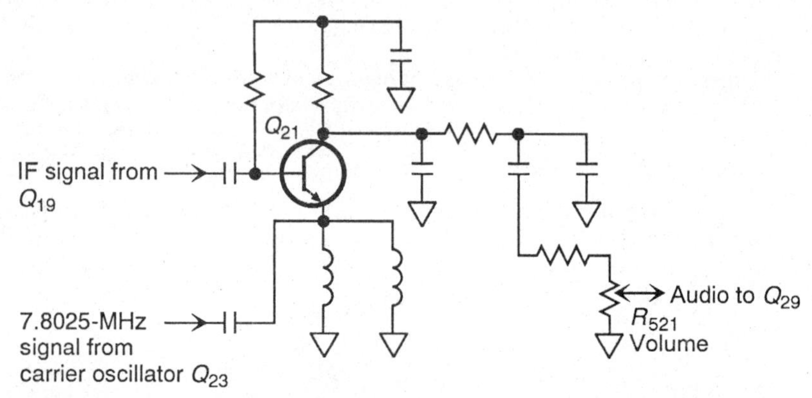

FIGURE 5.11 Typical SSB detector circuits.

5.13.4 SSB Transmitter RF-Power Check

This is normally the first check made on an SSB communications set. The test measures transmitter PEP to determine if the power level is normal. The test is similar to that for AM and FM transmitters, except SSB transmitters must be modulated (since SSB transmitters produce no power when not modulated), and a peak-reading instrument must be used to measure power.

1. Connect the equipment as shown in Fig. 5.12. Make certain that the RF wattmeter (Sec. 3.7) is capable of measuring the maximum PEP output power of the transmitter. If an RF wattmeter is not available, use a dummy load and a *peak-reading meter* to measure power, as described in Sec. 3.6.

2. Set the transmitter to the first channel (or lowest frequency) to be checked. Operate the transmitter controls to produce the maximum output power. In SSB operation, the carrier signal and one sideband are suppressed, and all RF power is carried on one sideband. Thus, there is no RF output when the transmitter is unmodulated.

3. Apply two simultaneous, equal-amplitude audio signals for modulation, such as 500 and 2400 Hz. The audio signals must be free from distortion, noise, and transients. The two audio signals *must not have a direct harmonic relationship,* such as 500 and 1500 Hz.

There are two alternatives for connecting the audio-modulation signal sources to the transmitter. You can connect the modulation sources directly to the transmitter at the microphone input (or to a modulation-input jack if the transmitter is

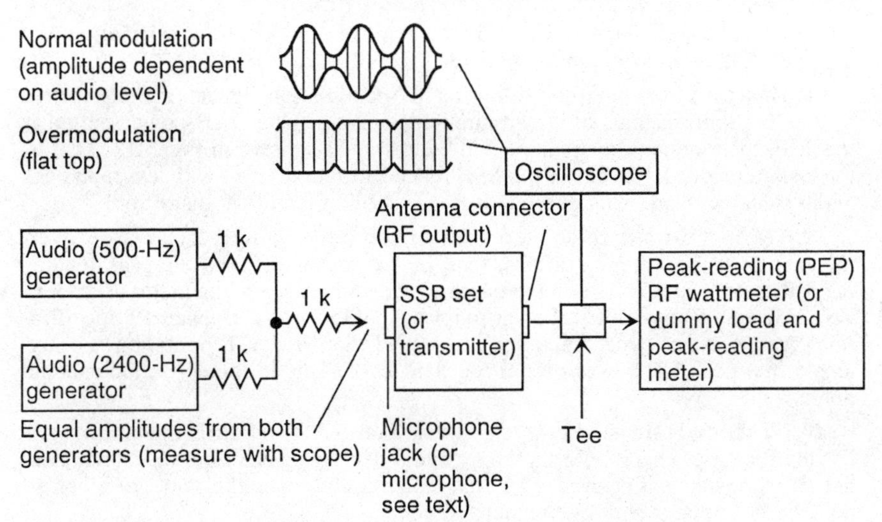

FIGURE 5.12 SSB transmitter test connections (PEP and modulation).

so equipped). This provides the most stable and uniform modulation source. The alternative is to place the microphone directly over a loudspeaker being fed by the audio generators.

Note that the alternative method, although somewhat less stable, provides for a check of the microphone, whereas the direct connection does not. Also, on some transmitters, the set must be keyed by a PTT switch on the microphone. Use whichever method is convenient.

4. If the transmitter is equipped with adjustable microphone gain, set the gain to midposition.

5. Secure the microphone over the speaker so that the speaker output drives the microphone with a constant tone. Adjust the modulation audio generators to frequencies of 500 and 2400 Hz, *equal amplitude,* unless other frequencies are specified in the service data.

6. Key the transmitter with the microphone PTT switch. Adjust the scope for a stable display of the modulation envelope. Note that if the vertical response of the scope is capable of handling the output frequency of the SSB transmitter (including the sidebands), the output can be applied through the scope vertical amplifier. If the scope is not capable of passing these frequencies, the transmitter output can be applied directly to the vertical-deflection plates of the scope display tube.

Note that, as in the case of AM modulation test (Sec. 3.2), there are two drawbacks when using direct connection to the vertical plates for SSB modulation measurement. First, the vertical plates may not be readily accessible. Next, the voltage output of the final RF amplifier may not produce sufficient deflection of the scope display.

7. Check the modulation-envelope patterns against the patterns in Fig. 5.12 or (preferably) against patterns shown in the SSB service data. Note that the typical SSB modulation envelope resembles the 100 percent AM-modulation envelope, except that the amplitude of the entire SSB waveform varies with the strength of the audio signal. Thus, the percentage of modulation calculation that apply to AM *cannot be applied to SSB.*

8. Increase the amplitude of both audio-modulation signals, making certain to maintain both signals at equal amplitudes. When peak SSB power output is reached, the modulation envelope will "flat-top," as shown in Fig. 5.12. That is, the instantaneous RF peaks of the SSB reach saturation, even with less than peak audio signal applied. This overmodulated condition results in distortion.

9. With the transmitter at maximum, or peak, power, but before overmodulation or flat-topping occurs, read the power in watts on the peak-reading RF wattmeter (Sec. 3.7) or on the dummy load and peak-reading meter (Sec. 3.6). This is the transmitter PEP. Power output should meet the transmitter specifications and must not exceed any applicable FCC limit for SSB operation. For example, the peak RF-power output for class D CB sets in the SSB mode should not exceed 12 W.

10. If the SSB transmitter is capable of both LSB and USB operation, as is usually the case, check both the upper and lower sidebands of at least one channel, or all channels if desired. The PEP readings should be the same for all channels, on both upper and lower sidebands.

5.14 SSB TRANSMITTER MODULATION CHECK

This check of modulation quality displays the SSB transmitter modulation envelope for examination. For ease of understanding, the check is described as a separate test. However, in practical work, the steps of the modulation check are often performed as part of the SSB transmitter RF power check (Sec.5.13.4). Since the transmitter must be modulated to generate RF output, both can be checked simultaneously.

1. Perform steps 1 through 10 of the SSB transmitter RF-power check, as described in Sec. 5.13.4.
2. Reduce the modulation inputs to zero (set the gain controls of both audio generators to zero). The RF wattmeter (or dummy load and peak-reading meter) connected to the transmitter output should indicate zero.
3. Slowly increase the amplitude of both audio-modulation signals, making certain to maintain both signals at equal amplitudes. When maximum power is approached, the amplitude of the waveform ceases to increase and the peaks flatten out. This is the overmodulation condition in Fig. 5.12 (flat-topping). There should be a smooth transition from no output to full, or peak, output, with no sudden jerks or dips.
4. If the set is equipped with adjustable microphone gain, decrease the amplitude of both audio-modulation signals to a point that produces one-half of the maximum measured RF power. Vary the microphone gain control from minimum to maximum, and again note there is a smooth transition from minimum power to full-power output.
5. The preceding tests are a convenient method for quickly verifying that the SSB performance is good. If a detailed evaluation of SSB operation is required, such as that performed with a spectrum analyzer (Sec. 3.11), the two-tone signals from the audio generators should be applied to a microphone jack

rather than through a loudspeaker and microphone. The PTT leads of the microphone jack can then be connected to a switch for keying the transmitter.

5.15 SSB TRANSMITTER FREQUENCY CHECK

This test measures the transmitter operating frequency in the SSB mode and should be performed immediately after the SSB transmitter RF power and modulation checks. Since the RF carrier is suppressed in SSB operation, the frequency of the sideband signal is measured. If a 1000-Hz modulating signal is applied, the frequency of the USB signal should equal the assigned carrier signal, plus 1000 Hz. A LSB signal should equal the assigned carrier signal, minus 1000 Hz. A stable single-frequency tone must be used for modulation during this check.

Many SSB communications sets are equipped with a *speech clarifier* adjustment (Sec. 5.13.1), which is a fine-frequency adjustment of the oscillator for clearest reception of SSB signals. Operation of the speech clarifier (sometimes called a *voice lock*) circuit is checked during this test. Although the speech clarifier circuit is used only while receiving, the circuit adjusts the oscillator (or frequency synthesizer), which is common to both the receiver and transmitter. It is easier to check operation of the speech clarifier circuit while measuring transmitter frequency.

1. Connect the equipment as shown in Fig. 5.13. As in the case of SSB modulation checks, you can connect the modulation source (audio generator) to the transmitter at the microphone input or through a loudspeaker and the microphone. The direct connection gives you the most stable modulating signal source, but the loudspeaker-microphone method requires no special wiring and provides a simultaneous check of the microphone.

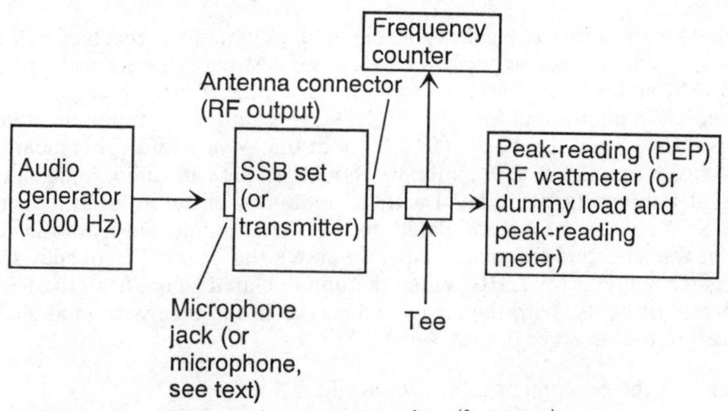

FIGURE 5.13 SSB transmitter test connections (frequency).

2. If the transmitter is equipped with adjustable microphone gain, set the gain to the midposition. Also set any speech-clarifier control to the midposition.
3. Select the desired channel on the transmitter. Select the USB mode on sets where both USB and LSB are available.

4. Adjust the modulation source to 1000 Hz. Use the frequency counter to monitor the audio-generator output to the transmitter. The audio generator should be at 1000 Hz, within ±100 Hz. An accuracy of ±10 Hz is preferable.
5. Key the transmitter with the PTT switch. Adjust the audio-generator output until the RF output (as measured on the RF wattmeter or dummy load and meter) is equal to one-half the maximum rated PEP (as measured in Sec. 5.13.4). However, you will not get accurate frequency measurements unless the PEP is at least 1 W (in most cases).
6. With the frequency counter connected to the RF output of the transmitter, read the frequency of the RF-output signal. The indicated frequency should be equal to the assigned channel frequency, plus the modulating frequency of 1 kHz. For example, if you are monitoring class D CB channel 9, the assigned carrier frequency is 27,065,000 Hz, and the USB frequency with 1000-Hz modulation should be 27,066,000 Hz.
7. Check the USB and LSB frequencies for each channel.
8. If the set is equipped with a speech-clarifier or voice-lock adjustment, it should be possible to adjust the frequency displayed on the counter during steps 6 and 7. The typical range of speech-clarifier adjustment should not exceed about ±1000 Hz (typically) from the assigned channel frequency.
9. To determine the adjustment range of the speech clarifier, get the USB frequency reading as outlined in steps 1 through 6 (with the speech-clarifier control centered). Set the speech-clarifier adjustment at the maximum counterclockwise position, and note the frequency reading on the counter. Set the speech-clarifier control to the maximum clockwise position, and note the frequency. The difference in readings is the total adjustment range of the speech clarifier.

5.16 SSB RECEIVER SENSITIVITY CHECK

This check measures the weakest usable level at which the receiver will receive SSB signals and should be performed after the AM checks (on receivers capable of both AM and SSB operation). As in the case of AM, SSB-receiver sensitivity is expressed in microvolts for a 10-dB (S + N)/N ratio at a minimum audio level: for example, 0.5-μV for 10 dB (S + N)/N at 0.5 W of audio. This means that a 0.5-μV signal into the receiver antenna should produce an audio voltage output at least 10 dB above the noise level with an audio-power output of at least 0.5 W.

For SSB checks, an unmodulated carrier signal is injected from the RF generator at the sideband frequency (slightly above the channel frequency for USB and slightly below it for LSB). When the unmodulated signal from the RF generator mixes, or beats, with the reinjected carrier in the receiver, an audio tone is produced in the receiver output.

1. Connect the equipment as shown in Fig. 5.14.
2. Select the desired channel on the receiver. Select the USB mode on sets where both USB and LSB are available.
3. Set the receiver volume control to maximum. If the receiver is provided with adjustable RF gain, adjust the gain for maximum.
4. Set the receiver squelch control so that it is fully unsquelched (fully counterclockwise on most sets). If the receiver is provided with accessory modes

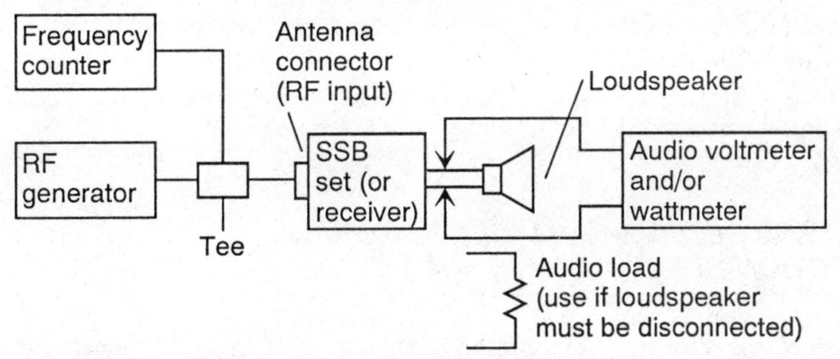

FIGURE 5.14 SSB receiver test connections.

such as an automatic noise limiter or ignition-noise blanker, turn all such modes off.

5. Set the RF generator to the unmodulated carrier mode. Adjust the RF-generator output to the 10-dB (S + N)/N level. Start with the receiver sensitivity specification for the SSB modes (typically 0.5 μV or less).
6. Tune the RF generator for peak reading on the audio meter and peak volume on the receiver loudspeaker. The frequency of the audio output will vary as the RF-generator frequency is changed. If no output is obtained on the meter and speaker, receiver sensitivity is poor, and a much greater RF-generator level may be required.
7. Read the audio-output level, in watts, from the audio meter. If an audio wattmeter is not available, use a voltmeter and calculate the power as described in Sec. 5.4, step 10. Adjust the receiver volume control for one-half of the receiver-rated maximum audio power. For example, if the receiver is rated at 2 W of audio, adjust for 1 W. If the audio output is less than one-half of the rated maximum, leave the volume control at maximum.
8. Note the audio-meter reading on the decibel scale.
9. Turn off the RF generator.
10. Again note the audio-meter reading on the decibel scale.
11. If step 10 is more than 10 dB below step 8, decrease the level of the RF-generator output and repeat steps 6 through 10. If step 6 is less than 10 dB below step 8, increase the level of the RF-generator output and repeat steps 6 through 10. Continue adjustment until there is a 10-dB difference in meter readings between steps 8 and 10.
12. Note the setting of the attenuator (or output control) on the RF generator. This setting, in microvolts, is the receiver sensitivity for 10 dB (S + N)/N.
13. It is not always necessary to measure the sensitivity in microvolts, but note whether or not the receiver meets specifications. In this case, set the RF generator to the specification level and note the audio-meter reading. Then turn off the RF generator and again note the meter reading. If there is a 10-dB or greater difference between the meter readings, the receiver meets the specification.
14. After making the sensitivity reading, note the audio output in watts. The audio output should equal or exceed the specifications.

15. Select the LSB mode and repeat the check. Sensitivity should be the same as for the USB mode.
16. Check USB and LSB sensitivity for each channel of operation. Sensitivity should be approximately the same for all channels.

5.17 SSB RECEIVER ADJACENT-SIDEBAND REJECTION CHECK

This check measures the ability of an SSB receiver to suppress signals received on the opposite sideband. When the receiver is set for USB operation, any LSB input signals should be suppressed at least 40 dB, and vice versa. The check should be performed after, or during, the SSB receiver-sensitivity check. The check can be performed on all channels or only on one, as desired.

1. Leave the connections as shown in Fig. 5.14. Repeat steps 1 through 6 in Sec. 5.16.
2. Note the audio-meter reading for reference when the RF generator is tuned for a peak on the audio meter.
3. Switch the receiver to the LSB mode on the same channel.
4. Increase the output level of the RF generator until the audio meter reading is the same as in step 2, if possible.
5. It should require at least a 40-dB increase in RF-generator output to produce the same audio-meter reading on the adjacent sideband. For example, if 1 μV is required in step 2, 100 μV (or more) should be required in step 4. If less than 40-dB suppression of the opposite sideband is measured, be sure the RF-generator frequency is not shifted toward the opposite sideband by the output-level adjustment. Recheck RF-generator tuning on the desired sideband.
6. Repeat the procedure, except adjust the RF generator for a peak on the LSB frequency, and measure USB suppression.

5.18 SSB RECEIVER SQUELCH-SENSITIVITY CHECK

This check measures the minimum amount of on-frequency RF carrier required to unsquelch the receiver when adjusted at squelch threshold (the point that barely suppresses receiver noise) and at tight squelch (the point that requires a large signal to overcome). A typical SSB receiver requires 0.5 μV for the squelch threshold and anything from about 30 to 500 μV for tight squelch. The receiver should not block strong signals, even when set at tight squelch.

The check should be performed after the sensitivity and sideband rejection checks, and it is very similar to the squelch-sensitivity checks for AM receivers, except that the receiver is operated in the USB or LSB modes, and an unmodulated signal is injected from the RF generator. The check can be performed on all channels or only on one, as desired.

1. Leave the connections as shown in Fig. 5.14.
2. Select the USB mode of operation on the desired channel.

3. Set the RF generator to the unmodulated-carrier mode. Tune the RF-generator frequency for maximum reading on the audio meter. Set the RF-generator output to the 10-dB (S + N)/N sensitivity level as determined in Sec. 5.16.
4. Turn off the RF generator. Adjust the receiver volume control as necessary so that noise is heard on the loudspeaker.
5. Adjust the receiver squelch control to the squelch threshold (to the point where noise is just squelched).
6. Set the RF-generator output to minimum, and turn on the RF generator (unmodulated carrier).
7. Slowly increase the RF-generator output until the receiver unsquelches. There should be at least a 20-dB difference in the audio-meter reading between the squelched and unsquelched condition. In the unsquelched condition, audio output should be at least one-tenth of the receiver-rated maximum audio output.
8. Note the setting of the attenuator (or output control) on the RF generator. This setting, in microvolts, is the SSB squelch threshold setting of the receiver. The reading should be equal to or less than the receiver specification for the SSB squelch threshold. Typically, this value is 0.5 μV or less.
9. Adjust the receiver squelch control for tight squelch.
10. Increase the output level of the RF generator until the receiver is unsquelched. To make sure that the RF generator remains on frequency when the output level is increased, temporarily reduce the squelch setting and retune the RF generator for peak reading on the audio meter, then return the squelch control to tight squelch.
11. Note the setting of the attenuator (or output control) on the RF generator. This setting, in microvolts, is the tight squelch sensitivity. The reading should be equal to or less than the receiver specification, which is typically in the range 30 to 500 μV.
12. Switch the receiver to the LSB mode on the desired channel, and repeat steps 1 through 11.

5.19 ANTENNA CHECK AND SWR MEASUREMENT

The antenna check is one of the most important performance checks that can be made when troubleshooting RF communications equipment. The quickest and most effective antenna check is an SWR measurement, since such measurements show just how well the communications set, antenna, and antenna cable are matched and tuned.

Note that, in many cases, what appears to be a defect in the antenna or antenna match is actually a problem in the antenna-switching circuits. Older communications sets use relays for antenna switching. Many present-day sets have "electronic switching" such as shown in Fig. 5.15. With electronic switching, the antenna is connected between the receiver and transmitter by diodes and a directional coupler. The coupler also provides for SWR measurements, as described in Sec. 3.9.

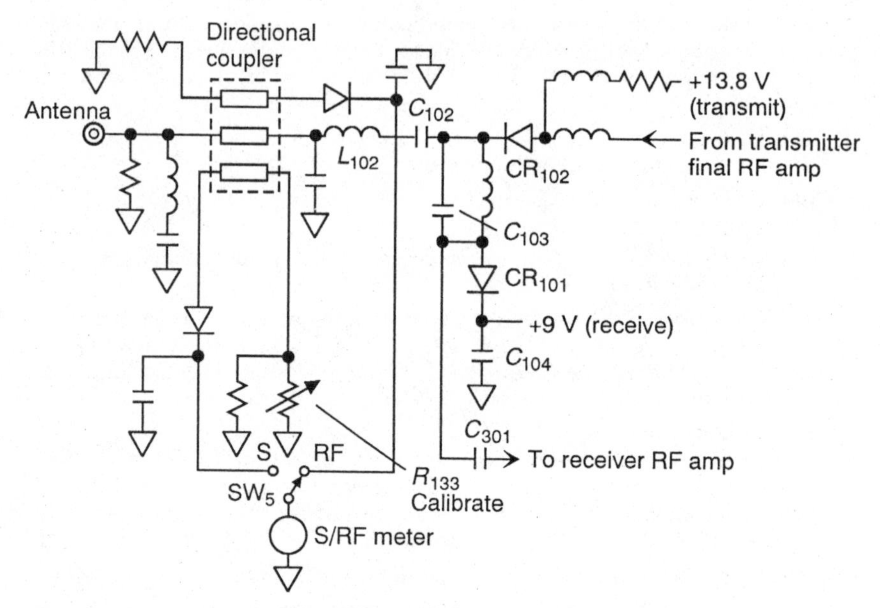

FIGURE 5.15 Antenna switching, SWR, and S/RF circuits.

In the receive mode, as shown in Fig. 5.15, +9 V is applied to the cathode of CR_{101}, thus reverse-biasing CR_{101}. At the same time, +13.8 V is removed from the anode of CR_{102}, reverse biasing CR_{102}. With CR_{102} reverse biased, the antenna is effectively disconnected from the transmitter output. Simultaneously, the receiver input is connected to the antenna through C_{301}, C_{103}, C_{102}, and L_{102}.

In transmit, when the PTT switch is pressed, +13.8 V is applied to the anode of CR_{102}, forcing CR_{102} into conduction. The path is then open from the transmitter output to the antenna though CR_{102}, C_{102}, and L_{102}. Simultaneously, +9 V is removed from the cathode of CR_{101}, forward biasing CR_{101}. With CR_{101} conducting, RF input to the receiver is shunted to ground through C_{301}, CR_{101}, and C_{104}, effectively disconnecting the antenna from the receiver input.

Note that the S/RF meter in Fig. 5.15 is controlled by switch SW_5. With SW_5 in the S position, the meter monitors incoming RF-signal strength to the receiver, passing through the directional coupler. In the RF position, the meter monitors outgoing RF-signal strength from the transmitter, passing through the directional coupler in the opposite direction. In many such circuits the S/RF meter can also be used to measure SWR instead of using an external SWR meter connected between the set and antenna. The following are a series of notes that apply to SWR measurements made either by built-in or external SWR meters.

As described in Sec. 3.9, the transmitter is keyed, and the SWR meter needle is aligned with some "set" or "calibrate" line (usually near the right-hand end of the meter scale). When the S/RF meter in Fig. 5.15 is used, the meter has two scales, and calibrate control R_{133} is used to set the needle. The switch is then set to the SWR measurement position, *with the transmitter still keyed,* and the SWR-meter needle drops back to read the SWR.

A low SWR reading is desired since this indicates that the transmitter is operating at maximum effectiveness, and receiver performance is optimum. A low

SWR results when the antenna is properly tuned to the operating frequency, and there is a close match of impedance between the transmitter output, antenna cable, and antenna.

Not all antennas are tunable. When the antenna is tunable, the SWR measurements can be used to find the optimum frequency. The SWR is measured and, with the transmitter still keyed, the antenna loading coil (or other tuning element) is adjusted for minimum SWR. However, always follow the antenna-tuning procedure recommended in the service literature. Antenna tuning is discussed further in Chap. 7.

The antenna is normally tuned for minimum SWR on the *center operating frequency* (for example, on channel 20 or 21 of a 1-to-40 channel CB). Compact mobile antennas and high-gain beam-type base-station antennas are frequency-sensitive and show variations in SWR as the transmitter is keyed on all channels (particularly when there is a wide frequency range for the lowest to the highest channel.

To be of any practical value, the SWR measurements must be made using the antenna and antenna cables that are normally used with the communications set. Be sure the SWR test includes all the antenna *cables and connectors* that are normally used. This is particularly important if you use an external SWR meter.

An SWR measurement is essential at the time of installation and should always be performed after repairs are completed. The check is also needed if damage to the antenna or antenna cable is suspected. A periodic SWR measurement will detect any gradual deterioration and assure continued high performance. A damaged antenna or cable, corroded connectors, or a similar problem can cause a very high SWR. In turn, a high SWR often causes premature failure of final RF-amplifier transistors.

5.20 RECEIVER FREQUENCY-RESPONSE CHECK

This check measures receiver audio-frequency response. An audio-frequency specification of 300 to 3000 Hz usually means that all audio frequencies from 300 to 3000 Hz at a given frequency level should produce audio outputs that are within 3 dB across the entire range. A 1000-Hz reference is often used to measure the 3-dB level. However, the point of reference can be the frequency within the response band at which maximum output level is developed.

The check is performed by applying a constant-amplitude modulated test signal to the receiver input. The modulation percentage is maintained at a constant value, and the receiver audio-output level is observed as the modulation frequency is varied.

1. Connect the equipment as described in Sec. 5.4. Use the audio generator connected to the external modulation input of the RF generator, as described in step 18 of Sec. 5.4.
2. Select the desired channel on the receiver. The check can be performed on any channel. Unsquelch the receiver. Adjust the receiver volume and RF gain (if adjustable) to the midposition. On AM/SSB/FM receivers, use the AM mode. Turn off *all accessory mode switches* (noise blankers, etc.) on the receiver.
3. Set the RF generator for external modulation. Set the audio generator to 1000 Hz. Adjust the audio-generator level and the modulation adjustment of the RF

generator for 30 percent modulation. Adjust the RF-generator output level for 1000 μV.

4. Adjust the receiver volume control for a convenient reference level on the audio meter (try 0 dB). Use a volume well below the maximum capability of the receiver to minimize distortion.
5. Tune the audio generator across the band of audio frequencies specified in the receiver service literature. In the absence of any specifications, tune across the audio range from 300 to 3000 Hz. Readjust modulation as required to maintain 30 percent modulation as the frequency of the audio generator is changed.
6. Read the audio meter as the frequency of the audio generator is changed. The meter reading should not change more than 3 dB over the entire range from 300 to 3000 Hz.

5.21 S-METER AND POWER-METER CHECK

Many communications sets are equipped with an S-meter that indicates received-carrier signal strength, and a power meter that indicates transmitted RF-output power. The S/RF meter shown in Fig. 5.15 provides both of these functions (by operating as an S-meter while receiving and as a power or RF meter while transmitting). Normally, these meters give *relative indications* rather than specific values in microvolts and watts.

The procedures in this test can be used to check proper operation of the S-meter and power-meter functions. This test can also establish a standard against which the set's meters can be calibrated, thus converting the relative indications to specific values (in microvolts and watts).

The receiver S-meter can be checked while performing the receiver AGC check (Sec. 5.9). The S-meter reading should vary as the RF-generator output level is changed. However, the RF-generator output signal does need not to be modulated since the S-meter responds to the carrier signal. If desired, note the RF-signal level required (in microvolts) for each increment on the S-meter scale.

The S-meter information can be recorded or plotted on a graph and retained for future use during troubleshooting. For example, if an input between 100 and 200 μV is required to produce an S_9 reading, and you find a receiver that requires 1000 μV for an S_9, the receiver has poor sensitivity (caused by improper adjustment, off-frequency synthesizer or PLL, weak RF-section transistors, or RF-tuner package problems).

When making comparisons of S-meter readings, make certain that the RF-generator is precisely on frequency (peak readings). If the RF generator is not *exactly* on frequency, it will take a higher output level from the RF generator to get the same reading.

The transmitter power-meter (RF) can be checked while performing the transmitter RF-power check (Sec. 5.1). The power meter of the transmitter (S/RF meter set to RF) should indicate the same reading as the RF wattmeter (or dummy load and meter) connected to the antenna when the transmitter is keyed.

If the set operates on both AM and SSB modes, the power meter will probably have a different range or scale for each mode. After the power meter is checked in the AM mode, check the SSB scales while performing the SSB transmitter RF power check (Sec. 5.13) and the SSB transmitter modulation check (Sec. 5.14). Note that the power meter and the RF-wattmeter indications vary as the modulation varies. Note the power (in watts) for each increment of the power meter (or as directed in the service literature).

5.22 EFFECTS-OF-VOLTAGE CHECK

All of the checks described thus far in this chapter should be performed with the communications set operated at the rated input voltage. For example, a typical mobile set requires an input or power-supply voltage of 13.8 V.

It is good troubleshooting practice to note the effects on test results when the power-supply voltage is varied, particularly the *effects of low voltage.* Where practical, repeat all tests and vary the power-supply voltage over the range of 11 to 15 V. Note the test results at various voltage levels. However, make certain not to exceed the maximum rated power-supply input voltage (typically about 15 V for a mobile communications set).

Power-supply voltage changes should have very little effect on transmitter frequency (theoretically no effect). However, transmitter power, receiver sensitivity, and audio power may be affected significantly, even by a 1-V change in the power supply. Testing at low and high voltage can occasionally reveal a fault that is undetected at normal voltage levels.

CHAPTER 6
BASIC RF TROUBLESHOOTING

This chapter is devoted to basic troubleshooting procedures for a cross section of RF circuits. The chapter emphasizes basic problems common to all RF equipment and concludes with specific examples of troubleshooting for two typical circuits.

6.1 TROUBLESHOOTING APPROACH

Troubleshooting can be considered a step-by-step logical approach to locating and correcting any fault in the operation of equipment. In the case of RF-circuit troubleshooting, seven basic functions are required.

First, you, the technician, must study the RF equipment, or set, using service literature, schematic diagrams, and so on, to find out how each circuit works when operating normally. In this way, you will know in detail how a given set should work. If you do not take the time to learn what is normal operation, you will never be able to distinguish what is abnormal.

Second, you must know the function of all controls, indicators, and adjustments and how to manipulate them. It is difficult to check out a set without knowing how to use the controls, even though many RF-equipment controls are fairly simple and relatively standard. Also, as the set ages, readjustments and realignment of critical circuits may be required.

Third, you must know how to interpret service literature and how to use test equipment. Along with good test equipment that you know how to use, well-written service literature is your best friend (unfortunately, a friend that is often missing).

Fourth, you must be able to apply a systematic, logical procedure in order to locate the trouble. Of course, a procedure that is logical for one type of set is quite illogical for another. For that reason, we discuss logical troubleshooting procedures for various sets, as well as basic procedures that apply to all sets.

Fifth, you must be able to analyze logically the information provided by improperly operating equipment. The information to be analyzed may be the equipment performance (transmission and reception on all channels of a communications set), indications taken from test equipment (voltage and resistance measurements), or indications taken from front-panel indicators (received signal strength and/or transmitted RF power and S/RF readings). No matter what form the information takes, it is your analysis of the information that makes for logical, efficient troubleshooting.

Sixth, you must be able to perform complete checkout procedures on the repaired equipment. Such a checkout may require only simple operation (switching through all channels, checking transmission and reception, squelch, ANL, and volume-control operation). At the other extreme, the checkout may involve complete realignment of the equipment. Either way, a checkout is always recommended after troubleshooting.

One reason for a checkout is that there may be more than one problem. For example, an aging part may cause high current to flow through a resistor, resulting in the burnout of the resistor. Logical troubleshooting may lead you quickly to the burned-out resistor, and replacement restores operation. However, only a thorough checkout can reveal the original high-current condition that caused the burnout.

Another reason for after-service checkout is that the repair may have produced a condition that requires readjustment. A classic example occurs when replacement of a part changes circuit characteristics. For example, a new transistor in an RF (or IF) stage may require complete realignment of the stage.

Seventh, you must be able to use the proper tools to repair the trouble. As a minimum for RF-circuit repair, you must be able to use soldering tools, wire cutters, long-nose pliers, screwdrivers, and socket wrenches. If you are still at the stage where any of these tools seem unfamiliar, you are not ready for RF service, even simplified service.

In summary, before starting any RF service work, ask yourself these questions: Have I studied all available service literature to find out how the particular equipment works? Can I operate the equipment properly? Do I really understand the service literature, and can I use all required test equipment properly? Using the service literature and/or previous experience on similar equipment, can I plan a logical troubleshooting procedure? Can I analyze the results of operating checks and procedures involving test equipment logically? Using the service literature and/or experience, can I perform complete checkout procedures on the equipment, including realignment, adjustment, and so on, if necessary? Once I have found the trouble, can I use common hand tools to make the repairs? If the answer to any of these questions is No, you are simply not ready for RF service. Start studying.

6.1.1 The Troubleshooting Sequence

There are four basic steps in the troubleshooting sequence: (1) determine the trouble symptoms, (2) localize the trouble to a functional area, (3) isolate the trouble to a circuit, and (4) locate the specific trouble, probably to a specific part. The following paragraphs are devoted to generalized examples of how the four steps may be used to troubleshoot a very basic (and theoretical) communications set (such as a CB, amateur radio, business radio, or other mobile-communications set).

6.2 TROUBLE SYMPTOMS

It is impractical to list all trouble symptoms that may occur in all communications sets. However, the list in Fig. 6.1*a* covers those problems most commonly found in a typical set. The troubles are grouped into functional areas (or circuits) of the set. These areas, or circuit groups, correspond to those in the block diagram in Fig. 6.1*b*.

Power supply
- Set dead, no LED channel indicator or front-panel lights
- No transmission, no reception on any channel

Receiver (Q_1 through Q_6)
- No reception on any channel, transmission normal
- Reception poor, transmission good

Transmitter (Q_9, Q_{10}, C_{11})
- No transmission on any channel, reception normal
- Transmission poor, reception good

Audio and modulation circuits (Q_{12}, Q_{13})

Audio clipper
- Poor modulation, sound good on reception

Frequency Synthesizer (Q_7, Q_8)
- Transmission and reception abnormal on certain channels
- Transmission off-frequency

Control circuits (switching relay)
- No transmission with PTT switch pressed

Low-pass filter
- Poor reception and transmission, carrier normal on S/RF meter

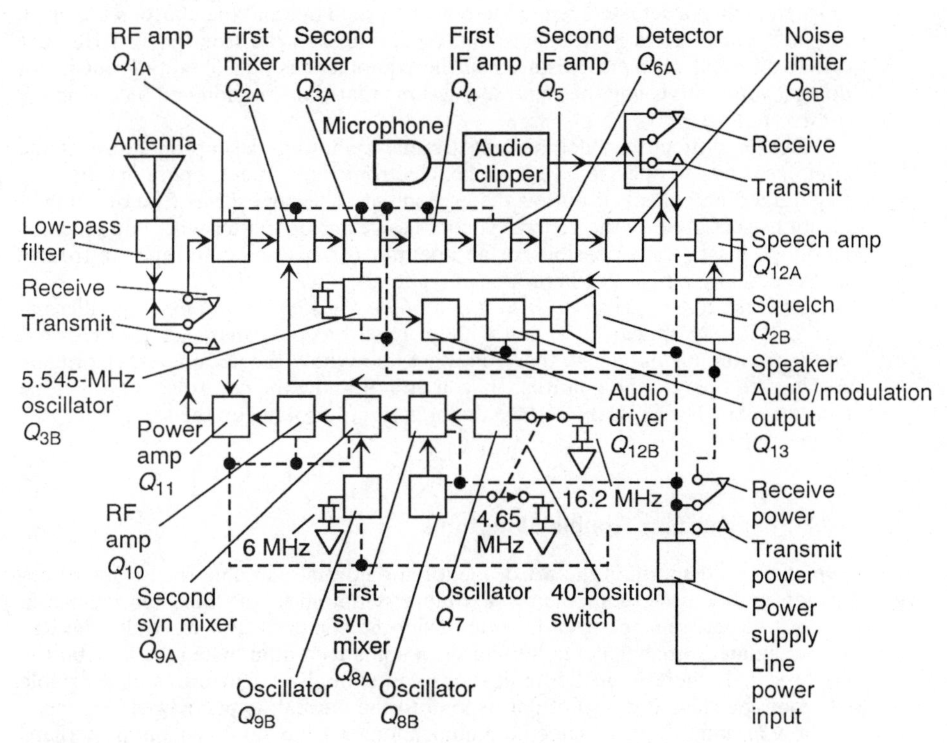

FIGURE 6.1 Basic communications set. (*a*) Trouble symptoms; (*b*) block diagram.

Note that the purpose of the Fig. 6.1 block is *to provide troubleshooting examples and to show how the troubleshooting sequence is applied* but not necessarily to represent a present-day set. Each stage in Fig. 6.1 is composed of a separate transistor, with related, discrete-component, parts. In present-day equipment, many of the stages are combined.

For example, RF stages Q_1 through Q_3 would probably appear as a receiver tuner package (Chap. 1), while IF Q_4 through Q_6 would appear as a single IC, as would audio Q_{12} and Q_{13}. The transmitter frequency-synthesizer Q_7 through Q_9 would probably be replaced by some form of PLL (Fig. 1.17), and the mechanical-relay functions would be performed by diodes and electronic switching, as discussed in Sec. 5.19.

Some of the symptoms listed in Fig. 6.1 point to only one area of the set as a *probable cause.* For example, if there is no reception on any channel but transmission is normal on all channels, the trouble is *most likely* in the receiver circuits. On the other hand, if there is some problem in both transmission and reception, the trouble *could be* in the low-pass filter, and audio/modulator circuits, the switching circuits, or possibly the power supply.

6.2.1 Determining Trouble Symptoms

It is obvious that trouble exists when electronic equipment does not operate, for example, when a set is connected to power, turned on, and the controls are properly set, but there is no transmission, reception, and the front-panel LEDs are off. A different problem exists when the equipment is still operating but is not doing a good job. Using the same set, assume that transmission and reception are *present but poor.*

Another difficulty in determining trouble symptoms is improper use of the equipment by the operator. In complex electronic equipment, operators are usually trained and checked out on the equipment. The opposite is true of communications sets used by the general public. However, no matter what equipment is involved, it is always possible for an operator (or customer) to report a trouble that is actually a result of improper operation.

For these reasons, *you must first determine the signs of failure,* regardless of how bad it may appear to be and without regard to the cause (set failure or operator trouble). This means that you must know how the set operates normally and how to operate the controls. (If you do not know the controls, try to get the *Operator's* or *User's Manual.* Use the pretext of "looking up the serial number" if necessary.)

6.2.2 Recognizing Trouble Symptoms

Symptom recognition is the art of identifying normal and abnormal signs of operation in electronic equipment. A trouble symptom is an undesired change in equipment performance or a deviation from the standard. For example, the RF-output indicator (panel meter) should show some RF output when the PTT button is pressed. If there is no RF indication, you should recognize this as a trouble symptom because it does not correspond to the normal, expected performance.

Now assume that the same communications set has poor reception, perhaps because of bad signal conditions in the area or a defective antenna. If the receiver circuits of the set do not have sufficient gain to produce good reception under these conditions, you could mistake this for a trouble symptom unless you were really familiar with the set. Poor reception (for this particular set operating under these conditions) is not abnormal operation nor is it an undesired change. Thus, it is not a true trouble symptom and should be so recognized.

6.2.3 Equipment Failure versus Degraded Performance

Equipment failure means that either the entire piece of equipment or some functional part of the equipment is not operating properly. For example, the total absence of any received signal when all controls are properly set is a form of equipment failure, even though there may be sound (background noise) from the speaker. Degraded performance occurs whenever the equipment is working but is not presenting normal performance. For example, the presence of hum in the speaker is degraded performance, since the set has not yet failed but the performance is abnormal.

6.2.4 Evaluation of Symptoms

Symptom evaluation is the process of obtaining more detailed descriptions of the trouble symptoms. The recognition of the original trouble may not in itself provide enough information to decide on the probable cause or causes of the trouble because many faults produce similar trouble symptoms.

To evaluate a trouble symptom, it is generally necessary to operate the controls associated with the symptoms and apply your knowledge of electronic circuits, supplemented with information gained from the service data. Of course, the mere adjustment of operating controls is not the complete story of symptom evaluation. However, the discovery of an incorrect setting can be considered a part of the overall symptom-evaluation process.

6.2.5 Examples of Evaluating Symptoms

When there is no sound of any kind coming from the loudspeaker of a communications set, there obviously is trouble. The trouble could be caused by defective ICs, shorted transistors, burned-out diodes, or any one of the several hundred components in the circuits (assuming that power is applied, the set is switched on, and the squelch is properly set). However, the same symptom may be produced when the gain control is turned down. Think of all the time you may save by checking the operating controls first, before you charge into the set with tools and test equipment.

To do a truly first-rate job of determining trouble symptoms, you must have a complete and thorough knowledge of the normal operating characteristics of the set. Your knowledge helps you decide if the set is doing the job for which it was designed. In most service literature it is called "knowing your equipment."

In addition to knowing the set, you must be able to operate all controls properly to determine the symptom (to decide on *normal* or *abnormal* performance). If the trouble is cleared by manipulating the controls, your analysis may or may not stop at this point. If you know the set, you should be able to understand why a specific control removes the apparent trouble.

6.3 LOCALIZING TROUBLE

Localizing trouble means that you must determine which of the major functional areas in a set are actually at fault. This is done by systematically checking each area selected until the faulty one is found. If none of the functional areas of your list shows improper performance, you must take a return path and recheck the symptom information (and observe more information, if possible). Several circuits could be causing the trouble, and the localizing step helps narrow the list to those in one functional area, as indicated by a particular block of the block diagram.

The problem of trouble localization is simplified when a block diagram and a list of trouble symptoms (such as those shown in Fig. 6.1) are available for the equipment being serviced. Remember that these illustrations apply to a "typical," or composite, communications set. However, the general arrangement shown in Fig. 6.1 can be applied to many sets. Thus, Fig. 6.1 serves as a universal starting point for trouble localization. More complex and sophisticated communications sets are discussed in Chap. 7.

6.3.1 Bracketing Technique

The basic bracketing technique makes use of a block diagram or schematic to localize the trouble to a functional area. Bracketing (sometimes known as the *good-input/bad-output* technique) provides a means of narrowing the trouble down to a circuit group and then to a faulty circuit. Symptom analysis and/or a signal-tracing test are used with, or are part of, bracketing.

Bracketing starts by placing brackets (at the good input and the bad output) on the block diagram or schematic. Bracketing can be done mentally, or it can be physically marked with a pencil, whichever is most effective for you. No matter what system is used, with the brackets properly positioned, you know that the trouble exists somewhere between the two brackets.

The technique involves moving the brackets, one at a time (either the good input or the bad output) and then making tests to find if the trouble is within the newly bracketed area. This process continues until the brackets localize a circuit group.

The most important factor in bracketing is to find where the brackets should be moved in the elimination process. This is determined from your deductions based on you knowledge of the set and on the symptoms. All moves of the brackets should be aimed at localizing the trouble with a minimum of tests.

6.3.2 Examples of Bracketing

Bracketing may be used with or without actual measurement of voltages or signals. That is, sometimes localization can be made on the basis of symptom evaluation alone. In practical service, both symptom evaluation and tests are usually required, often simultaneously. The following examples show how the technique is used in both cases.

Assume that you are serving a base-station business-radio set and that you find a "no reception and no-transmission" symptom. That is, the power is applied, the set is turned on, the front-panel LEDs or lamps are on, but there is no RF indication on the panel meter during transmission and no signal-strength or

S-meter indication during receive. The power supply is a logical suspect as the faulty circuit group.

You place a good-input bracket at the 115-V line input and a bad-output bracket at the dc output, as shown in Fig. 6.2*a*. To confirm the symptom, you measure both the dc output voltage (or voltages) and the ac input voltages. If the input is normal but one or more of the output voltages is absent or abnormal, you have localized the trouble to a specific circuit in the power supply.

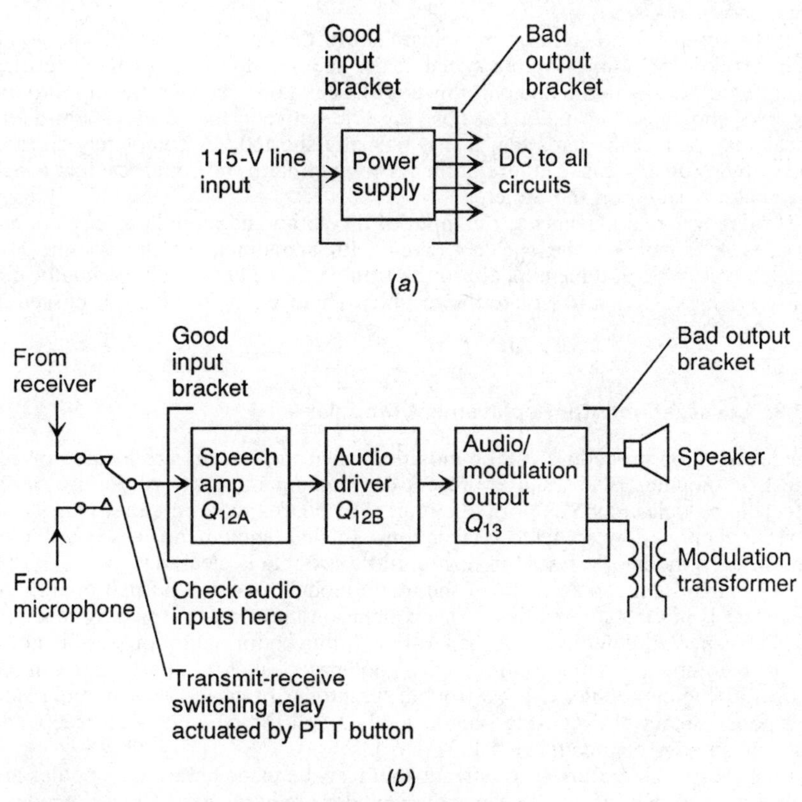

FIGURE 6.2 Examples of bracketing in RF troubleshooting.

From a practical troubleshooting standpoint, it is possible that the power-supply output voltages are normal, but you still have a "no-reception and no-transmitting" symptom. For example, the lines carrying the dc voltages to other circuit groups could be open, or the switching-relay circuits could be malfunctioning to interrupt the dc voltages. This can be checked by measuring the voltages at the circuit end of the lines as well as at the power-supply end.

Also, it is possible that a failure in another circuit could cause the power-supply voltages to be abnormal. For example, if there is a short in one of the circuits on the dc power supply line, the dc output voltage will be low. Of course, this will show up as an abnormal measurement and can be tracked down during the "isolate" step of troubleshooting (Sec. 6.4).

As another example of bracketing, assume that the "no-reception and no-transmission" condition still exists, but the symptoms are somewhat different. Now, there is an RF indication on the panel meter during transmission and an S-meter indication during reception, but there is no sound in the loudspeaker.

You could start by placing a good-input bracket at the input to the audio and modulation circuits and a bad-output bracket at the output of these circuits (at the speaker or modulation transformer), as shown in Fig. 6.2*b*. However, from a practical standpoint, your first move should be adjustment of the volume and squelch controls.

If the trouble is not cleared by adjustment of the controls, confirm the good-input bracket by monitoring the signal at the audio and modulator input (during reception there should be audio at this point). Make this check at the input to the audio as shown in Fig. 6.2*b*. Possibly the line between the receiver and audio circuits is open, or perhaps that line is partially shorted. (A completely shorted line would probably cause failure of the receiver circuits and could result in a lack of signal indication on the S-meter.)

If there are audio signals at the input of the audio and modulator circuits but there is no sound on the speaker (even with adjustment of the volume and squelch controls), you have localized the trouble to the audio and modulator circuits. The next step is to isolate the trouble to a specific circuit, as discussed in Sec. 6.4.

6.3.3 Localization with Replacement Modules

The localization procedure can be modified when the circuits are located on replaceable modules. The trend in present-day equipment is toward the use of replaceable modules, such as printed-circuit (PC) boards that are either plugged in or require only a few soldered connections. In such equipment, it is possible to replace each module or board in turn until the trouble is cleared.

For example, if replacement of the audio module restores normal operation, the defect is in the audio module. This conclusion may be confirmed by reinserting the suspected defective module. Although this confirmation process is not a part of theoretical troubleshooting, it is a good practical check, particularly in the case of a plug-in module. Often a trouble symptom of this sort may be the result of a poor contact between the plug-in module and the connector or receptacle (such as an edge connector on a PC board).

Some service literature recommends that tests be made before all modules are arbitrarily replaced, usually because the modules are not necessarily arranged according to functional area. Thus, there is no direct relationship between the trouble symptom and the modules. In such cases, always follow the service-literature recommendations. Of course, if modules are not readily available in the field, you must make tests to localize the trouble to a module (so that you can order the right module, for example). Also, operating controls and connectors are not usually found on replacement modules, so they must be tested separately.

6.3.4 Which Circuit Group to Test First

When you have localized trouble to more than one circuit group, you must decide which group to test first. Several factors should be considered in making this decision.

As a rule, if you can run a test that eliminates several circuits, or circuit groups, use that test first, before making a test that eliminates only one circuit. This requires an examination of the diagrams (block and/or schematic) and a knowledge of how the set operates. The decision also requires that you apply logic.

Test-point accessibility is the next factor to consider. A test point can be a special jack, or terminal, located at an accessible spot on the PC board. The jack, or terminal, is electrically connected (directly or by a switch) to some important operating voltage or signal path. At the other extreme, a test point can be any point where parts (or wires) are connected together.

Another factor (although definitely not the most important) is your past experience and a *history of repeated failures.* Past experience with identical or similar sets and related trouble symptoms, as well as the probability of failure based on records of repeated values, should have some bearing on the choice of a first test point. However, all circuit groups related to the trouble symptom should be tested, no matter how much experience you may have had with the set. Of course, the experience factor may help you decide which group to test first.

Anyone who has had any practical experience in troubleshooting knows that all the steps of a localization sequence rarely (if ever) proceed in textbook fashion. Just as true, many troubles listed in service data may never occur in the set you are servicing. These troubles are included in the literature as a guide and are not meant to be hard and fast rules.

6.3.5 Universal Trouble Localization for Communications Sets

The following paragraphs describe a universal trouble-localization process for a theoretical communications set or transceiver. The procedures are based on the assumption that the set circuit arrangement is as shown in Fig. 6.1*b*; thus it is possible to group the troubles as shown in Fig. 6.1*a*. Refer to Chap. 7 for trouble localization of more complex communications-set circuits.

If the set is completely dead (no LEDs or panel lights, no transmission, no reception), check the input and output of the power supply. Also check the power-supply fuse (if any). If one or more power-supply voltages are absent or abnormal, you have localized the problem to the power-supply circuits. If all power-supply voltages are normal, check for proper distribution of voltages at the remaining circuits (receiver, transmitter, etc.). If the receiver voltages are normal but not the transmitter voltages (or vice versa), check the relay or switching circuits.

If there is no reception or reception is poor, but transmission is normal (with all controls properly set), check for a signal at the receiver output with the receiver tuned to an active channel. The volume control is generally a convenient test point, as shown in Fig. 6.3*a.* The audio signal can be monitored (or traced) at this point (to check operation of the audio receiver circuits Q_1 through Q_6). As an alternate, an audio signal can be injected at the volume control (to check operation of the audio and modulation section Q_{12} and Q_{13}), as shown in Fig. 6.3*b.*

These same tests can be made at the relay contacts that switch the audio and modulation circuit input between the receiver output and audio clipper. Either way, if an audio signal is present at the volume control or relay contacts, the receiver circuits are cleared. Using the alternate test, if the audio signal passes to the speaker, the audio and modulator circuits are cleared.

Instead of signal tracing through the receiver circuits, the receiver can also be checked by injecting a modulated RF signal at the antenna connector (the RF sig-

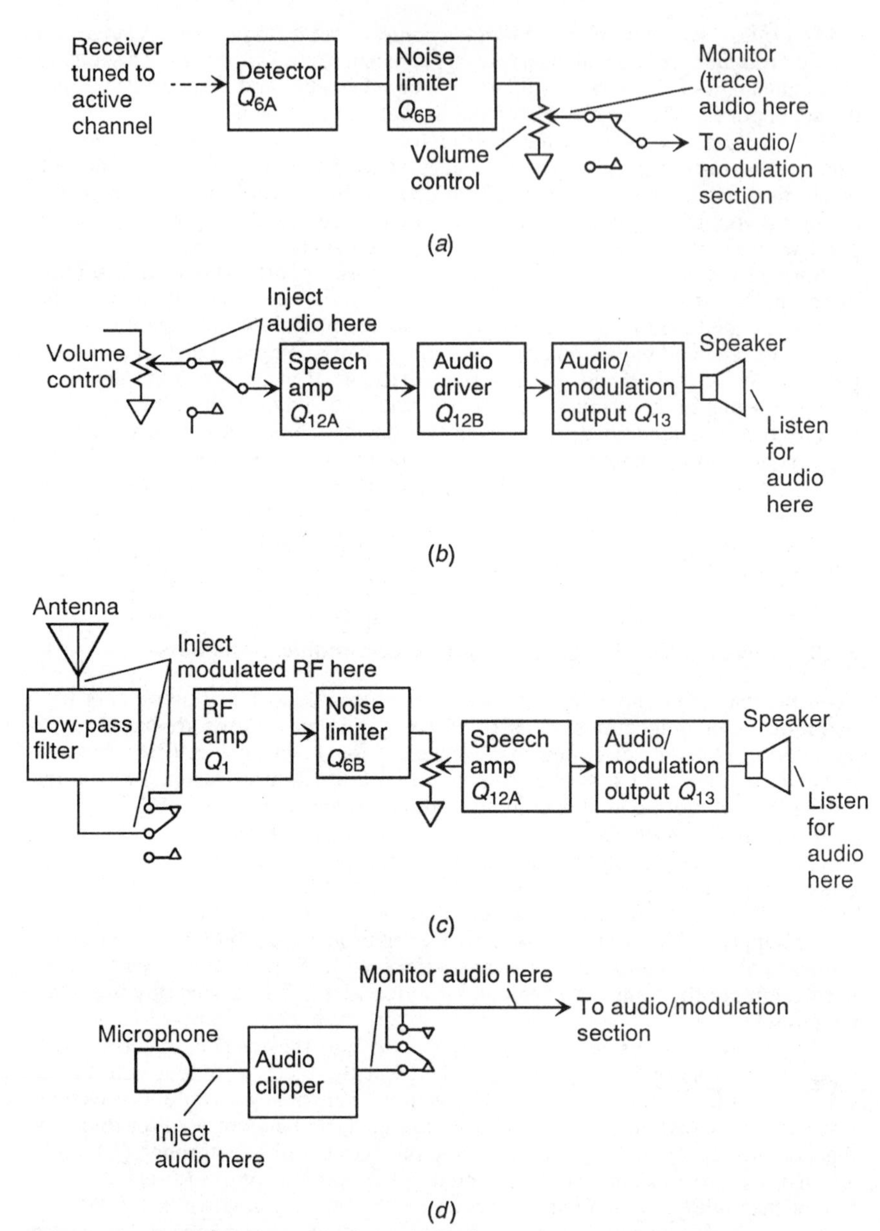

FIGURE 6.3 Examples of signal tracing and injection.

nal must be modulated by an audio tone) and listening for the modulated tone in the speaker, as shown in Fig. 6.3*c*. Since this symptom may be caused by a defect in the switching circuits, inject the modulated RF signal on both sides of the relay contacts as shown.

To eliminate the AGC circuits as trouble suspects, apply a fixed voltage to the AGC line and check operation of the receiver. (This is known as *clamping* the AGC line.) If operation is normal with the AGC line clamped but not when the clamp is removed, you have localized trouble to the AGC circuit.

AGC circuit problems are often difficult to localize because the AGC circuit uses *feedback.* For example, if the IF amplifiers are defective, the detector Q_{6a} and AGC circuits do not receive proper IF signals. (The AGC voltage is developed by the detector Q_{6a} circuit.) In turn, the lack of proper AGC voltage may cause the IF amplifiers to operate improperly. Conversely if the AGC circuits are defective, the IF amplifiers do not receive proper AGC voltages and do not deliver a proper IF signal to the AGC circuits.

As a guideline, if clamping the AGC line eliminates a "no-reception" or "poor reception" symptom, the trouble is probably localized to the AGC circuits (but could be in an IF or tuner package, in present-day sets).

If there is no transmission or transmission is poor, but reception is normal, check for an RF indication on the front-panel meter (with the PTT button pressed). If the RF indication is absent or abnormal (very low), you have localized trouble to the transmitter circuits (Q_9, Q_{10}, and Q_{11}).

This symptom may also be caused by a defect in the switching circuits. Check that the relay operates when the PTT button is pressed and that *all sets of contacts* switch from receive to transmit (antenna from receiver input to transmitter output, power from receiver to transmitter, audio input from receiver to microphone or audio-clipper input, and the audio output to the speaker disabled). If any of these functions is not normal, trouble is localized to the switching circuits. (If the set uses electronic switching, such as shown in Fig. 5.15, check that the voltage at *all switching diodes* changes each time the PTT button is pressed.)

If there is no modulation on transmission and no background noise in the speaker during reception (with the squelch properly set), the problem is most likely in the audio and modulation circuits. These circuits can be checked quickly by injecting an audio signal at the input (volume control or corresponding relay contacts) and listening for a tone in the speaker.

If there is no modulation on transmission but there is background noise in the speaker during reception, the problem is probably in the audio-clipper circuits. Three circuits may be checked by injecting an audio signal at the microphone input and checking at the corresponding relay contacts, as shown in Fig. 6.3*d*.

This same symptom can be produced by a *defective microphone.* Generally substitution is the most practical means of checking microphones. Finally, it is possible that the symptom is caused by bad relay contacts (at the audio and modulator circuit input). On sets with electronic switching (Fig. 5.15), it is possible that the switching diodes are leaking (but this is rare).

Note that some communications sets are provided with a modulation indicator. Often this indicator is a front-panel lamp that glows when modulation is applied (strong modulation produces a bright glow, and weak modulation produces a dull glow). Such an indicator makes it easy to localize modulation problems. For example, if the modulator lamp is not glowing (to indicate modulation) but the panel meter indicates proper RF output, the problem is between the microphone and the modulation-output circuit.

If transmission and reception are absent or abnormal only on certain channels, the problem is likely to be in the frequency-synthesis (crystal-control) circuits, or in the channel-selector switching circuits (or in the PLL circuits of present-day sets). Either way, troubles producing these symptoms are usually easy to localize. However, the trouble may not be easy to isolate once you are into the circuits, particularly on those sets using a PLL for channel selection. (Such circuits are described in Sec. 7.1.)

If transmission is poor but the RF indication is normal (on the panel meter) and reception is poor (weak S-meter indication) but background noise appears normal, the low-pass filter is a likely suspect. For example, the low-pass filter could have shorted capacitors or open coils (or there could be poor solder connections where the capacitors and coils are mounted on the PC board).

The same symptoms may be caused by a *defective antenna system.* Needless to say it is possible to spend hours trying to localize problems in a perfectly good communications set if the antenna or lead-in is bad (shorted coaxial, improper connections, mismatching, etc.). When you are confronted with some mysterious "poor reception and poor transmission" symptom, always try the set on a known-good antenna. Also, check the antenna as described in Sec. 7.7.

6.4 ISOLATING TROUBLE TO A CIRCUIT

The first two steps (symptoms and localization) of the troubleshooting procedure give you initial information about the trouble and describe the method of localizing the trouble to a *probably faulty* circuit group. Both steps involve a minimum of testing. In the isolate step, you will do extensive testing to track the trouble to a specific faulty circuit.

6.4.1 Isolating Trouble in ICs and Replaceable Modules

ICs and replaceable modules are common in present-day equipment. For example, all parts of a PLL except the crystals are contained in a replaceable module (tuner, etc.). Likewise, the entire IF circuit is replaced by a single IC. These modules may be plug-ins, but they usually require some solder connections.

In sets with ICs and replaceable modules, the trouble can be isolated to the IC or module input and output but not to the circuits (or individual parts) within the IC. No further isolation is necessary, since parts within the IC cannot be replaced on an individual basis.

This same condition is true for some sets where groups of circuits are mounted on *sealed, replaceable* boards or cards. Note that not all modules are sealed; many have replaceable parts.

6.4.2 Using Diagrams in the Isolation Process

No matter what physical arrangement is used, the isolation process follows the same reasoning you have already used: the continuous narrowing down of the trouble area by making logical decisions and performing logical tests. Such a process reduces the number of tests that must be performed, thus saving time and reducing the possibility of error.

A block diagram is a convenient tool for the isolation process since a block shows circuits already arranged in circuit groups. Unfortunately, as discussed, you may or may not have a block diagram supplied with your service literature. Often, you must work with nothing but a schematic diagram.

With either diagram, if you can recognize *circuit groups* as well as *individual circuits,* the isolation process is much easier. For example, if you can subdivide (mentally or otherwise) the schematic into circuit groups rather than individual circuits, you can isolate the group by a single test at the input or output for the group.

The block diagram in Fig. 6.1*b* is arranged into individual circuits with each block representing a transistor and the related circuit parts. You can arrange these blocks into circuit groups, as is done in Fig. 6.1*a*. For example, the blocks representing Q_1 through Q_6 form the receiver circuit group. Blocks Q_7 and Q_8 form the frequency synthesizer, Q_9 through Q_{11} form the transmitter, and Q_{12} and Q_{13} form the audio and modulation group. All communications sets have some similar (but not identical) arrangement. Make it a practice to group the circuits mentally on the block or schematic as a first step in isolating trouble.

No matter what diagram you use, or what equipment arrangement is found, you are looking for three major bits of information: the *signal path* (or paths), the *signal form* (waveform, amplitude, frequency, etc.), and the *operating and adjustment controls* in the various circuits along the signal paths. If you know what signals are supposed to go where and how the signals may be affected by controls, you can isolate trouble quickly in any electronic equipment.

No waveforms are shown in Fig. 6.1. This is standard for most RF diagrams (both block and schematic). The signals in a communications set are typically sine waves, and the shape or form is not critical (with the possible exception of the audio and modulation section). Instead, you are interested in *amplitude and frequency* of the signals. Although there is no standardization, most RF service literature shows the amplitude and frequency of critical signals. Usually, this information is found in the alignment and adjustment instructions rather than on the diagrams. About the only waveforms found in RF literature are those pertaining to the *modulation envelope.*

No operating and adjustment controls are shown in Fig. 6.1. This is typical for most RF block diagrams. The controls are shown on the schematic diagrams. Once you have arranged the individual circuits into groups, your next step is to locate *all operating and adjustment controls* in the group. This is just as important as localizing all inputs and outputs for the circuit group.

6.4.3 Comparison of Signals

In the simplest form, the isolation step involves comparing the actual signals produced along the circuit paths against the signals given in the service data. This is known as *signal tracing.* The isolation step may also involve *injection* or *substitution* of signals usually found along the signal paths. Signal tracing and injection are discussed throughout the rest of the book. With either technique, you check and compare inputs and outputs of circuit groups (and circuits) in the signal paths.

Typically, the input signal is injected at the base, and the output signal is traced at the collector (or possibly the emitter) of transistors, as shown in Fig. 6.4. For a circuit group, the input is at the first base of the path, whereas the output is at the last collector (or emitter). The input signal is injected at one point, and the output signal is obtained simultaneously at a point several stages further along in the same signal path.

Before studying the subject of signal paths, remember the following points. The symptoms and relation information obtained in the previous steps (symptoms and localization) should not be discarded now or at any time during the troubleshooting process. From this information you will be able to identify those circuit groups which are *probable* trouble sources.

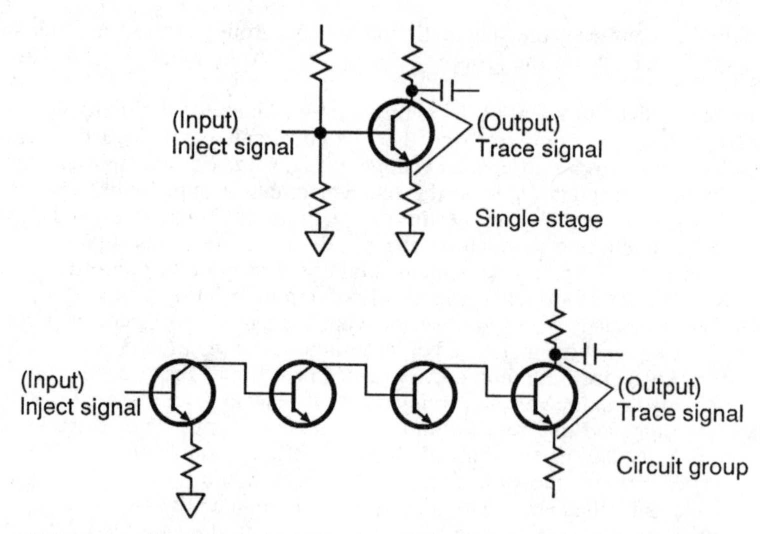

FIGURE 6.4 Signal tracing and injection at RF circuit-group inputs and outputs.

Note that the physical location of the circuit groups within the set has no relation to the representation on the diagrams (block or schematic). You must consult *part-placement* diagrams or photos (if any) to find physical location.

6.4.4 Signal Tracing versus Signal Substitution

Both signal-tracing and signal-substitution (or signal-injection) techniques are used frequently in troubleshooting all types of RF equipment. The choice between tracing and substitution depends on the test equipment used and the circuits involved. *Signal tracing* is generally used for the oscillator, frequency synthesizer, and transmitter RF circuits because these circuits generate or amplify signals that are readily traced and need not be substituted. *Signal substitution* is generally used for the receiver RF, IF, noise, squelch, and audio circuits. Both tracing and substitution are used for the PLLs found in RF tuners, as discussed in Sec. 7.1.

Signal tracing is done by examining the signals at test points with a monitoring device such as a frequency counter, scope meter, or speaker. The monitoring-device probe is moved from point to point with a signal applied at a fixed point. Signal substitution is done by injecting an artificial signal (from a generator) into a circuit or group. The injected signal is moved from point to point, with a monitoring device remaining fixed at one point. Both signal tracing and substitution are often used simultaneously in RF troubleshooting. For example, it is common practice to inject a modulated RF signal at the antenna and monitor the output of the receiver circuits with a meter or scope.

6.4.5 Half-Split Technique

The half-split technique is based on using any test that eliminates the maximum number of circuits simultaneously. Unless the trouble symptoms point definitely to one circuit that might be the trouble source, the first test is made at a convenient point halfway between the good-input and bad-output brackets.

Example of the Half-Split Technique. Figure 6.5 is a simplified version of the receiver circuit group for the set shown in Fig. 6.1*b*. Note that in Fig. 6.5, the first and second mixers are combined into one block, as are the first and second IF amplifiers.

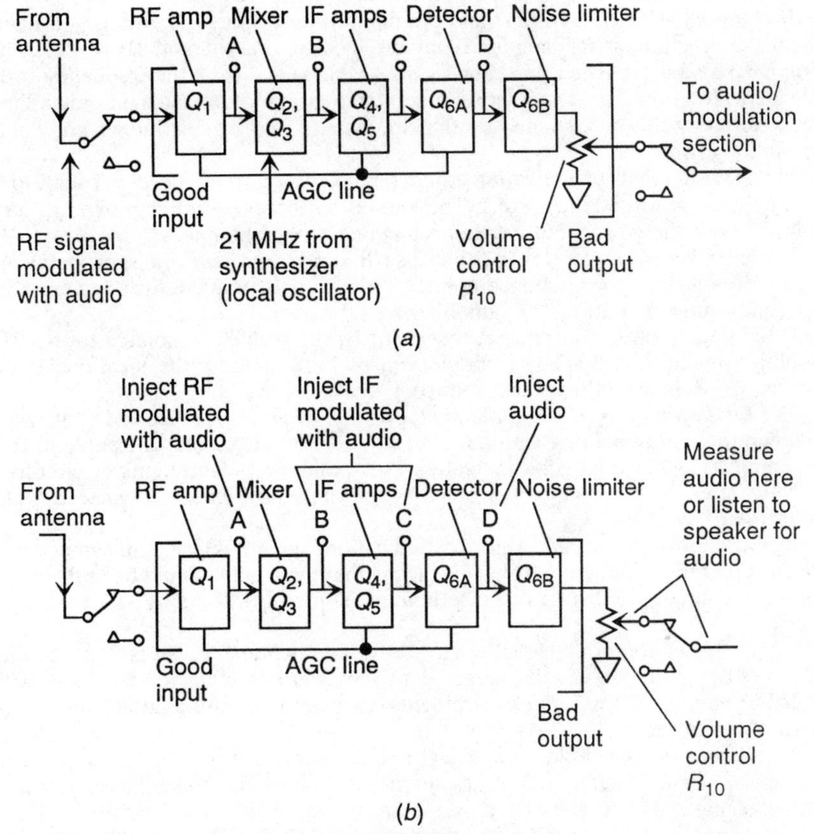

FIGURE 6.5 Half-split technique in RF-circuit troubleshooting.

The brackets placed at the antenna (good-input) and volume control (bad-output) show the trouble being localized to the receiver circuit group. Brackets should be placed at these points as a result of a "no reception on any channel, transmission normal" or "reception poor, transmission good" trouble symptom.

The next phase of troubleshooting is to isolate the trouble to one of the circuit groups (mixers or IF amplifiers) or to one of the individual circuits (RF amplifier, detector, or noise limiter) in the signal path. The selection of test points during this phase depends on the *accessibility* and *method of troubleshooting* (signal tracing or signal injection).

Assuming that test points A, B, C, and D are equally accessible (and that there are no special symptoms that would point to a particular circuit or group), test point C is the most logical point for the first test if signal tracing is used. Test point B is the next most logical choice. When using signal injection, however, test points C and D are the most logical choices. Here are the reasons for this thinking.

Half-Split Technique Using Signal Tracing. With the equipment connected as shown in Fig. 6.5*a,* each of the test points (A through D) can be monitored in turn (with meter, scope, or counter), as described in Chap. 4.

If test point C is chosen first and the monitoring display is normal, you have cleared four circuits (RF amplifier, mixer, local oscillator, and IF amplifiers). You have also established that there is a 21-MHz signal from the frequency synthesizer. However, there may still be defective circuits (detector and noise limiter). Note that this process divides the circuits into two groups (known good and possibly bad).

Now assume that the indication at test point C is abnormal (totally absent, low in amplitude, off-frequency, etc.). The bad-output bracket can be moved to test point C, with the good-input bracket remaining at the antenna.

The next logical test point is B because B is near the *halfway point* between the two brackets. (If you choose point A instead of B, you confirm or deny the possibility of trouble in the RF amplifier *only.*)

If the indication is abnormal at test point B, the trouble is isolated to the RF amplifier or the mixers. (The term *mixer* used here includes the local oscillator and the 21-MHz signal from the frequency synthesizer.)

The final step is to monitor the signal at test point A because A is half way between the two remaining circuits. A bad indication at A isolates trouble to the RF amplifier, whereas a good indication at A points to defective mixer circuits.

Now let us see what happens when a *test point other than C is monitored first, using signal tracing.*

If you choose A and get a normal indication, the trouble is somewhere between A and the volume control. This means that the trouble could be in the mixers, IF amplifiers, detector, or noise limiter. You still have many test points to check.

If you get an abnormal signal at A, the trouble is immediately isolated to the RF amplifier, but this is a lucky accident, not good troubleshooting. You will probably have as many unlucky accidents as lucky ones throughout your troubleshooting career.

The same condition holds true if test point D is chosen first (if you use signal tracing). A normal signal at D clears all circuits except the noise limiter. An abnormal signal at D still leaves the possibility of trouble in many circuits.

If test point B is chosen first and the signal is normal, this clears three circuits (RF amplifier, mixers, local oscillator) but leaves four circuits possibly defective (both IF amplifiers, detector, and noise limiter). The opposite is true if the signal is abnormal at B.

From a practical standpoint, B is not a bad choice for a first test using signal tracing. If test point B is more accessible than C, use B.

Half-Split Technique Using Signal Injection. With the equipment connected as shown in Fig. 6.5*b,* appropriate signals are injected at each of the test points (A through D), and the results are monitored at the volume control or speaker. (Note that signal injection requires several different types of signal sources at different frequencies (RF, IF, audio, etc.), whereas signal tracing requires only one signal at the input.

Using signal injection with half-split, the first signal is again injected at C. However, *a normal response now clears the final circuits* (detector and noise limiter). Under these conditions, the next logical points for signal injection are B and A, in that order.

An absent or abnormal response at the volume control or speaker (with a signal injected at C) isolates the trouble to the detector or noise limiter. Under these conditions, the next logical test point is D.

A normal response with a signal injected at D, but not at C, isolates the trouble to the detector. An absent or abnormal response with a signal at D isolates the trouble to the noise limiter.

Keep in mind that these examples using the half-split technique, signal tracing, signal injection, and bracketing to isolate the trouble to a circuit group by no means cover all the possibilities that may occur. The examples simply illustrate the basic concepts involved when following a systematic, logical troubleshooting procedure in servicing typical RF equipment.

6.5 LOCATING A SPECIFIC TROUBLE

The ability to recognize symptoms and to verify them with test equipment will help you to make logical decisions regarding the selection and localization of the faulty circuit group. You will also be able to isolate trouble to a faulty circuit. The final step of troubleshooting—locating the specific trouble—requires testing of the various branches of the faulty circuit to find the defective part.

The proper performance of the locate step enables you to find the cause of trouble, repair it, and return the set to normal operation. As a follow-up, record the trouble so that, from the history of the set, future troubles may be easier to locate. Also, such a history may point out consistent failures that might be caused by a design error.

6.5.1 Locating Troubles in Replaceable Modules

Because the trend in modern electronic equipment is toward IC and replaceable-module design, technicians often assume it is not necessary to locate specific troubles to individual parts. They assume that all troubles can be repaired by replacement of ICs, packages, or boards.

Although the use of replaceable modules often minimizes the number of steps required in troubleshooting, it is still necessary to check branches to parts outside the module. Front-panel operating controls are a good example of this since such controls are connected to the terminal of an IC, board, or other module.

6.5.2 Inspection Using the Senses

After the trouble is isolated to a particular circuit, the next step is to conduct a preliminary inspection using your physical senses. For example, burned or charred resistors may be observed visually or by smell. Overheated parts, such as hot transistor cases, may be located quickly by touch. The sense of hearing can be used to listen for arcing between wires and PC wiring, for "cooking," or overloaded or overheated transformers, or for hum or lack of hum. Although all of the senses are involved, the procedure is usually referred to as a visual inspection.

6.5.3 Testing to Locate a Faulty Part

Most transistors, ICs, and diodes are not easily replaced, so the old electronic troubleshooting procedure of replacing parts at the first sign of trouble is not applicable to present-day RF equipment. Instead, *circuits are analyzed by testing* to locate faulty parts.

Testing the Active Device. For service purposes, the transistor, IC, and diode may be considered the active devices in any electronic circuit. Because of their key position in the circuit, these devices are useful in evaluating operation of the entire circuit (through signal, voltage, and resistance tests). Preliminary tests at the terminals of the active device often locate the trouble quickly.

Signal Testing. Usually, the first step in circuit testing is to analyze the output signal of the active device (the circuit output), typically at the collector of a transistor (but possibly at the emitter). Of course, in some circuits (such as power supplies) there is no output signal, as such.

In addition to checking for the presence of a signal, the signal must be analyzed in detail for proper amplitude and frequency. As discussed throughout this book, a careful analysis often pinpoints the branch of a circuit most likely to be defective.

Transistor and Diode Testers. It is possible to test transistors and diodes in circuit, using in-circuit testers. Such tests are usually adequate for transistors used at lower frequencies, particularly in the audio range. However, most in-circuit tests do not show the high-frequency characteristics of transistors. (The same is true for out-of-circuit transistor and diode testers.) For example, it is possible for a transistor to perform well in the audio section of a communications set but be hopelessly inadequate in the RF, IF, oscillator, or frequency-synthesizer sections.

Voltage Testing. After signal analysis and/or in-circuit tests, the next logical step is voltage measurement at the active device terminals or leads. Always pay particular attention to those terminals which show an abnormal signal since those terminals are most likely to show an abnormal voltage (but not always).

When properly prepared service literature is available (with signal, voltage, and resistance information), the actual voltage measurements may be compared with the normal voltages listed in the service literature. This test often helps isolate the trouble to a *single branch of the circuit.*

Relative Voltages. It is often necessary to troubleshoot circuits without benefit of adequate voltage and resistance information. This can be done using the schematic diagram to make a logical analysis of the relative voltages at the transistor terminals. For example, in an NPN transistor, the base must be positive in relation to the emitter if there is to be emitter-collector current flow. The problem of troubleshooting RF circuits on the basis of relative voltages is discussed further throughout this chapter.

Resistance Measurements. After signal and voltage measurements are taken, it is often helpful to make resistance measurements at the same point on the active device where an abnormal signal and/or voltage is found. Of course, other points (for example, suspected parts) in the circuit may also be checked by resistance measurement, or a continuity check can be made to find point-to-point resistance of the suspected branch. Considerable care must be used when making resistance measurements in solid-state circuits since the junctions of transistor act like diodes. When biased with the right polarity (by the ohmmeter battery), the diodes conduct and produce false readings. This condition is also discussed throughout the rest of this chapter.

Current Measurements. In rare cases, the current in a particular circuit branch can be measured directly with an ammeter. However, it is usually simpler and more practical to measure the voltage and resistance of a circuit and then calculate the current.

6.5.4 Signal Measurements

When testing to locate trouble, the signals are measured with the circuit in operation and usually with an input signal (or signals) applied. The signals can originate at a test generator or from another communications set, whichever is convenient for the particular measurement. The signal tracing and signal injection techniques found in Sec. 6.4 are typical. If you use signal information found in the service literature, follow all notes and precautions described in the literature. Usually, the literature specifies the position of operating controls, typical input signal amplitude, and so on.

Note that there is a relationship between signals and trouble symptoms. Complete failure of a circuit usually results in the absence of a signal, whereas a poorly performing circuit usually produces an abnormal signal. For example, an improperly adjusted circuit could produce an off-frequency signal, or a defective circuit could produce a low-amplitude signal.

6.5.5 Voltage Measurements

In troubleshooting, voltages are measured with the circuit in operation but usually with no signals applied. In some cases, the voltage measurements are made with the equipment operating. If you use the voltage information found in service data, follow all notes and precautions. Usually the service literature specifies the position of operating controls, typical input voltages, and so on.

If you have had any practical experience in troubleshooting, you know that voltage (as well as resistance and signal) measurements are seldom identical to those listed in the service literature. This brings up an important question concerning voltage measurements: How close is good enough? Several factors must be considered in answering this question.

The tolerances of resistors, which greatly affect the voltage readings in a circuit, may be 20, 10, or 5 percent. Resistors with 1 percent (or better) tolerances are used in some critical RF circuits. The tolerances marked or color-coded on the parts are thus one important factor. The wide range of transistor, IC, and diode characteristics can also cause variations in voltage readings.

The accuracy of test instruments must also be considered. Most voltmeters have accuracies of a few percent (typically 5 or 10 percent). Precision laboratory meters (generally not used in routine troubleshooting) have a much greater accuracy.

To operate properly, the voltages of critical RF circuits must fall within a narrow range of tolerances (at least 10 percent and probably closer to 2 or 3 percent). However, many RF circuits can operate satisfactorily within tolerance ranges of 20 to 30 percent.

In troubleshooting, the most important factors to consider when measuring voltages are the symptoms and the output signal. If no output signal is produced by the circuit, you should expect a wide variation of voltages in the trouble area. Trouble that results in circuit performance that is just out of tolerance may cause only a slight change in circuit voltages.

6.5.6 Resistance Measurements

Resistance measurements must be made with no power applied. However, in some cases, various operating controls must be in certain positions to produce resistance readings similar to those found in the service literature. This is particularly true of controls with variable resistances.

Always observe any notes or precautions found in the service literature. In any circuit, always check that the filter capacitors are discharged before making resistance measurements (but after the power has been turned off). Unless otherwise directed by the service notes, measure the resistance from the terminals of the active device to a ground point on the PC board or between any two points that are connected by wiring or parts.

Do not be surprised if you find service literature with little or no resistance data. Often, the only resistances given are the values of resistors and the dc resistance of coils and transformers. In well-prepared literature, you will find the resistance from all terminals of each active device, including all IC terminals.

The reasoning for the omission of resistance values from various terminals has some merit. If there is a condition in any active-device terminal circuit that produces an abnormal resistance (say, an open resistor or a resistor that has changed drastically in value), the voltage at that terminal is abnormal. If such an abnormal voltage reading is found, it is necessary to check out each resistance in the terminal circuit on an individual basis.

Because of the *shunting effect* of other parts connected in parallel, the resistance of an individual part or circuit may be difficult to check. In such cases, it is necessary to disconnect one terminal of the part being checked from the rest of the circuit. This leaves the part open at one end, and the value of resistance measured is that of the part only.

6.5.7 Duplicating Signal, Voltage, and Resistance Measurements

If you are responsible for service of one type or model of RF equipment, it is strongly suggested that you duplicate all signal, voltage, and resistance measurements found in service literature with your own test equipment. This should be done with RF equipment that is known to be good and operating properly.

With a good set of duplicate measurements, you can spot even slight variations during troubleshooting. Always make the initial measurements with test equipment normally used in troubleshooting. If more than one set of test equipment is used, make the initial measurements with all available test equipment, and record any variations.

6.5.8 Using Schematics in RF Troubleshooting (Example)

Regardless of the trouble symptom, the actual fault can be traced eventually to one or more of the circuit parts (transistors, ICs, etc.). The checks of signals, voltage, and resistance then indicate which branch within a circuit is at fault. Finally, you must locate the particular part that is causing the trouble in that branch.

To do this, you must be able to read a schematic diagram. These diagrams show what is inside the blocks on a block diagram and provide the final picture of the equipment. Often, you must service electronic equipment with the aid of nothing more than a schematic. If you are fortunate, the diagram may also show some voltages.

Figure 6.5 shows the block diagram for the receiver circuits of a communications set. The following example shows how to use schematics to localize trouble in one of the receiver circuits. Although only one receiver circuit (the mixer) is discussed here, the same basic troubleshooting principles apply to all circuits in the set.

Assume that the receiver circuits in Fig. 6.5 are being serviced, using signal injection as shown in Fig. 6.5*b*. The trouble is initially isolated to the receiver circuits by a "no reception, good transmission" trouble symptom. An IF signal modulated with audio is injected at test point B, and a proper response is noted at the volume control (receiver-circuit output).

These symptoms indicate that the 21-MHz signal from the frequency synthesizer to the mixer is normal (on-frequency and of correct amplitude). However, there is no response when an RF signal (at the selected channel frequency) modulated with audio is applied to the mixer input (test point A). Now you have localized trouble to the mixer circuit.

The next step is to measure the voltages at the terminals of the active device in the mixer. These are shown as the collector, emitter, and base of Q_2 in Fig. 6.6*a*. If any of the Q_2 elements show an abnormal voltage, the resistance of that element should be checked first. Note that the collector voltage is specified as -7 V. Neither the base and emitter voltages nor any of the resistance values are given. This lack of information is typical for most RF equipment. Thus, you must be able to interpret schematic diagrams to *estimate approximate voltages.*

For example, the voltage at the junction of R_3 and R_4 is given as -8.5 V. This is logical because the source is 9 V. The value of R_2 is approximately 25 percent of the value of R_3. Thus, the drop across R_2 is about 25 percent of -8.5 V, or approximately -2 V, and the base should be about -2 V. If Q_2 is silicon, the emitter is about 0.5 V different from the base, or about -1.5 V (emitter more positive or less negative, in this case). If Q_2 is germanium, the base-emitter differential is about 0.2 V, and the emitter should be about -1.8 V.

Remember that this method of interpreting the schematic gives you approximate voltages only. In practical troubleshooting, the voltage differentials between circuit elements and transistor terminals are the most important factor. Troubleshooting based on voltage differentials is discussed fully in the rest of this chapter.

Now assume that there is no voltage at the collector but that the base and emitter show what appears to be normal or logical voltages. The next step is to remove power, discharge C_{14} and C_{15}, and measure the collector resistance (to ground), as shown in Fig. 6.6*b*. Because no resistance values are given for the elements of Q_2, you must use the schematic to estimate the approximate values.

Of course, a zero resistance at the collector indicates a short. For example, capacitor C_5 could be shorted. On the other hand, an infinite resistance indicates

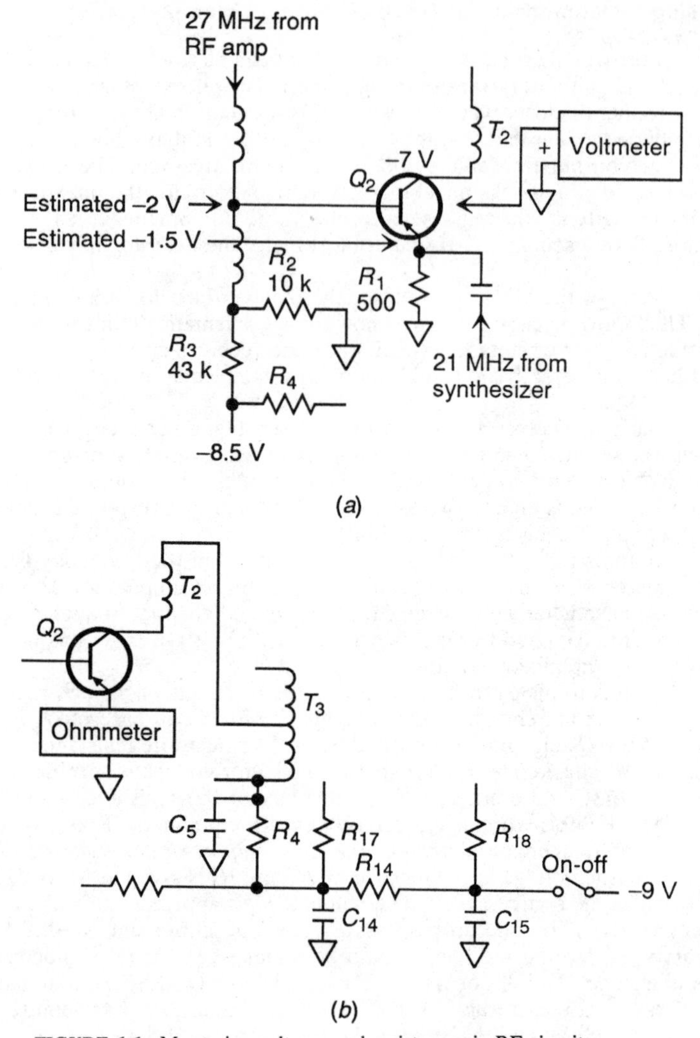

FIGURE 6.6 Measuring voltages and resistances in RF circuits.

an open circuit. For example, the coil winding of T_2 and T_3 could be open, or R_4 could be burned out and open. It is usually easy to locate the fault when you find such extreme resistance readings.

Unfortunately, a resistance reading that falls between these two extremes does not provide a really sound basis for locating trouble. To make the problem worse, the effect of solid-state devices in the circuit can further confuse the situation. For example, assume that the fault is an open T_3 winding, as shown in Fig. 6.6*c*. This results in a no-voltage reading at the collector of Q_2. However it is still possible to measure a resistance to ground from the Q_2 collector, if the following conditions are met.

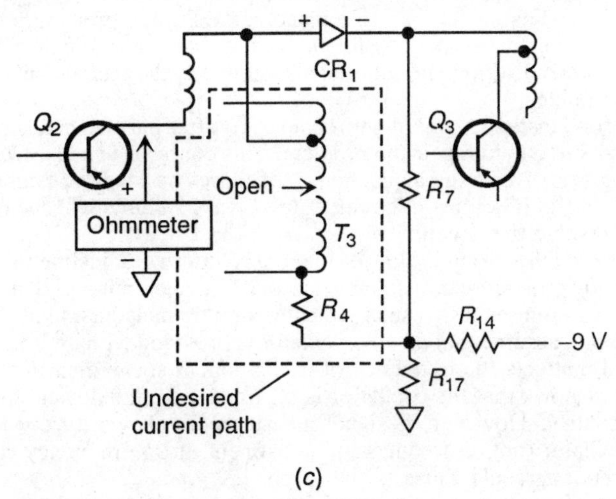

(c)

FIGURE 6.6 *(Continued)* Measuring voltages and resistances in RF circuits.

Assume that the ohmmeter leads are connected so that the positive terminal of the ohmmeter battery is connected to the Q_2 collector. This forward biases CR_1 and causes the ohmmeter to measure the resistance across R_7 (the collector supply of Q_3) and R_{17}. Also, if the collector of Q_2 is made positive in relation to the base, the Q_2 base-collector junction is forward biased, resulting in possible current flow.

The problem illustrated in Fig. 6.6*c* can be eliminated by reversing the ohmmeter leads and measuring the resistance both ways. *If there is a difference in the resistance values with the leads reversed* (in any solid-state circuit), check the schematic for possible forward-bias conditions in diodes and transistor junctions in the associated circuit.

6.5.9 Internal Adjustments During Trouble Localization

Remember that adjustment of controls (both internal-adjustment and operating controls) can affect circuit conditions and may lead to false conclusions during troubleshooting. For example, the amplitude of the signal from Q_6 (Fig. 6.5) is set directly by the volume control R_{10}, which is an operating control.

If the signal from Q_6 is very low, it could be that the volume control is set to the minimum position. Of course, because the volume control is an operational control, a run-through of the operating sequence at the beginning of troubleshooting will pinpoint such an obvious condition. However, a low output from the detector can be caused by poor alignment of the IF transformers. Because the IF transformers require internal adjustments, poor alignment is not detected through the use of operating procedures. This condition, or a similar one, may lead to one of two unwise courses of action.

Following the first course, you might launch into a complete alignment procedure (or whatever internal adjustments are available) once you have isolated the trouble to a circuit and are trying to locate the specific defect. No internal control, no matter how inaccessible, is left untouched. You may reason that it is easier to make the adjustments than to replace parts. Such a procedure does eliminate improper adjustment as a possible fault, but it may also create more trouble than is repaired. Indiscriminate internal adjustment is the technician's equivalent of operator trouble.

In the second instance, you might replace part after part when a simple screwdriver adjustment would repair the problem. This course is usually taken because of an inability to perform the adjustment procedures or a lack of knowledge concerning the control functions in-circuit. Either way, a study of the service literature should resolve the situation.

There is a middle ground. Do not make any internal adjustments during the troubleshooting procedure until trouble is isolated to a circuit and then only when the trouble symptom or test results indicate possible maladjustment.

For example, assume that an RF oscillator is provided with an internal adjustment control that sets the frequency of oscillation. If signal measurement at the circuit output shows that the oscillator is off-frequency, it is logical to adjust the frequency control. However, if signal measurement shows a very low output from the oscillator (but on-frequency), adjustment of the frequency control during troubleshooting could cause further problems.

An exception to this rule occurs when the service literature recommends alignment or adjustment as part of the troubleshooting procedure. Generally, alignment and adjustment are checked after testing and repairs are completed. This assures that the repair procedure (replacement of parts) has not upset circuit adjustments.

6.5.10 Trouble Resulting from More Than One Fault

A review of all symptoms and test information obtained thus far helps you to verify that the part located is the *sole trouble* or to isolate other faulty parts. This is true whether the malfunction of these parts is caused by the isolated part or by some entirely unrelated problem.

If the isolated malfunctioning part can produce *all the symptoms and indications* that you have accumulated, you may logically assume that it is the sole cause of trouble. If the part cannot, you must use your knowledge of electronics and the particular equipment to determine what other part (or parts) could have become defective and produced all the symptoms.

When one part fails, the failure often causes abnormal voltages or currents that could damage other parts. Trouble is often isolated to a faulty part that is a result of the original trouble, rather than to the trouble source. For example, assume that the troubleshooting procedure thus far has isolated a transistor as the cause of trouble and that the transistor is burned out.

Start by asking What could cause this burnout? Excessive current can destroy the transistor by causing internal shorts or by altering the characteristics of the semiconductor material, which is sensitive to temperature. Thus, the problem becomes a matter of finding how such excessive current was produced.

Excessive current in a transistor can be caused by an extremely large input signal (which overdrives the transistor). Such an occurrence indicates a fault somewhere in the circuits ahead of the transistor input. Power surges (intermittent excessive outputs from the power supply) may also cause the transistor to burn out. Such conditions should be checked before a new transistor is placed in the circuit.

Some other typical malfunctions, along with their common causes, include: burned-out transistor caused by *thermal runaway* (Chap. 2), power-supply overload caused by a short circuit in some portion of the voltage-distribution network, shorted blocking capacitors, and blown fuses caused by power-supply surges or shorts in filtering networks.

It is obviously impractical to list all the common malfunctions and the related causes that you may find in troubleshooting RF equipment. Generally, when a part fails, the cause is an operating condition that exceeded the maximum ratings of the part. However, it is quite possible for a part simply to "go bad" in any equipment, no matter how well designed.

The operating condition that causes a failure can be temporary and accidental, or it can be a basic design problem, as a history of repeated failures would indicate. No matter what the cause, your job is to find the trouble, verify the source or cause, and then repair the trouble.

6.5.11 Repairing Troubles

In a strict sense, repairing the trouble is not part of the troubleshooting procedure. However, repair is an important part of the total effort involved in getting the equipment back in operation. Repairs must be made before the set can be checked out and declared ready for operation.

Never replace a part if the part fails a second time, unless you make sure that the cause of trouble is eliminated. Actually, the cause of trouble should be pinpointed before you replace a part the first time. However, this is not always practical. For example, if a resistor burns out because of an intermittent short and you have cleared the short, the next step is to replace the resistor. However, the short could recur and burn out the replacement resistor. If this happens, you must recheck every element and lead in the circuit.

When replacing a defective part, an *exact replacement* should be used if available. If the part is in any circuit that can affect the transmitter output (frequency, output power, percentage of modulation, etc.), *you must use an exact replacement recommended by the manufacturer.* If not, the set may no longer be *type accepted,* and you will be in violation of FCC regulations. Type acceptance is discussed further in Sec. 6.5.13.

In noncritical circuits, such as those of the receiver, if an exact replacement is not available and the original part is beyond repair (the usual case with most electronic parts), an *equivalent or better part* should be used. Never install a replacement part that has characteristics or ratings inferior to those of the original.

Another factor to consider during repair is that the replacement part be *installed in the same physical location* as the original, with the *same lead lengths,* if at all possible. This precaution is optional in most low-frequency circuits but must be followed for high-frequency applications. In any RF, IF, and frequency-synthesizer circuits (all high-frequency circuits), changing the location of parts or the length of leads may detune the circuit or otherwise place the circuit out of alignment.

6.5.12 Operational Checkout

Even after the trouble is found and the faulty part located and replaced, the troubleshooting effort is not necessarily completed. An operational check is necessary to verify that the equipment is free of all faults and is performing properly again. Never assume that simply because a defective part is located and replaced, the set will automatically operate normally again. In practical troubleshooting, never assume anything.

Run the set through the complete operating sequence, on all channels. In this way you make sure that one fault has not caused another. Follow the procedure found in the service literature (when available). When you are servicing a communications set for someone else (a customer), have that person go through the entire sequence (if practical), but verify operation yourself.

When the operational check is completed and the set is again operating normally, make a brief record of the symptoms, the faulty part, and the remedy. This is particularly helpful when you must service the same set on a regular basis or when you must troubleshoot similar sets. Even a simple record of troubleshooting gives you a valuable history of the set for future reference.

If the set does not perform properly during the operational checkout, you must continue troubleshooting. If the symptoms are the same as, or similar to, the original trouble symptoms, retrace your steps, one at a time. If the symptoms are entirely different, you may have to repeat the entire troubleshooting procedure from the beginning. However, this is usually not necessary.

For example, assume that the receiver circuits do not check out because a replacement IF-amplifier transformer has detuned the circuit. In this case, you should repair the trouble by IF alignment rather than by returning to the first troubleshooting step and repeating the entire procedure. Remember that you have arrived at the defective circuit or part by a systematic procedure. Thus, retracing your steps—one at a time—is the logical course of action.

6.5.13 Type Acceptance of Transmitters

Most transmitters used in the United States (amateur radio, CB, business radio, cellular telephone, cordless telephone, etc.) are type accepted by the FCC. We will not discuss type acceptance in any detail here, except as it applies to service. In brief, the manufacturer (or an independent testing laboratory acting in behalf of the manufacturer) makes comprehensive tests of a transmitter (frequency stability, modulation, etc.) and submits the full details to the FCC. If the FCC approves, a type acceptance number is assigned to the transmitter design.

From a service standpoint, once a type-acceptance number is assigned, any change in the design not made with FCC permission, or any modifications made to the equipment, voids the type acceptance and makes operation of such a transmitter illegal. In brief, if you replace any part with another part that is not an exact duplicate of the original or a part not recommended by the manufacturer, type acceptance can be voided, and operation of the transmitter is illegal.

Obviously, type acceptance can raise some service problems. For example, some older communications sets may no longer be manufactured, and spare parts may no longer be available. (Most CB manufacturers are no longer in the CB business, for example.)

The RF coils and transformers (and other parts that affect frequency stability, modulation, etc.) are a particular problem. Of course, if you have full technical details of such a part (winding data, impedances, etc.), it is possible to duplicate the part, but *you must take full responsibility* for the part replacement.

Another problem may arise if you are called upon to modify or test a non-type-accepted transmitter or to modify a type-accepted transmitter. The author does not recommend accepting such a job unless you have had extensive experience in communications service and laboratory testing. Remember that any modification of a type-accepted set automatically voids the original type acceptance. A possible exception may occur if the manufacturer provides a *modification kit* or unit that has been type accepted.

6.6 RF TROUBLESHOOTING NOTES

The following notes summarize practical suggestions for troubleshooting all types of RF equipment.

Transient voltages: Be sure that power to the equipment is turned off or that the line cord is removed when making in-circuit tests or repairs (except for voltage measurements and/or signal tracing, or course). Components can be damaged from the transient voltages developed when components are changed (plugging in a new board for example). Remember that certain circuits may be "live" even when the power switch is off.

Disconnected parts: Do not operate the equipment with any parts, such as loudspeakers, disconnected. This can result in damage to transistors and/or ICs.

Intermittent conditions: If you run into an intermittent condition and find no fault using routine checks, try tapping (not pounding) the components. If this does not work, try rapid heating and cooling. (A small portable hair dryer and a spray-type circuit cooler make good heating and cooling sources.) Apply heat first, then cool. The quick change in temperature normally causes an intermittently defective component to go bad permanently (the component opens or shorts). Never hold a heated soldering tool directly on a component case.

Operating control settings: If transistor or IC pins appear to have a short, check the setting of any operating or adjustment controls associated with the circuit. For example, the gain control between amplifier stages can show what appears to be a short to ground (from a transistor element or IC pin) simply because the control is set to zero or minimum.

Making a record of gain readings: If you must service any particular make or model of RF equipment, record the transistor/IC gain readings of a unit that is working properly for future reference. Compare these readings with the values listed in service literature.

Shunting capacitors: Do not shunt suspected capacitors with known-good capacitors when troubleshooting RF equipment. This technique is good only if the suspected capacitor is open; it can cause damage to other circuits (because of the voltage surge). Use the procedures described in Sec. 6.15 to check capacitors.

Test connections: Many metal-case transistors (and a few ICs) have their cases tied to the collector (or to some point within the IC). Avoid using the case as a test point unless you are certain as to what point or circuit element is connected to the case. Avoid clipping onto some of the subminiature resistors used in certain RF equipment. Any subminiature component can break if handled roughly.

Injecting signals: Make sure that there is a blocking capacitor in the signal-generator output when injecting a signal into an RF circuit. If there is not, connect a capacitor between the generator output and the point of signal injection (transistor base, IC pin, etc.).

6.7 MEASURING VOLTAGES IN RF CIRCUITS

One of the basic troubleshooting techniques for RF equipment is to measure the voltages at all pins of the ICs and at all elements of the transistors. This tells you instantly if any voltages are absent or abnormal and provides a good starting point for troubleshooting.

6.7.1 Basic RF Transistor Connections

Figure 6.7 shows the basic connections for both PNP and NPN transistor RF circuits (with capacitors, coils, etc. removed for clarity). The purpose of this figure is to establish normal RF transistor relationships. With a normal pattern established, it is relatively simple to find an abnormal condition.

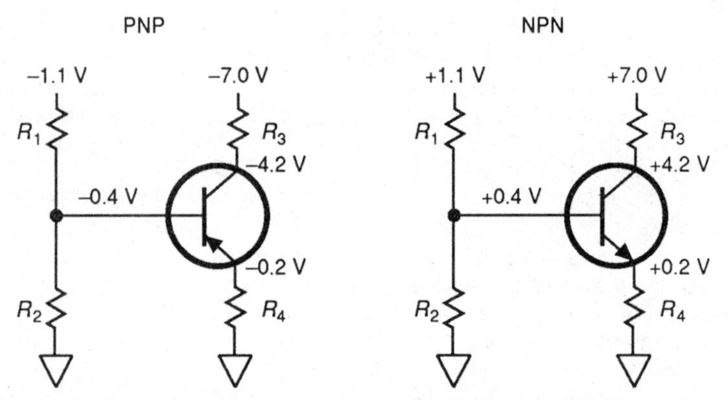

FIGURE 6.7 Basic connections for transistors in RF circuits.

In many RF circuits, the transistors are biased so that the transistor is cut off, and current flows only when there is a signal applied to the base. This is true for RF amplifier and buffer circuits found in transmitters. In other cases, the emitter-base junction is forward-biased to get current flow, with or without an input signal, as shown in Fig. 6.7.

In a PNP transistor, this means that the base is made more negative (or less positive) than the emitter. Under these conditions, the emitter-base junction draws current and causes electron flow from the collector to the emitter. In an NPN, the base is made more positive (or less negative) than the emitter to cause current flow from emitter to collector.

6.7.2 Rules for Practical Analysis of RF Transistor Voltages

The following general rules can be applied to practical analysis of transistor voltages in troubleshooting RF circuits:

1. The middle letters in PNP and NPN always apply to the base.
2. The first two letters in PNP and NPN refer to the *relative bias* polarities of the emitter with respect to either the base or the collector. For example, the letters PN (in PNP) show that the emitter is positive in relation to both the base and emitter. The letters NP (in NPN) show that the emitter is negative in relation to both the base and collector. An exception is found in transmitter amplifier and buffer circuits where the emitter and base are often at 0 V with respect to ground.
3. The collector-base junction is always reverse biased.
4. The emitter-base junction is usually forward biased in RF circuits, except for transmitter amplifier and buffer circuits.
5. A base input voltage or signal that aids or increases the forward bias also increases the emitter and collector currents.
6. A base input voltage that opposes or decreases the forward bias also decreases the emitter and collector currents.
7. The dc electron flow is always against the direction of the arrow on the emitter.
8. If electron flow is into the emitter, electron flow is out from the collector.
9. If electron flow is out from the emitter, electron flow is into the collector.

Using these basic rules, normal RF transistor voltages can be summed up this way:

1. For an NPN, the base is positive, the emitter is not quite as positive, and the collector is far more positive. For an NPN in a transmitter amplifier and buffer, the collector is positive, with the base and emitter at or near zero.
2. For a PNP, the base is negative, the emitter is not quite as negative, and the collector is far more negative. For a PNP in a transmitter amplifier and buffer, the collector is negative, with the base and emitter at or near zero.

6.7.3 Practical Measurement of RF Transistor Voltages

There are two schools of thought on how to measure RF transistor voltages. The most common method is to measure from a *common or ground to the element.* RF service literature generally specifies transistor voltage this way. For example, all of the voltages for the PNP in Fig. 6.7 are negative with respect to ground. This method may be confusing, since the rules appear to be broken. (In a PNP, some elements should be positive, but all PNP elements in Fig. 6.7 are negative.) However, the rules still apply.

In the case of the PNP in Fig. 6.7, the emitter is at −0.2 V, whereas the base is at −0.4 V. The base is *more negative* than the emitter. Thus the emitter is *positive in relation to* the base, and the base-emitter junction is forward biased (normal). On the other hand, the base is at −0.4 V, whereas the collector is at −4.2 V. The base is *less negative* than the collector. As a result, the base is *positive with respect* to the collector, and the base-collector junction is reverse biased (normal).

Some troubleshooters prefer to measure transistor voltages from *element to element* and note the *difference* in voltages. For example, in the circuits in Fig. 6.7, a 0.2-V differential exists between base and emitter. The element-to-element method of measuring transistor voltages quickly establishes forward to reverse bias.

6.8 TROUBLESHOOTING WITH RF TRANSISTOR VOLTAGES

This section presents an example of how voltages measured at the elements of a transistor can be used to analyze failure in RF circuits.

Assume that an NPN transistor circuit is measured and that the voltages found are similar to those shown in Fig. 6.8*a*. Except for transmitter amplifier and buffer circuits, these voltages show a defect. It is obvious that Q_1 is not forward biased because the base is less positive than the emitter (reverse bias for an NPN). The only RF circuit in which this might be normal is one that requires an input signal to turn Q_1 on.

The first troubleshooting clue in Fig. 6.8*a* is that the collector voltage is almost as large as the collector source at R_3. This means that very little current is flowing through R_3 in the collector-emitter circuit. Q_1 could be defective. However, the trouble is more likely to be one of bias. The emitter voltage depends mostly on the current through R_4. Unless the value of R_4 has changed substantially (this is unusual), the problem is incorrect base bias.

The next step in this case is to measure the bias-source voltage at R_1. If the voltage is at 0.7 V, instead of the required 2 V (as shown in Fig. 6.8*b*), the problem is obvious; the external bias voltage is incorrect. The condition should show up as a defect in the power supply and should appear as an incorrect voltage in other circuits.

If the source voltage is correct, as shown in Fig. 6.8*c*, the trouble is probably a defective R_1 or R_2 (or Q_1).

The next step is to remove all voltage from the RF circuit and measure the resistance of R_1 and R_2. If either value is incorrect, the corresponding resistor must be replaced. If both values are good, it is reasonable to check the value of R_4. However, it is more likely that Q_1 is defective. This can be established by testing and/or replacement.

6.8.1 Practical In-Circuit Resistance Measurements

Be careful when measuring resistance values in transistor circuits with the resistors still connected. One reasons for this is that the voltage produced by the meter could damage some transistors. More important, there is a chance for error because transistor junctions pass current in one direction. This can complete a circuit through other resistors and produce false indications.

For example, assume that a battery is connected across R_2 with the negative terminal of the meter (internal battery) connected to ground, as shown in Fig. 6.8*d*. Because R_4 is also connected to ground, the negative terminal is connected to the end of R_4. Further, because the positive terminal is connected to the base of Q_1, the base-emitter junction is forward biased and there is current flow. In effect, R_4 is in parallel with R_2, and the meter reading is incorrect. To prevent the

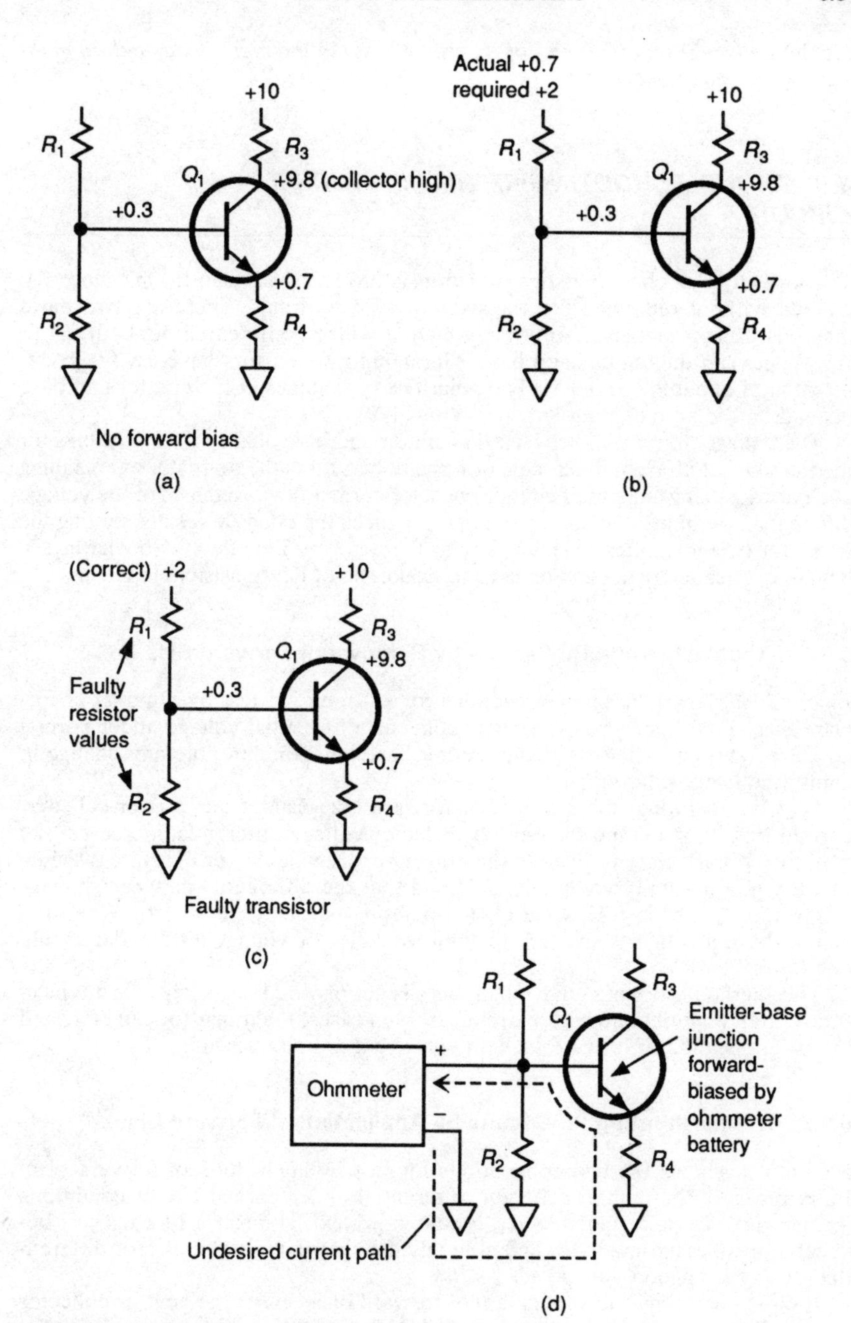

FIGURE 6.8 Troubleshooting with transistor voltages.

problem, *disconnect either end of* R_2 *and/or reverse the meter leads before making the measurement.*

6.9 TROUBLESHOOTING RF TRANSISTORS IN CIRCUIT

The forward-bias characteristics of a transistor can be used to troubleshoot RF circuits without removing the transistor from the circuit. There are two basic methods. Silicon transistors normally have a voltage differential of about 0.4 to 0.8 V between the emitter and base. Germanium transistors have a voltage differential of about 0.2 to 0.4 V. The polarities of voltages at the emitter and base depend on the type of transistor (NPN or PNP).

The voltage differential between the emitter and base acts as a forward bias for the transistor. Sufficient differential or forward bias turns the transistor on, resulting in a corresponding amount of emitter-collector current flow. Removal of the voltage differential, or an insufficient differential, produces the opposite results, cutting the transistor off (no emitter-collector flow or there is very little flow). Now let us see how these characteristics can be used to troubleshoot RF transistors in circuit.

6.9.1 Troubleshooting RF Circuits by Removal of Forward Bias

Figure 6.9*a* shows the test connections for an in-circuit test by removal of forward bias. First, measure the emitter-collector differential voltage under normal circuit conditions. Then short the emitter-base junction and note any change in emitter-collector differential.

If Q_1 is operating, the removal of forward bias causes the emitter-collector current flow to stop, and the emitter-collector voltage differential increases (the collector voltage rises to or near the power supply value). For example, assume that the power supply voltage is 12 V and that the differential between the collector and emitter is 6 V when Q_1 is operating normally (no short). When the emitter-base junction is shorted, the emitter-collector voltage differential should rise to (or near) 12 V.

This method is not effective when there is no forward bias, such as in a typical transmitter amplifier and buffer circuit. In such cases, the transistor can be tested by the application of forward bias (in some, but not all, circuits).

6.9.2 Troubleshooting RF Circuits by Application of Forward Bias

Figure 6.9*b* shows the test connections for test by application of forward bias. First, measure the emitter-collector differential under normal circuit conditions (or measure the voltage across R_4, as shown). Next, connect a 10-k resistor between the collector and base, and note any changes in emitter-collector differential (or any change in voltage across R_4).

If Q_1 is operating, the application of forward bias causes the emitter-collector current flow to start (or increase), and the emitter-collector voltage differential decreases (or the voltage across R_4 increases). Note that this method does not always provide a significant indication on those RF circuits where there are no load resistances in the collector and/or emitter.

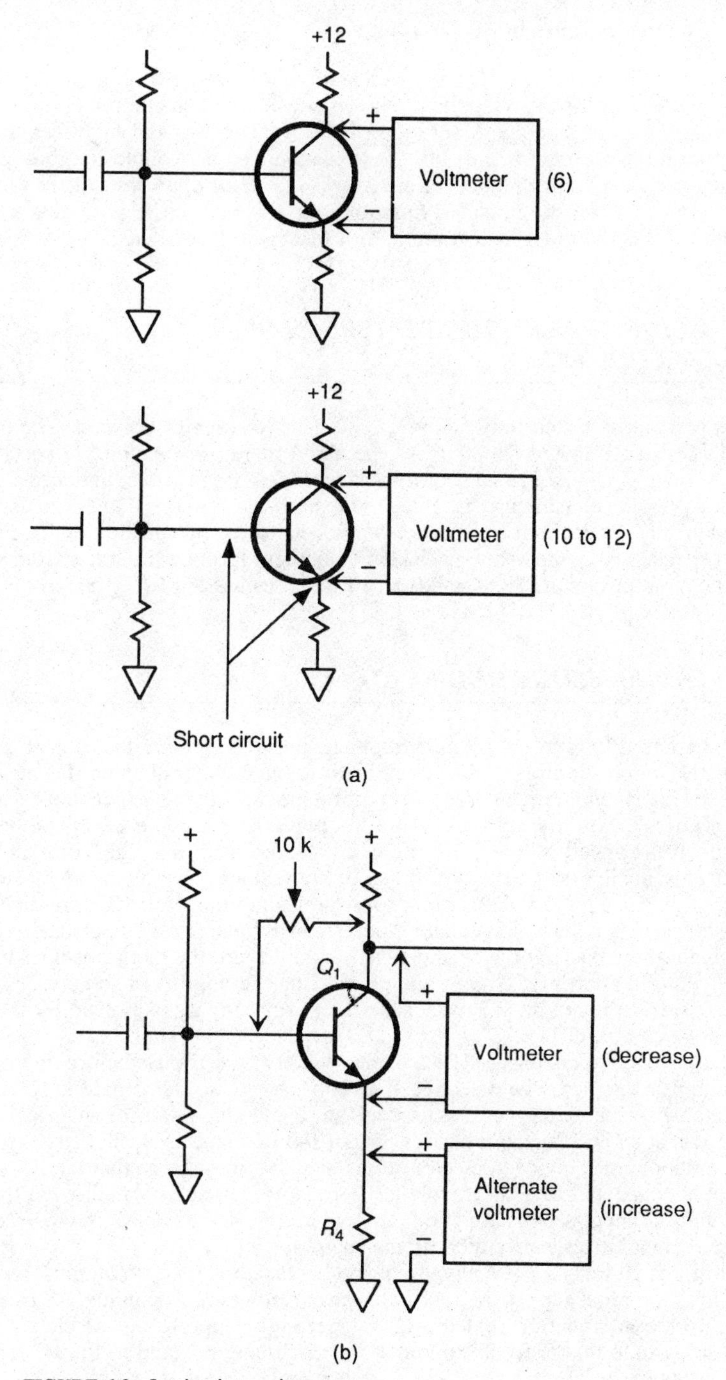

FIGURE 6.9 In-circuit transistor tests.

6.9.3 Go/No-Go Tests in RF Circuits

The tests in Fig. 6.9 show that the transistor is operating on a go/no-go basis, which is sufficient for many applications. However, the tests do not show gain or leakage and do not establish operation of the transistor at the high frequencies usually found in RF equipment. For these reasons, some troubleshooters reason that the only satisfactory test of a transistor is *in-circuit operation.* If the transistor does not perform the intended function in a given circuit, the transistor must be replaced. So the most logical method of test is replacement.

6.10 USING TRANSISTOR TESTERS IN RF TROUBLESHOOTING

Transistors can be tested in or out of circuit using commercial testers. The use of such testers in RF troubleshooting is generally a matter of opinion. At best, such testers show the gain and leakage of transistors at low frequencies, under one set of conditions (fixed voltage, current, etc.). In the author's opinion, transistors used for high-frequency applications in RF circuits are best tested by substitution (in circuit) or with special test equipment (special lab equipment for measuring transistor characteristics) out of circuit. The same is true for diodes used in RF circuits.

6.11 TROUBLESHOOTING RF ICs

There is some difference of opinion about testing ICs in circuit or out of circuit during RF troubleshooting. An in-circuit test is the most convenient because the power source is available, and you need not unsolder the IC (which can be a real job). Of course, you must first measure the power-source voltages (plus, minus, ground, etc.) applied at the IC terminals to make sure that the voltages (and grounds) are available and correct. If any of the voltages or grounds are absent or abnormal, this is a good starting point for troubleshooting. If the ICs are digital or microprocessors, reset voltages and clock signals must also be checked.

With all of the basic voltages and signals established, the in-circuit IC is tested by applying the appropriate input and monitoring the output. In some cases, it is not necessary to inject an input because the normal input is supplied by the circuits ahead of the IC.

One drawback to testing an RF IC in circuit is that the circuits before (input) and after (output) the IC may be defective. This can lead you to think that the IC is bad. For example, assume that the IC is used as the IF and video-detector stages of a TV set. To test such an IC, you inject a signal at the IC input (typically, from the TV tuner) and monitor the video-detector output signal. Now assume that the IC output terminal is connected to a short circuit. There is no output indicated, even though the IC and the input signals are good. Of course, this shows up as an incorrect resistance reading (if resistance measurements are made).

Out-of-circuit tests for ICs have two obvious disadvantages. You must remove the IC and you must supply the required power. However, if you test a suspected IC after removal and find that the IC is operating properly out of circuit, it is logical to assume that there is trouble in the circuits connected to the IC. This is very convenient to know *before* you install a replacement IC.

6.11.1 RF IC Power-Source Measurements

Although the test procedures for RF ICs are essentially the same as those used for discrete-component circuits, measurements of the power-source voltages applied to the IC may not be the same. Some ICs require connections to *both* a positive and a negative power source. This is particularly true when the IC contains an op-amp or similar balanced circuits. Also, in some older equipment where the ICs are in metal cases, the case may be "hot" (at a voltage above or below ground or common). However, in most present-day RF equipment, the ICs have plastic or other nonconducting cases.

6.12 EFFECTS OF CAPACITORS IN RF TROUBLESHOOTING

During the troubleshooting process, suspected capacitors can be removed from the circuit and tested on bridge-type checkers. This establishes that the capacitor value is correct. With a correct value, it is reasonable to assume that the capacitor is not open, shorted, or leaking. From another standpoint, if the capacitor shows no shorts, opens, or leakage, it is also fair to say that the capacitor is good. So, from a practical troubleshooting standpoint, a simple test that checks for shorts, opens, or leakage is usually sufficient.

There are two basic methods for a quick check of capacitors during RF troubleshooting: one with circuit voltages and one with an ohmmeter.

6.12.1 Checking Capacitors with RF-Circuit Voltages

As shown in Fig. 6.10*a,* this method involves disconnecting one lead of the capacitor (the ground or cold end) and connecting a voltmeter between the disconnected lead and ground. In a good capacitor, there should be a momentary voltage indication (or surge) as the capacitor charges up to the voltage at the hot end.

If the voltage indication remains high, the capacitor is probably shorted. If the voltage indication is steady, but not necessarily high, the capacitor is probably leaking. If there is no voltage indication whatsoever, the capacitor is probably open.

6.12.2 Checking RF-Circuit Capacitors with an Ohmmeter

As shown in Fig. 6.10*b,* this method involves disconnecting one lead of the capacitor (usually the hot end) and connecting an ohmmeter across the capacitor. Make certain that all power is removed from the circuit. As a precaution, short across the capacitor (after the power is removed) to make sure that no charge is retained. In a good capacitor, there should be a momentary resistance indication (or surge) as the capacitor charges up to the voltage of the ohmmeter battery.

If the resistance indication is near zero and remains so, the capacitor is probably shorted. If the resistance indication is steady at some high value, the capacitor is probably leaking. If there is no resistance indication (or surge) whatsoever, the capacitor is probably open.

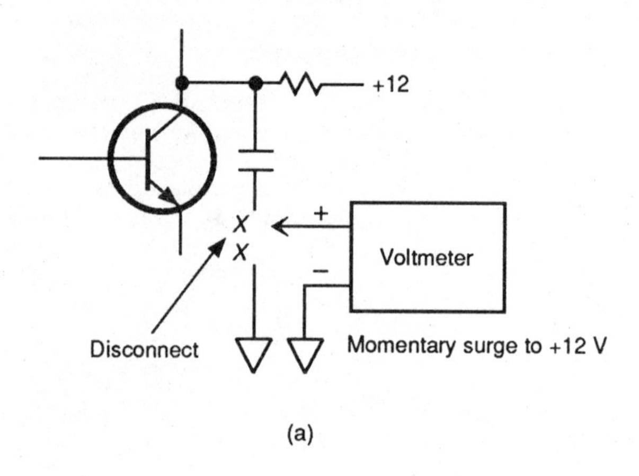

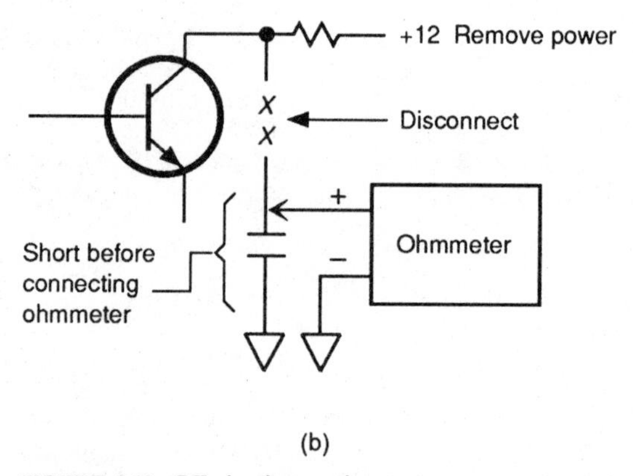

FIGURE 6.10 RF-circuit capacitor tests.

6.13 EXAMPLE OF RECEIVER RF-CIRCUIT TROUBLESHOOTING

This step-by-step troubleshooting problem involves locating a defective part in the receiver section of a communications set. The problem has been localized to the receiver section using the basic procedures described thus far in this chapter (Fig. 6.1). The trouble symptoms are: "no reception on any channel, transmission normal" or "reception is poor and transmission is good."

Before presenting a step-by-step example of receiver troubleshooting, we shall discuss basic test procedures for the receiver section.

6.13.1 Oscillator Injection Signals to the Receiver

Remember that the receiver section must have injection signals from other sections of the set to operate properly. In the simplest communications receivers, such as those found in walkie-talkie sets, there is a single local oscillator signal that is mixed with the incoming signal to produce the IF signal. In dual-conversion sets (and most communications sets are dual-conversion) there are two oscillators. In SSB receivers there is another oscillator that replaces the missing carrier. In sets with PLL, the signals are provided by the PLL IC (as discussed in Sec. 7.1).

As a first step in troubleshooting the receiver, check that all oscillator signals are available and are of the correct frequency and amplitude. The frequencies may be found from the schematic or other part of the service literature (it is hoped). Service literature usually lists the crystal and/or synthesizer frequencies. However, you may have to guess at the oscillator signal amplitudes. Typically, communications-set oscillator voltages are 1 V or less and rarely above 1.5 V.

6.13.2 ANL Circuit Troubleshooting

Most communications receivers are provided with some form of automatic noise limiter (Sec. 5.11). Usually, the ANL is a simple series diode between the detector output and the audio section or volume control. In other sets, there is a separate noise-signal channel that controls the audio. Either way, a failure in the ANL circuits can make it appear that the receiver section is malfunctioning. For this reason, always check the ANL circuits before launching into the receiver section.

First, disable the ANL circuit and check operation. If this restores operation, the trouble is probably in the ANL circuit rather than in the receiver section. On some sets the ANL may be disabled by a switch. In other sets, it is necessary to short across the ANL series diode. If audio passes through with the ANL diodes shorted, check the diodes and associated parts by substitution.

6.13.3 Squelch Circuit Troubleshooting

Most communications sets are provided with some form of squelch circuit. Although there are many types of squelch circuits, they all set the receiver input-signal level at which the audio will pass. If the squelch is set too high, the receiver section appears to be dead, or if the squelch circuits are defective, the receiver section may seem to be defective.

Before starting any receiver-section troubleshooting, try resetting the squelch control. If the receiver appears to be dead with the squelch set fully counterclockwise, check the squelch circuit, as described in Secs. 5.7 and 5.8.

6.13.4 General Troubleshooting Instructions

Figure 6.11 shows the schematic diagram of the receiver-section RF and IF circuits (of a CB set in this case). Note that an IC is used for the IF stages. Thus, if a fault is traced to any circuit within the IC, it is necessary to replace the entire IC package. Also note that the mixer/RF and oscillator stages are shown as discrete-component circuits (to illustrate the troubleshooting process). In present-day equipment, these circuits are usually part of a PLL (as discussed in Sec. 7.1).

The test points shown in Fig. 6.11 do not appear in the service-literature schematic. When test points are not shown, it is sometimes helpful to assign test points as a guide for troubleshooting. In general, use collectors and bases of transistors for test points. It is entirely possible to troubleshoot the receiver section (and most other sections of the set) without service-data test points, using only the minimum data shown in Fig. 6.11. Consider the following.

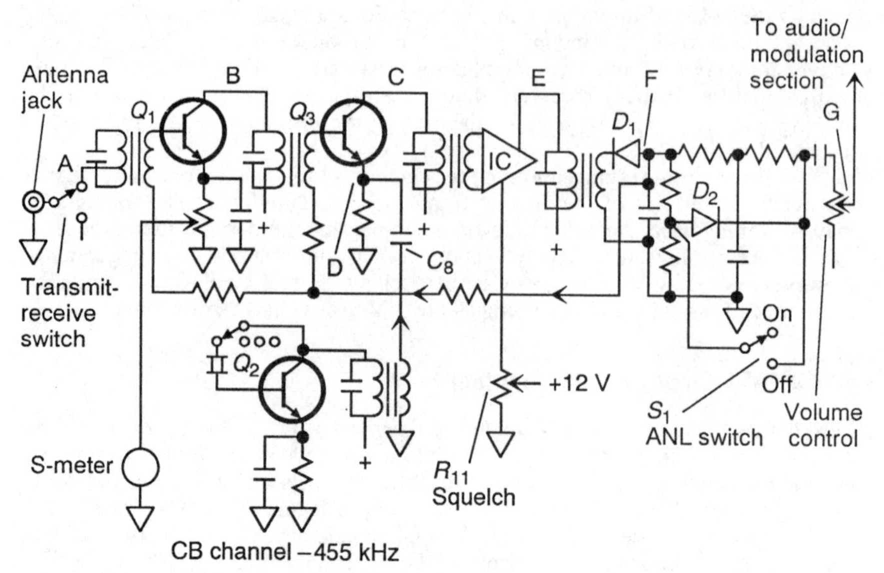

FIGURE 6.11 Receiver-section RF/IF circuits.

You know that the input RF signal is at the selected channel frequency (one of the 40 CB channels in this case). The schematic shows that the IF is at 455 kHz. From a troubleshooting standpoint, this establishes certain conditions.

You know that an RF signal may be introduced at test point A. This signal must be at the channel frequency. The same is true for test point B. However, if you inject a signal at B, it must be greater in amplitude because Q_1 acts as an RF amplifier and usually supplies considerable voltage gain. If you are monitoring the signal at B, it should be identical to that at A, except for amplitude.

The signal at test point D, the local-oscillator injection signal, is at the channel frequency, plus or minus 455 kHz. The schematic indicates that the local-oscillator frequency is at the channel frequency, minus 455 kHz. Since the receiver is single-conversion, only one oscillator signal need be considered.

Signals at test points C and E are at 455 kHz. From a monitoring standpoint, the signal at E should be much larger in amplitude than that at C because of the gain produced in the IC amplifier stages. The signal at test point C may also include the local-oscillator frequency developed by Q_2.

The signals at test points F and G are in the audio range. Thus, from the standpoint of signal injection, you must inject an audio signal at F. There is little point in injecting a signal at G since this is the input to the audio and modulation section, and the trouble symptom stated that transmission was normal. As a possible exception, the output from G (the volume control) passes through contacts of a switch or relay to the audio and modulation section input. It may be that the contacts are bad.

If test points F and G are used for signal tracing (instead of signal injection), you must use a meter or scope with a probe to monitor the signals. Also, if the incoming RF signal (either a CB station signal or a signal from an RF generator connected to some point ahead of E) is unmodulated, there is no signal at E, even though the receiver is operating properly. (If an RF generator is connected to A, and you expect to monitor the signal at E with a scope or meter, you must modulate the RF signal generator output.)

With this limited information scrounged from the schematic (and your vast knowledge of CB receivers), you are now ready to start the troubleshooting.

6.13.5 Determine the Symptoms

The user reports that no signals can be heard on any channel. You confirm this symptom, but you then check further.

The schematic shows that the ANL circuit can be cut out when the ANL switch S_1 is set to off (switch closed to short D_2). You find that with the ANL switch off, there is still no improvement.

You rotate the squelch control to fully off and recheck for signals at known active channels. There is still no improvement.

You check the S-meter indication on all channels or on known active channels. If the S-meter is working in the normal manner (S-meter indication varies as channels are changed), it is fair to assume that all stages up to the S-meter are good (at least operating).

Note that an S-meter check is of real value when the meter is operated from one of the last stages in the receiver (such as the last iF stage) since you can clear several circuits at once if you get a good S-meter indication. Unfortunately, in our example, the S-meter monitors only the first stage Q_1. A good S-meter indication proves only that Q_1 is probably operating properly (in our case).

You rotate the volume control to fully on and notice that some noise can be heard at the speaker. You then put your finger on test point G (the volume-control arm) and notice that there is considerable hum in the speaker. What is the most logical test point at which to begin the troubleshooting?

To simplify this example, you can eliminate test points A and G. In effect, you just checked G when you touched the volume-control arm. The speaker hum, coupled with the original trouble symptom of "good transmission" clears the audio and modulation section and the wiring between G and the audio input (through relay or switch contacts, etc.). Of course, if you insist, you can inject an audio signal at G (about 15 or 20 mV should be sufficient) and positively clear the audio and modulation section. Test point A is eliminated as a first test point to check since A is for signal injection only.

You could choose test point F. If you monitor at F and get a good signal, this points to a bad ANL circuit. However, this same condition would probably show up when you set the ANL switch to off. If you get no signal at F, it proves very little. The problem is still somewhere between A and F.

If you inject a signal at F (the signal must be audio, probably at 1 V or less, at 1 kHz) and the signal passes, this proves only that the ANL circuits are probably good. However, it is possible that a shorted or badly leaking ANL diode D_2 can pass audio, so the test is not conclusive. Try again.

You could choose test point D. This is not a bad choice, if you intend to align the receiver section. One of the first steps in alignment is to monitor the local-oscillator signal. However, as a first choice for troubleshooting, a good signal at D proves very little (only that the local-oscillator signal is at the correct frequency and is of the correct amplitude). A bad signal at D pinpoints the problem, but this is a lucky guess. Try again.

You could choose test point B. If you inject a signal at B and the response is good (that is, if you inject a modulated RF signal and hear the tone on the speaker), you have isolated the trouble to the RF voltage amplifier Q_1, and you are also very lucky. (The good S-meter indication tells you just about as much.) On the other hand, if the response is bad with a signal injected at B, you still have many circuits to check. You could also monitor the signal at B, but this requires two test instruments (a signal generator at A and a monitoring meter or scope at B). Try again.

You could choose test point E. If you inject a signal at E and the response is bad (that is, if you inject a modulated RF signal and do not hear the tone on the speaker), you have isolated the trouble to the detector D_1 circuit, and you are again very lucky. If the response is good with a signal injected at E, you still have many circuits to check. Again, you could monitor at E, but this also requires two test instruments.

There is another problem with test point E (if signal injection is used). It is possible that a shorted or badly leaking D_1 could pass an audio signal but might not pass the audio portion of the modulated RF signal. Try again.

You should choose test point C. By choosing C, after you have checked the ANL, squelch, S-meter, and volume control, you have effectively split the receiver section in half. You are on the right track. But do you inject a signal at C or monitor the signal at C? Many technicians will argue either way. Since injection is simpler (in this case), you should choose to inject a 455-kHz modulated signal at C. Now assume that you hear a tone on the speaker and there is plenty of volume.

The trouble is now isolated to three circuits, the RF amplifier Q_1, the local oscillator Q_2, or the mixer Q_3. To isolate the trouble further, there are two obvious (and equally logical) steps. You can inject a B or monitor at D. Assume that you choose to inject at B, using a modulated RF signal at the channel frequency. If the response is good, you have isolated the trouble to the RF amplifier Q_1. But let us say that you are not lucky this time. The response is bad (there is no tone on the speaker).

Your next step is to monitor at D. Now you are lucky, and there is no local-oscillator signal at D. You quickly conclude that Q_2 is bad. You are wrong. The real problem is that C_8 is open, which you locate by moving the monitoring device to both sides of C_8. The signal is present and normal at the oscillator side of C_8 but not at D. You replace C_8 and restore normal operation to the receiver section.

6.14 EXAMPLE OF TRANSMITTER RF-CIRCUIT TROUBLESHOOTING

This step-by-step troubleshooting problem involves locating a defective part in the transmitter section of a communications set. The problem has been localized to the transmitter section using the basic procedures described in Sec. 6.1. The trouble symptoms are: "no transmission on any chanel, reception normal" or "transmission is poor, reception good."

6.14.1 General Troubleshooting Instructions

Figure 6.12 shows the schematic diagram of the transmitter circuits (of the same CB set described for receiver-section troubleshooting). The test points shown do not appear in the service-literature schematic but have been arbitrarily assigned

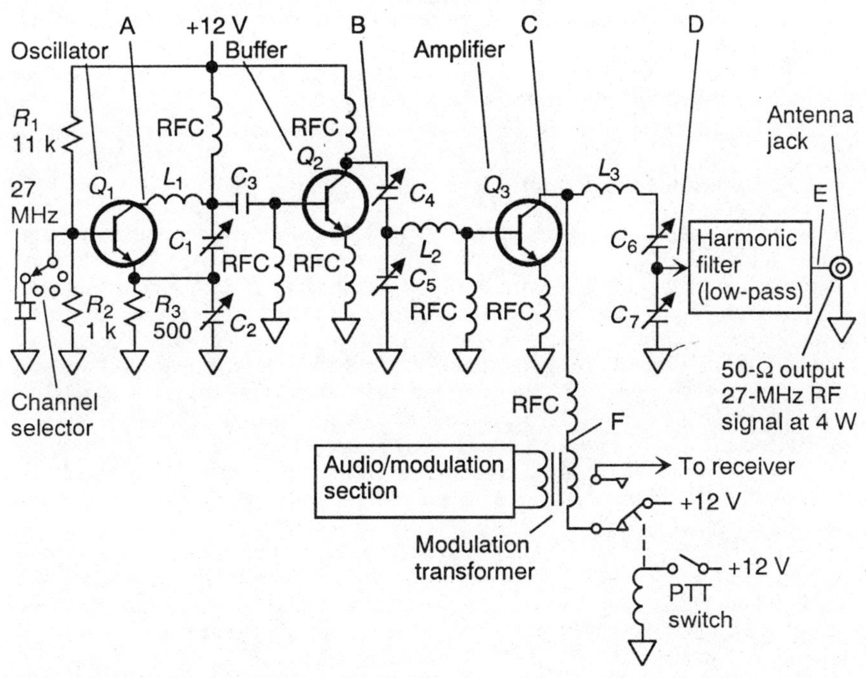

FIGURE 6.12 Transmitter-section RF circuits.

to illustrate the troubleshooting process. As discussed, it is entirely possible to troubleshoot the transmitter section using only the minimum information found on the schematic diagram (and this may be all you get).

No voltage or resistance information (at transistor terminals) is available except for the +12-V power supply voltage found on the schematic. However, you should be able to calculate the voltages found at the terminals. The collectors of the three transistors Q_1 through Q_3 are all connected to the +12 V through RF chokes. Such chokes generally have very little dc resistance and thus produce very little voltage drop (usually a fraction of 1 V). Thus, it is reasonable to assume that the dc voltage at the collectors is about +12 V (or slightly less).

The emitters of Q_2 and Q_3 are connected to ground through RF chokes. Thus, the dc voltage of the emitters should be zero, and the resistance to ground should be a few ohms at most. Possibly some dc voltage might be developed across the chokes, but is is not likely. The same is true for the Q_2 and Q_3 bases.

The base of Q_1 has a fixed dc voltage applied through the voltage-divider network of R_1 and R_2. The ratio of R_1 to R_2 indicates that the voltage drop across R_2 is about 1 V. Thus, the base of Q_1 is about 1 V. Typically, the emitter of Q_1 is within 0.5 V of the base (assuming that Q_1 is a silicon transistor). The resistance to ground from the emitter of Q_1 should be equal to R_3, or about 500 Ω.

The schematic shows that the output is 4 W into a 50-Ω antenna. The crystals are in the 27-MHz range, so all three stages are tuned to the channel frequency, and there is no frequency multiplication in any stage.

Capacitors C_1, C_2, and C_4 through C_7 are tuning adjustments. There are no other operating or adjustment controls. Once you press the PTT switch, the transmitter section should perform its function and transmit AM-modulated signals to the antenna.

With this great wealth of information to draw upon, you are ready to plunge into the troubleshooting by determining the symptoms.

6.14.2 Determine the Symptoms

In some troubleshooting work, you will find that the set was never properly tuned, adjusted, or otherwise correctly put into operation. Some troubleshooters approach each new problem with this assumption. However, in most communications problems, you will find that the set had been working satisfactorily for a time before the trouble occurred. This is the case in our example. The set had been operating properly for many months, and the operator now reports to you that there is no output. The transmission cannot be heard by any CB stations tuned to the same channel. What is your first step?

You could start by checking each circuit to determine which one is not operating. This procedure rates a definite no, even though you would eventually locate the faulty circuit. It is not a logical approach and would probably require several unnecessary tests before the trouble was isolated to a circuit. Try again.

You could perform the tuning procedure to determine which circuit group does not perform properly. This would be even more illogical than checking each stage. Although you might locate the trouble area using this procedure, it would probably require many unnecessary steps. Each step should provide the most information with the least amount of testing. Try again.

You should check the output. Notice that we do not tell you how to check the output at this time, but you have two methods for doing so: with or without test equipment. You may also check the output with the antenna connected or disconnected.

The first place to look is the S/RF meter on the set front panel. A good RF output indicates that the transmitter section is operating. The power may be low and the signal may be off-frequency, but there is RF output. Unfortunately, as you can see in Fig. 6.12, there is no S/RF meter in our example.

With the antenna still connected, the output may be checked by tuning a receiver to the same channel, keying the suspected transmitter, and trying to transmit a signal (voice into the microphone). Although this test is quick and easy, it does not provide all the information you need.

Even if you hear the voice transmission on the test receiver, it is possible that the transmitter output is low. A weak transmission might be picked up by a nearby receiver but could not be heard over the normal communications range. If you do not hear the transmission on the receiver, it proves that the operator was right, but you have no clue as to which section of this set is at fault. The problem could be in the audio and modulation section.

By using a field-strength meter (Sec. 3.8) or one of the special test sets described in Chap. 3 that includes a field-strength function, you can leave the antenna connected and check the output signal.

It is possible that the antenna or lead-in is defective. To verify this, check the output with a dummy load and meter or scope (including an RF probe) or with an RF wattmeter (Secs. 3.6 and 3.7). The wattmeter is best for checking the RF output, since a wattmeter provides an immediate answer to the question Is there RF output and is the power correct? With the dummy load and meter or scope, you must measure the voltage and calculate the power. For a typical CB set, you will get about 14 V of RF voltage with a 50-Ω antenna or load. Power is equal to E^2/R, or $14^2/50 = 4$ W.

Even if you get the correct power output, it is still possible that the transmitter section is not being modulated by the audio and modulation section. Thus, the most positive test of the transmitter output is to monitor the output on a scope, using one of the modulation tests described in Sec. 3.2. The most accurate results are obtained by modulating the set with an audio tone. However, for our example, you are more concerned that the set performs the normal function of transmitting voice signals. (Typically, you should get between 85 and 100 percent modulation with normal voice.)

Now assume that you disconnect the antenna, connect a dummy load, and check the percentage of modulation using one of the methods described in Sec. 3.2. Further assume that there is no RF indication on the scope (no vertical deflection whatsoever) with the transmitter keyed (PTT button pressed) and someone speaking directly into the microphone. In which section of the set do you think the trouble is located?

Modulation section. If you choose the audio and modulation section, you will probably not succeed as a troubleshooter. You could try being a computer consultant or writing short stories. Or you are not paying attention. Or you simply do not understand transmitters. The vertical deflection on the scope is produced by the RF signal; the shape of the signal is determined by the modulation. Thus, *with no vertical deflection, there is no RF signal,* and the trouble is in the transmitter. Try again.

Transmitter section. You have made a logical choice. The next step is to isolate the trouble to one of the circuits (oscillator Q_1, buffer Q_2, or power amplifier Q_3) in the transmitter section. What is the next most logical test point?

You could check the power-supply voltage for the transmitter section. This approach does have some merit. Note that Q_1 and Q_2 receive collector voltage directly from the +12 V power-supply line, whereas Q_3 gets collector voltage through a secondary winding on the modulation transformer and through relay contacts (which are operated by the PTT switch). Thus, to make a complete check of power-supply voltage for the transmitter RF circuits, you must measure the voltage at each of the collectors. Of course, if the correct voltage is present at any one of the collectors, the power supply itself is good. Try again.

You could check for RF signals at test point A. This is not a bad choice. If there is no RF signal at A, the trouble is traced to the oscillator (Q_1 and associated parts). If there is RF at A, you know the oscillator is operating, but you eliminate only one circuit as a possible trouble area. Thus testing point A is not the most logical choice. Try again.

You could check for RF signals at test points C, D, and E. This proves very little. In effect, you are checking at these points when you monitor the output signal at the antenna. Try again.

You should check for RF signals at test point B. This is the most logical choice, since it is essentially a half-split of the transmitter circuits (Sec. 6.4.5). As shown in Fig. 6.13, you can check with a meter or scope and an RF probe. If the scope is capable of passing the 27-MHz signals, you can use the alternate test method shown in Fig. 6.13*b*. However, the probe is usually the most practical method.

If there is a good RF-signal indication at B, both Q_1 and Q_2 are operating properly, and the trouble is traced to the power amplifier Q_3 or possibly to the low-pass TV interference filter. You also eliminate the power supply (which must be good if Q_1 and Q_2 are functioning normally). However, you do not eliminate the

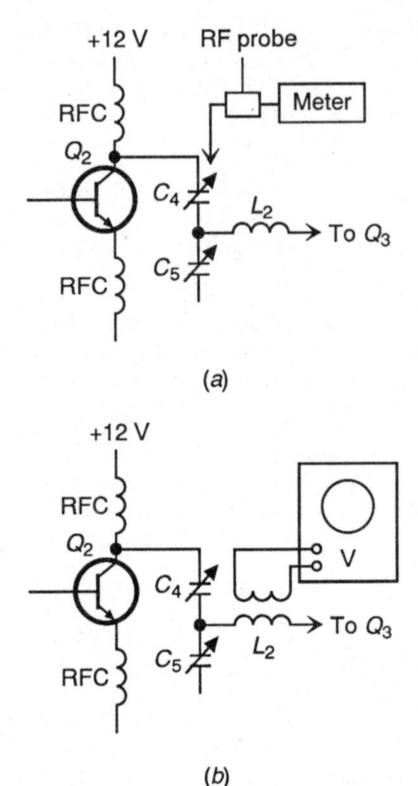

FIGURE 6.13 Checking for RF at test point B.

line between the power supply and the collector of Q_3 (which is a separate path from that between the power supply and the Q_1 and Q_2 collectors).

If there is no RF indication at B, the trouble is traced to Q_1 or Q_2. The next step is to check for RF signals at A.

Now assume that there is a good RF indication at B. The power amplifier (Q_3 and associated parts) may be bracketed as the faulty circuit because the input signal is good and the output is bad. To confirm this, check for RF signals at C, D, and E on the off chance that there is a good signal at C and D but not at E (because of a bad low-pass filter).

6.14.3 Locate the Specific Fault

Now that the trouble is isolated to the circuit, the problem must be located. First, perform a visual inspection of the power amplifier. Note that Q_3 is probably provided with a heat sink or is mounted on metal that acts as a heat sink. Assume that there is no apparent sign of where the trouble is located. There is no sign of overheating, and all components as well as the PC board or wiring appear normal. What is your next step?

You could make an in-circuit test of Q_3. This is difficult since Q_3 is operated class C and will not respond to in-circuit forward-bias tests (as described in Sec. 6.9). No forward bias is applied to Q_3, so you cannot remove the bias. If you apply bias, the voltage relationship between emitter and collector will probably not change because there is no dc load in either the collector or emitter. An in-circuit transistor tester might prove that the transistor is good (at low frequencies). However, in-circuit testers are generally useless at radio frequencies. Try again.

You could make a substitution test of Q_3. This would be more satisfactory than any in-circuit test. Likewise, it is possible that you may have to substitute Q_3 before you locate the fault. However, more convenient tests may be made at this time. Try again.

You could check resistance at all elements of Q_3. Although it will probably be necessary to check resistance and continuity before you are through, resistance checks at this time prove very little. The resistance to ground at the emitter or base of Q_3 should be a few ohms at most. Only a high resistance to ground at the base and emitter (or a very low resistance at the collector) would be significant. Try again.

You should check the voltage at all elements of Q_3 *first.* The voltage at the base and emitter should be zero (we are speaking of dc voltage, not RF-signal voltage). The dc voltage at the collector should be about +12 V. If the voltages are all good, you may skip the resistance-to-ground measurements. However, you may still have to make continuity checks if the voltages are abnormal. Let us examine possible faults indicated by abnormal voltages:

Large dc voltages at base or emitter: Large dc voltages at the base or emitter of Q_3 indicate that the elements are not making proper contact with ground. For example, a high-resistance solder joint between the Q_3 emitter and ground could produce a dc voltage at the emitter. Also, if the emitter-ground connection is completely broken, the emitter will "float" and show a dc voltage.

No dc voltage at collector: If the collector shows no dc voltage, the fault is probably in the RF choke, the modulation-transformer secondary winding, or the relay contacts, indicating the need for a continuity check. Check the dc voltage at test point F. If the voltage is correct at F but not at C (the collector of Q_3), the RF choke is at fault. If the voltage is absent at F, the fault is in the transformer winding or the relay contacts.

Now, to summarize this troubleshooting example, let us assume that the trouble is caused by an open L_3 coil winding. Assume that the transmitter has been subjected to excessive vibration, and the L_3 winding is broken from the coil terminal, underneath where you cannot see the break (that is where they always break). This trouble will not affect dc voltage or resistance. Substitution of Q_3 will not cure the problem. These are the kinds of problems you will find in real RF troubleshooting: Everything appears to be good, but the set will not work.

To solve such problems, *you must make point-to-point continuity checks* through the PC wiring. In this case, if you checked from point C to the top of C_6, you would have found the open coil winding.

CHAPTER 7
ADVANCED RF TROUBLESHOOTING

This chapter is devoted to advanced troubleshooting procedures for RF equipment. Here, the emphasis is on sophisticated RF devices such as digital tuning and mobile communications.

7.1 TROUBLESHOOTING PLL AND DIGITAL-TUNING CIRCUITS

As discussed in Chap. 1, PLL circuits, or so-called digital tuning, are used in most present-day RF equipment. This is true for AM/FM tuners as well as for TV tuners. PLLs are also used in present-day communications sets for both the receiver and transmitter tuning, in place of the frequency-synthesis tuning discussed in Chap. 6.

Operation and circuit functions of PLL AM/FM and TV tuners are discussed in Chap. 1. In this chapter, we are concerned with troubleshooting PLL circuits. Before we get into troubleshooting, let us review the PLL circuits found in typical communications sets. Again, we use a CB set as an example.

7.1.1 Typical PLL Used in RF Communications Sets

There are two definite advantages for PLL used in communications sets. First, the oscillator is not only crystal-controlled but is locked in frequency to the crystal by a feedback circuit that detects even slight shifts in both frequency and phase between oscillator and crystal. This maintains frequency accuracy far better than is possible with a conventional crystal oscillator, even with well-designed circuits. Second, only two or three crystals are required for PLL to cover many channels (40 CB channels for example), instead of the 12 or 14 crystals required for a frequency synthesizer.

Figure 7.1*a* is the block diagram of a typical PLL system used in communications circuits. The VCO produces a signal at the desired frequency. In some systems, the VCO signal is at the channel frequency and is applied directly to the transmitter amplifiers, thus substituting for the signal usually obtained from a frequency synthesizer.

In the most popular systems, the VCO output is mixed with other signals to produce the transmitter frequency and receiver local-oscillator frequencies. With

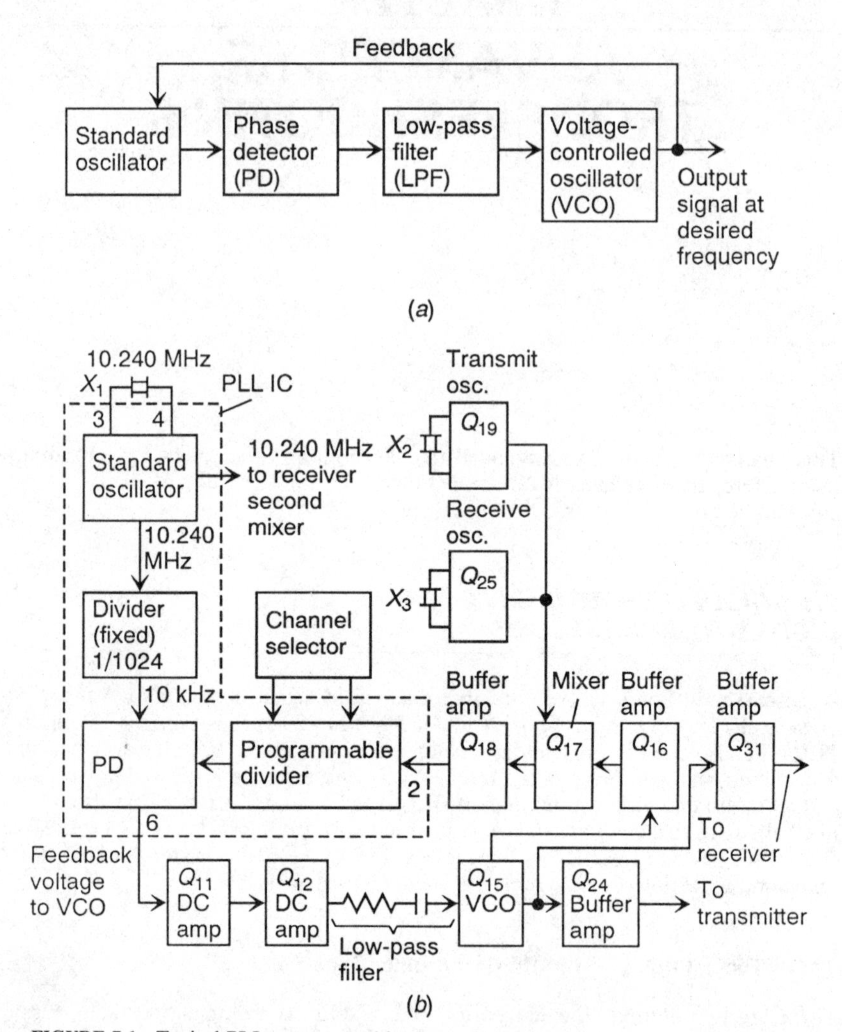

FIGURE 7.1 Typical PLL system used in communications circuits.

any system, the VCO output is fed back and compared with a standard signal as to frequency and phase. In a similar manner, with any system, the VCO frequency is determined by a voltage from the phase detector (PD) applied through a low-pass filter (LPF). In most systems, the PD produces a pulsed output, which is converted to a voltage by the LPF.

The PD pulse width is determined by the frequency and phase difference of the standard oscillator and the VCO. If both oscillators are locked in exact frequency and phase, the PD pulse width is zero (or at some fixed width, depending on the system), the LPF output is zero (or fixed), and the VCO output frequency remains fixed. If the VCO frequency deviates from that of the standard oscillator (even a fraction of one cycle), the PD pulse width changes, as does the LPF voltage to the VCO, and the VCO frequency also changes.

This change opposes the initial change in frequency. For example, if the VCO output frequency is lowered from that of the standard oscillator (perhaps because of a change in temperature, power-supply voltage, etc.), the PD senses this change and produces a wider pulse. In turn, the wider pulse produces an increase in voltage from the LPF and increases the VCO frequency so as to offset the undesired deviation from that of the standard. Note that in most present-day systems, the PD and standard oscillator (except for the crystal) are contained in an IC, whereas the LPF and VCO are usually separate circuits.

Figure 7.1*b* is the block diagram of a PLL system used in a CB set. In this system, the VCO output frequency becomes equal to the transmitting frequency during transmit and is equal to the receiver local-oscillator frequency during receive.

Note that the IC contains two dividers in addition to the phase detector and standard oscillator. The fixed divider divides the standard-oscillator frequency by 1/1024 to convert the 10.240-MHz crystal frequency to 10 kHz. The programmable divider is controlled by the channel selector and divides the mixer output frequency as necessary to produce 10 kHz.

For example, if CB channel 23 is selected, the desired VCO output must be 27.255 MHz during transmit. The VCO signal is mixed with the 29.515-MHz signal from the transmitter oscillator to produce a differential frequency of 2.260 MHz. With the channel selector set at 23, the programmable divider divides the difference frequency of 2.260 MHz by 1/226 to produce 10 kHz.

Note that the standard frequency of 10.240 MHz is used directly for the receiver's second local-oscillator signal to produce the desired 455 kHz IF. Using CB channel 23 again, the VCO produces 16.560 MHz, which combines in the receiver's first-mixer circuit with the incoming 27.255 MHz to produce an IF of 10.695. This is combined in the receiver's second mixer with the 10.240 MHz standard frequency to produce 455 kHz. In a similar fashion, all 40 CB-channel frequencies (including the transmitter frequency and both receiver oscillator frequencies) are produced with only three crystals.

Troubleshooting Approach. The first step in troubleshooting the circuits in Fig. 7.1 is to check that each oscillator signal is present and on frequency. If the transmit-oscillator signal is absent or off frequency, suspect Q_{19} and/or X_2. If the receive-oscillator signal is absent or off frequency, suspect Q_{25} and/or X_3. If the standard-oscillator signal (at pins 3 and 4 of the PLL IC) is absent or off frequency, suspect X_1 and/or the PLL IC.

If all three oscillator signals are good, monitor the input (pin 2) and output (pin 6) of the PLL IC while rotating the channel selector through each channel position. If there is no change at pins 2 and 6 at any position of the channel selector, suspect the PLL IC. If there are changes at the input and output of the PLL IC but not at one or two channels, suspect the wiring between the selector and PLL IC.

As an alternate, check that the VCO Q_{15} changes frequency each time the channel selector is moved from one position to another. If it does not, suspect the PLL IC, Q_{11}, Q_{12}, and Q_{15} through Q_{18}. Finally, check that both the receiver and transmitter circuits receive signals from Q_{15}. If they do not, suspect Q_{24} and/or Q_{31}.

7.1.2 Troubleshooting TV Digital-Tuning (PLL) Circuits

Figure 7.2 shows the PLL circuits for a typical TV tuner, such as described in Sec. 1.6. The following outlines a troubleshooting approach for such circuits.

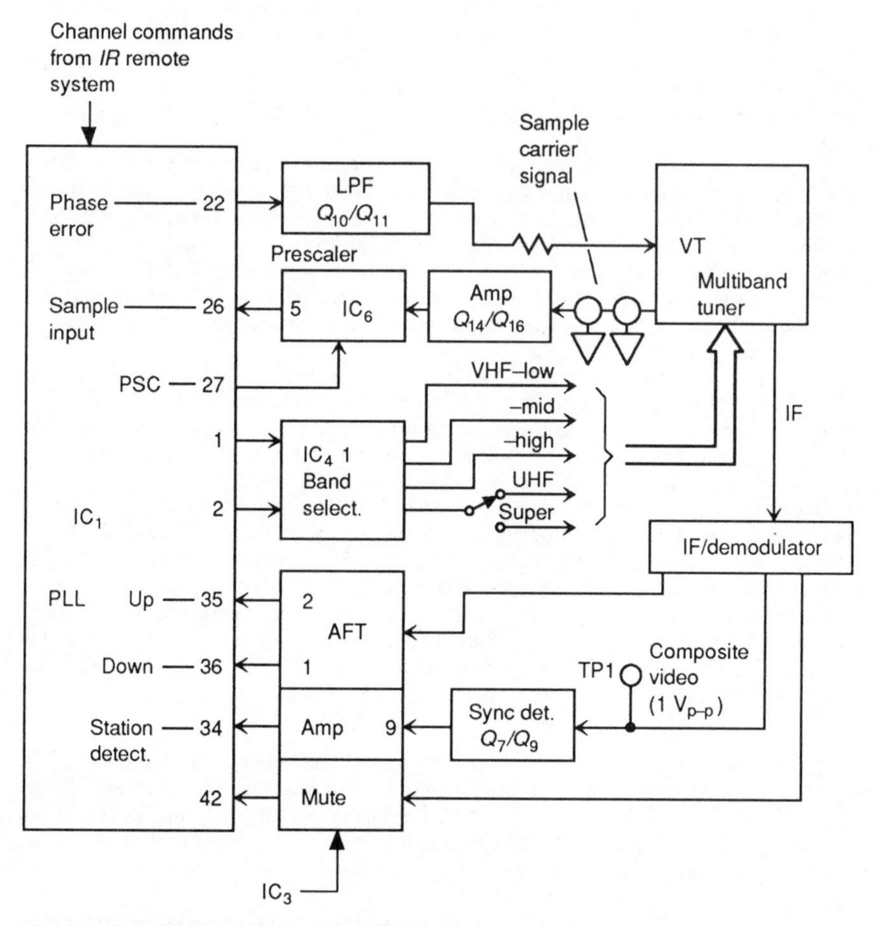

FIGURE 7.2 PLL circuits for a typical TV tuner.

The most common symptoms for failure of any TV PLL tuning systems are a combination of *no stations received, failure to lock on channels, picture snowing, audio noise,* and *color dropping in and out.* The author experienced all of these symptoms with a recently purchased TV set. The cause was traced to poorly soldered terminals on both the tuner and PLL IC. However, the following steps lead to the defective solder junctions.

The first troubleshooting step is to isolate the problem to the tuner and IF demodulator or the PLL system. Start by checking for power to all ICs and components. For example, the tuner requires +12 V. Once you are satisfied that power is available to all components, start the isolation process.

Select a channel and confirm that the band-switching signal for that particular channel appears at the tuner input and band-select output. For example, if Channel 4 is available locally, select Channel 4 and check that the VHF-low band-switching signal is present (high). This signal appears at pin 1 of IC_4 and at the tuner band-switch input. If the band-switching signal is not at pin 1 of IC_4, suspect IC_4 (or possibly IC_1).

Next, apply a *substitute tuner-control voltage* to the tuner oscillator (at the terminal marked VT). Then monitor the video signal at test point TP_1 with a scope. (The TP_1 signal is the typical composite video output, with both video and sync information, and is about 1 V peak to peak.)

If the video signal does not appear at TP_1, suspect a problem in the tuner and demodulator assembly. In most present-day TV and VCR equipment, this means replacing the complete assembly or possibly replacing individual tuner and IF-demodulator ICs.

If you can tune in stations using the substitute tuner-control voltage, the problem is most likely in the PLL components rather than in the tuner components.

If the video appears at the test point with a substitute voltage, check for a high at the station-detect input of IC_1 (pin 34). If the station-detect input is missing, suspect IC_3 or the sync-detector circuits Q_7 and Q_9.

If the station-detect signal at $IC_{1\text{-}34}$ *is normal,* monitor the AFT up and down inputs to IC_1, pins 35 and 36, while changing the substitute tuning voltage. (Note that the substitute tuning voltage can come from any external source but must match the normal voltage range at terminal VT of the tuner).

If there are no changes at pins 35 and 36 as you tune through a station (as you vary the voltage at VT), suspect the window-detector circuit within IC_3 (or possibly the AFT detector within the IF demodulator).

If the inputs at pins 35 and 36 appear to be normal, check for a sample oscillator signal at pin 26 of IC_1. If the oscillator signal is missing, suspect IC_6, the oscillator amplifier Q_{14} and Q_{16}, or possibly the shielded cable from the tuner. (It is also possible that IC_6 is not receiving proper PSC pulses from pin 27 of IC_1, although you should still get sample signals from IC_6 to $IC_{1\text{-}26}$.)

If the input at pin 26 of IC_1 *appears to be normal* but the outputs at pins 1 and 2 are absent or abnormal, suspect IC_1.

7.1.3 Troubleshooting AM/FM Tuner PLL Circuits

Figure 7.3 shows the PLL circuits for a typical AM/FM tuner, such as described in Sec. 1.6. The following outlines a troubleshooting approach for such circuits.

A failure in the PLL IC of an AM/FM tuner, or in the circuits controlled by the PLL, can cause many trouble symptoms. Unfortunately, a failure in other circuits can cause the same symptoms. For example, assume that you operate the tuning up and down buttons and see that the front-panel frequency display varies accordingly, but the AM or FM section does not tune across the corresponding band (no stations of any kind are tuned in). This can be caused by a PLL failure or by a failure of the commands to reach the PLL IC_{503} (even though the commands are displayed), which creates special troubleshooting problems for AM/FM tuners with PLL.

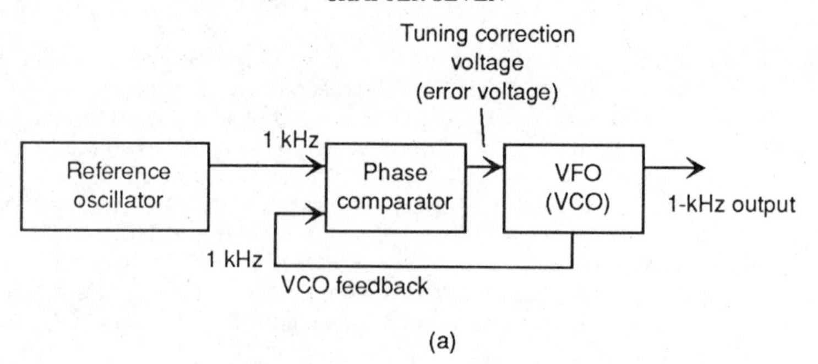

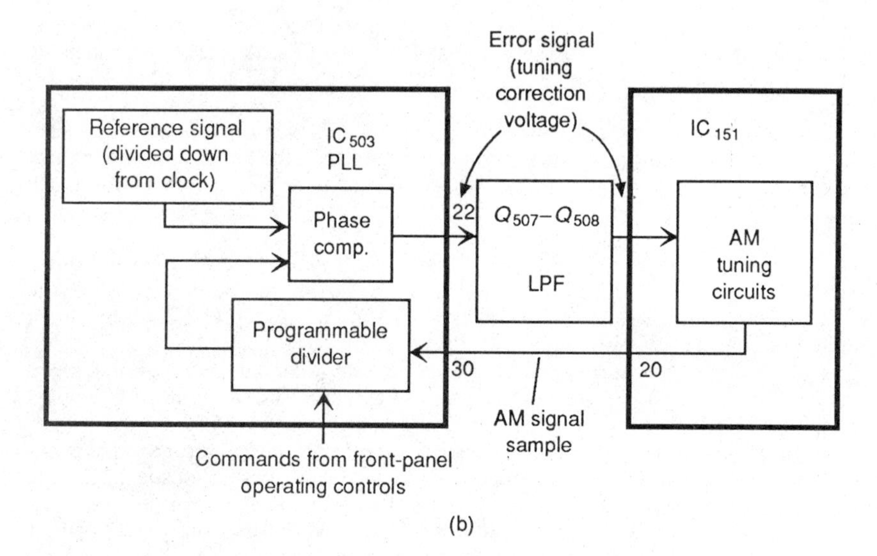

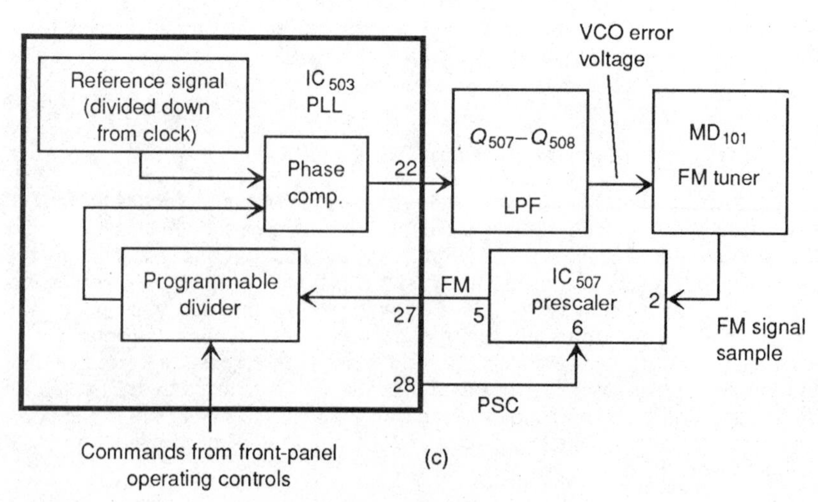

FIGURE 7.3 PLL circuits for a typical AM/FM tuner.

Common AM/FM-Tuner Failure Symptoms. The most common symptoms for failure of the PLL are a combination of *no stations received* (on both AM and FM) and *noisy audio* (audio not muted when you tune across the broadcast band). Of course, not all tuners have both auto and mono modes, as does our tuner, and not all muting circuits operate in exactly the same way. However, the following approach can be applied to the basic PLL problem without regard to the exact method of PLL tuning.

Operating Control Checks. First, make certain that the operating controls are properly set for a particular PLL function. For example, on our tuner, you must be in the auto mode (auto mode button pressed, auto mode indicator on) before the PLL seeks FM stations as you tune across the FM band with the tuning up and down buttons.

If you have selected mono (auto mode indicator off) or the circuits have gone into mono because of a failure, the PLL tunes across the FM band in 200-kHz increments whether the stations are present or not. Always look for some similar function on the tuner you are servicing.

When you are certain that the controls are set properly and that there is a true malfunction, the next step is to isolate the problem to the tuning circuit of the PLL circuit. Two basic approaches are discussed in the following sections.

Error Voltage Checks. The first approach is to apply a frequency-change command to the PLL and see if the error voltage and/or sample voltage changes accordingly. For example, to check the AM section of our tuner, press the tuning up and down buttons and see if the voltage at pin 22 of IC_{503} changes as the frequency display changes. If it does not, suspect IC_{503} or the circuits between the front-panel tuning up and down buttons and IC_{503}.

If the error voltage at pin 22 of IC_{503} changes, check that the frequency of the signal at pin 30 of IC_{503} (or pin 20 of IC_{151}) also changes as the frequency display changes. If it does not, suspect the tuning circuit D_{153}, Q_{151}, and/or low-pass filter Q_{507} and Q_{508} (Fig. 1.13).

Next, press the tuning up and down buttons while in the FM mode and see if the error voltage at the VCO input of MD_{101} changes. If it does not, suspect IC_{503} and/or IC_{507}. You can also check that the PSC pulses at pin 28 of IC_{503} change, but this is usually more difficult to monitor.

If the error voltage applied to the VCO input of MD_{101} changes, check that the frequency of the sample-frequency output of MD_{101} and pin 27 of IC_{503} change as the front-panel frequency display changes. If it does not, suspect MD_{101}. If the frequency does change at MD_{101} and IC_{507} but not at pin 27 of IC_{503}, suspect IC_{507}. (It is also possible that IC_{507} is not receiving proper PSC pulses from pin 28 of IC_{503}.)

Substitute Tuning Voltage. An alternative technique is to apply a substitute tuning, or error, voltage to the tuning circuits and see if the circuits respond by producing the correct frequency. Although this sounds simple, here are some considerations.

First, you must make certain that the substitute tuning voltage is in the same range as the error voltage. For example, the error voltage in our tuner varies from about 1 to 20 V (at the tuning circuits). You can cover this range with a typical shop-type variable power supply. However, the shop power supply can possibly load the tuning circuit with unwanted impedance, reactance, and so on.

Remember that if you apply a lower voltage, the circuits will not respond properly. If you apply a voltage higher than the tuning-circuit range, the circuits can be damaged.

Although a number of tuners use circuits similar to our tuner (the MD_{101} package is quite common in present-day AM/FM tuners), the tuning circuits are not the same for all tuners. Some tuners combine AM and FM functions in a single package. Of course, if you are lucky, you can find the error-voltage range in the service literature, often in the adjustment chapter.

Station-Detect. If the tuner has a station-detect function (most tuners do, at least in the FM section), you can use this feature together with a substitute tuning voltage to isolate PLL problems. Simply vary the substitute tuning voltage across the range and see if stations are detected.

For example, in our tuner, you can check at pin 12 of IC_{201} (Fig. 1.13) and/or pin 26 of IC_{503} for a change of status each time a station is tuned in and out. Pin 26 of IC_{503} should go high and pin 12 of IC_{201} should go low each time an FM station of sufficient strength is tuned in. The status of the pins should reverse when the station is tuned out (in between stations).

7.2 COMMUNICATIONS-SET TROUBLESHOOTING APPROACH

The remainder of this chapter is devoted to some practical tips on using the procedures described in Chaps. 5 and 6 to isolate trouble in communications sets. The procedures are presented in a logical sequence based on an analysis of the set's operation. The techniques presented here demonstrate a guideline for developing a logical, systematic approach to RF troubleshooting rather than step-by-step instructions for a specific RF equipment or set.

Although each manufacturer (and, in fact, each model) has its own design variations, there is a basic similarity among most communications sets. For example, Fig. 7.4 shows a block diagram of an AM-only transceiver, while Fig. 7.5 is the block diagram of an AM/SSB transceiver. Again, we use discrete-component circuits, rather than PLLs and ICs, to illustrate the troubleshooting approach.

Note that both sets use a dual-conversion receiver with 7.8-MHz and 455-kHz IF circuits, synthesizer-type transmit and receive oscillator, and an audio section that is common to both transmit, receive, and public address (PA) modes of operation. The SSB receiver IF circuits and transmitter RF circuits are independent of the AM circuits, but several SSB circuits are common to transmit and receive operations.

The troubleshooting procedure described here is tailored to the circuit designs shown in Figs. 7.4 and 7.5 to demonstrate the analyzing technique. However, the basic technique can be modified to troubleshoot communications sets of almost any design. As discussed in Chap. 6, by carefully observing the symptoms, it is often possible to go directly to a specific checkout procedure and bypass unrelated checks.

The procedures described here isolate the defect to a small area consisting of only a few parts. The voltage and resistance measurements described in Chap. 6 can then be made within the suspected circuit to locate the defective part (or the ever-popular break in PC wiring).

7.3 INITIAL CHECKS AT MOBILE UNIT AND BASE STATION

When troubleshooting mobile communications sets, it is advisable to perform a few checks to eliminate all items external to the set before removing the set from

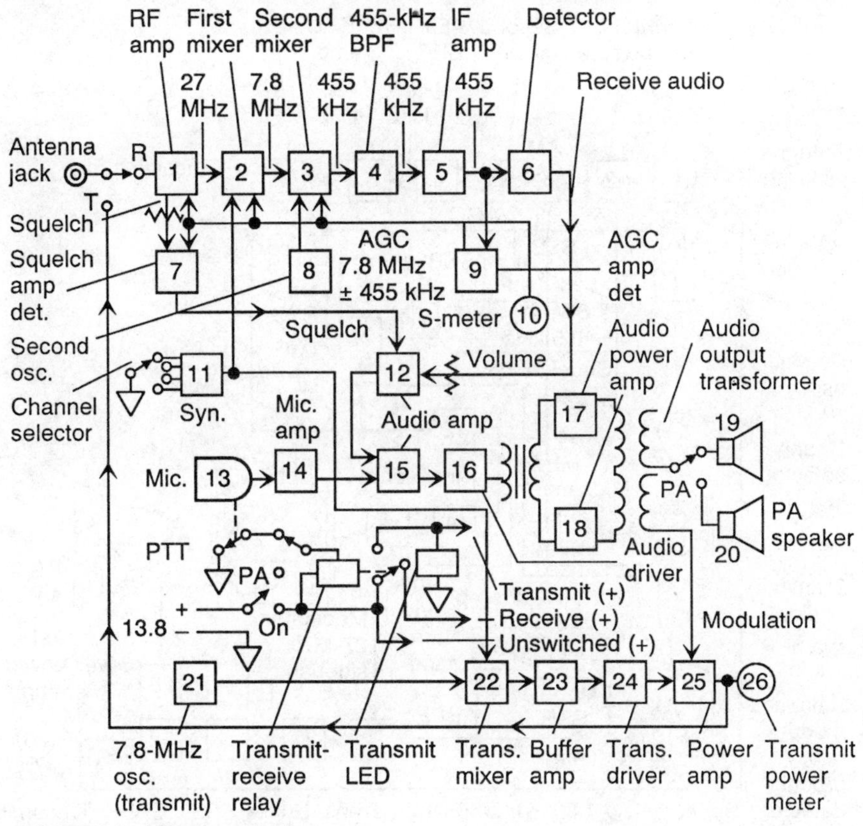

FIGURE 7.4 AM-only transceiver.

the vehicle for a bench test. External items such as the power cable, antenna, antenna cable, microphone, and external speaker are often more subject to physical damage than the set and must always be considered as a likely source of trouble. The following checks verify if external items are good or bad.

If the set is already removed from the vehicle and is in the shop for checkout, proceed with the bench check. However, if the set indicates normal operation on the bench, these in-vehicle checks may be necessary when the set is returned to the vehicle. For AM/SSB sets, perform all of the following checks in the AM mode (unless there has been a specific report of SSB-mode-only problems).

7.3.1 Power-Cable Check

1. Turn on the set. Be sure that the vehicle ignition switch is also on, if required for radio operation.
2. On most sets, there is some type of indicator (usually an LED or possibly a channel or meter light) which turns on when power is applied. Note whether the indicator is on or off (with all switches on).

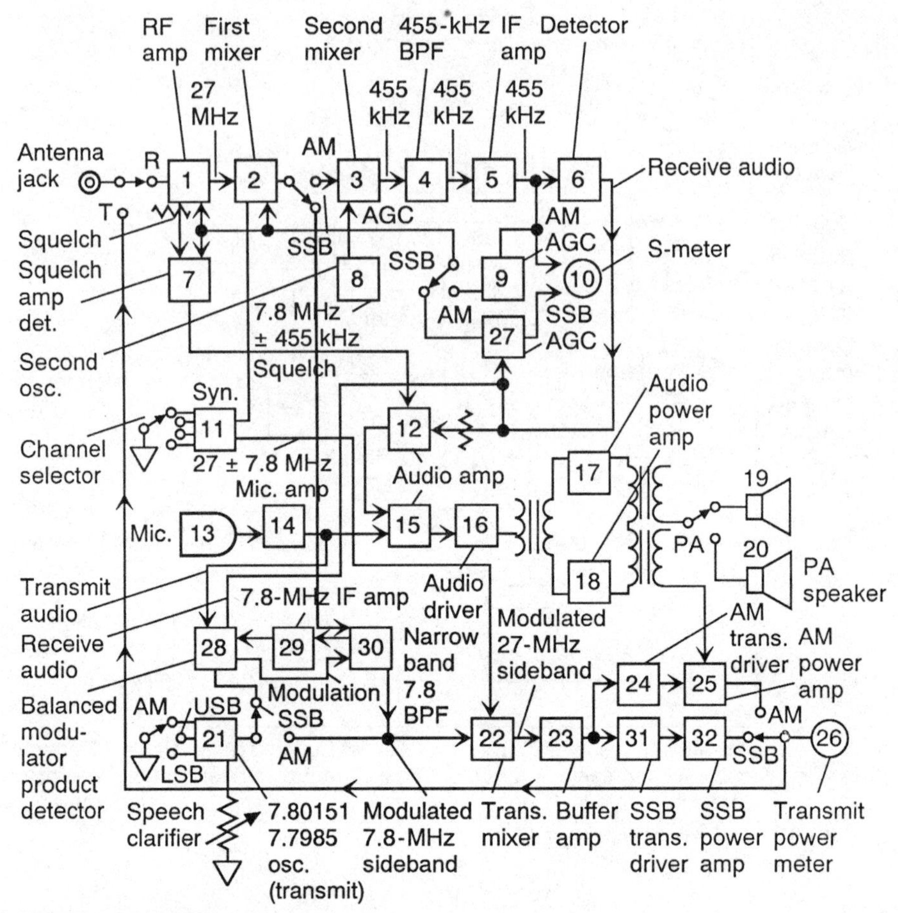

FIGURE 7.5 AM/SSB transceiver.

If the indicator is on, the power cable is probably good. If the indicator is not on when power is applied, check the fuse.

If the fuse is good, power is not getting through the power cable to the set. Make voltage and continuity checks on the cable and the related connectors. If power is available to the set power-input connector, the problem is within the set.

If the fuse is open, replace it with a new fuse of the proper rating. If the new fuse blows, there is an apparent overload or short (probably in the receiver section).

If the new fuse does not blow, key the transmitter (probably press the PTT button). If this causes the fuse to blow, there is an apparent overload or short (probably in the transmitter section).

If the new fuse does not blow when the transmitter is keyed, normal operation is restored, at least temporarily. If practical, operate the set in both the receive and transmit condition while driving over a rough street. If the fuse does not blow, the likelihood of future fuse failure is greatly reduced. If the new fuse blows during the test drive, there is an apparent intermittent short in the set.

7.3.2 External Speaker Check

If the installation does not include an external speaker, skip to the antenna check (Sec. 7.3.3).

1. Unsquelch the receiver and turn up the volume. If receiver background noise or received signals are heard on the external speaker, the speaker is operating.

2. If received signals are heard on the external speaker, but severe audio distortion is noted (as could be produced by a torn speaker cone), disconnect the external speaker and listen to the internal speaker (or connect a substitute speaker). Never operate a receiver without any speaker (or without a dummy load of appropriate impedance). This could destroy the transistors in the final audio section. If no distortion is noted on the internal or substitute speaker, replace the external speaker. If distortion is noted on both speakers, the trouble is within the set.

3. If nothing is heard from the external speaker, either the set or the speaker could be the cause. Disconnect the external speaker and listen for receiver noise or signals on the internal speaker. (If nothing is heard, and you do not know if the set has an internal speaker, connect the substitute speaker to the external speaker jack or connector. Use the PA speaker if the vehicle is so equipped.)

If noise or received signals are heard on the internal or substitute speaker, the trouble is in the external speaker or speaker cable and connectors. If nothing is heard on the internal or substitute speaker, it is unlikely that both speakers are defective. Take a wild guess that the trouble is within the set.

7.3.3 Antenna Check

This check cannot be completed unless there is a transmitter power output. If any step shows the need for troubleshooting the set, the antenna check should be delayed until the set is returned to the vehicle. Also note that antenna-check procedures are described, in boring detail, in Sec. 7.7.

1. If the set is equipped with a transmitter power indicator of any kind (S/RF meter, etc.), key the transmitter and observe the power indicator. (If the transmitter cannot be keyed, skip to the microphone check, Sec. 7.3.4.) If abnormal power output is indicated, you can generally omit the antenna check because the problem is probably in the set. However, an extreme problem in the antenna (shorted cable, etc.) can cause an abnormal power indication.
2. If the transmitter power-output reading is normal, or if the set is not provided with a transmitter power-output indicator, connect an external test SWR/power-meter between the set and antenna.
3. Key the transmitter and check power to the antenna. If abnormal power output is indicated on the external meter, the trouble is probably in the set (or there is an extreme antenna problem).
4. Perform the antenna SWR measurements as described throughout various chapters of this book, and note the following:

   If the SWR is less than 1.2, the antenna and antenna cable are excellent. If the SWR is from about 1.2 to 1.5, the antenna and cable are probably satisfactory.

   Higher SWR readings indicate a mismatched condition which may be caused by a damaged antenna, corroded or improperly fitted connectors, excessive moisture, or crushed antenna cables.

An extremely high SWR indicates an open-circuit or short-circuit condition, such as an antenna cable cut in two (or disconnected) or the inner and outer conductors of the antenna cable shorted together.
If you suspect antenna problems of any kind, follow the procedures described in Sec. 7.7

7.3.4 Microphone Check

A microphone check is required only if one or more of the following symptoms is indicated: *transmitter cannot be keyed, no transmitter modulation, and/or no output in the PA mode.*

Disconnect the microphone and connect a known-good substitute microphone. If no change in symptoms is noted, the trouble is within the set. If the symptom is corrected when using the substitute microphone, the original microphone is defective. If no substitute is available, or the microphone is the wired-in type, remove both the set and microphone from the vehicle for bench service.

7.3.5 PA Speaker Check

This check applies only if the installation includes a PA speaker and the set includes a PA mode.

1. Operate the set in the PA mode. If the announcement is heard on the PA speaker (without excessive distortion), the PA speaker is good.
2. If there is excessive distortion in the PA speaker, such as could be produced by a torn cone, disconnect the PA speaker and connect a substitute. If there is no distortion in the substitute, the PA speaker is defective. If there is distortion in both speakers, the trouble is in the set (and there will probably be distortion in the receiver audio and transmitter modulation).

7.3.6 Conclusion of Mobile-Set Initial Checks

If all items external to the communications set are good, the fault is in the set. Remove the set from the vehicle for bench service.

When repairs are complete and the set is reinstalled in the vehicle, always perform an antenna SWR check. The installation should be adjusted for minimum SWR by tuning the transmitter to match the antenna (or vice versa or both). Not all antennas are tunable. Always follow the procedures recommended in the service manual for tuning the transmitter, after you have studied Sec. 7.7.

If you cannot get an SWR below about 1.5 on all channels, and preferably better on the middle channels, look for problems in the antenna installation. A high SWR can cause burnout of the transmitter final RF amplifier, so *if you find final-stage burnout, look for antenna problems.*

If a bench check indicates normal performance, yet the set performs poorly in the vehicle, and you have tuned for minimum SWR and maximum power output, check the following:

1. Poor ground connection.
2. Too much resistance in the power cable. Voltage into the set substantially below 13.8 V (or below the recommended power-supply voltage).
3. Vehicle battery voltage low or poor voltage regulation from the vehicle regulation system.
4. Too much ignition interference and electrical disturbance because of inadequate noise-suppression on the vehicle. Refer to Sec. 7.10.
5. Antenna poorly located for good propagation characteristics.

7.3.7 Base-Station Initial Checks

The initial checks for a communications base station are essentially the same as for a mobile station. Often, base-station equipment has more meters and indicators for monitoring performance. Also base stations usually operate from a 120-V ac supply. In many cases, this power is converted to 13.8 V dc, which is the same as the vehicle battery voltage for most mobile communications sets.

As with mobile sets, eliminate all items as possible troubles before starting a bench check on a base station:

1. Connect the base station to the power source and antenna, if it is not already connected.
2. Turn on the set. If the front-panel LEDs are on, assume that the power-supply voltages are correct.
3. If the LEDs are not on, check the fuse and replace it if necessary. If a new fuse restores operation, proceed to step 4. If a new fuse does not restore operation, remove the base-station set, including the power supply, for bench check.
4. Key the transmitter. If the transmitter RF power-output is normal, check the antenna SWR (with a built-in or external SWR meter). Generally, base-station SWR is better than that found in most mobile stations.
5. Check the base-station microphones by substitution, or include them in the bench check.

7.4 CURRENT-DRAIN CHECKS

When a mobile communications set is brought in for service, the set is connected to a bench power supply. Most bench supplies include (or should include) output voltage and current meters. Typically, voltage should be adjusted to 13.8 V, and current limiting (a feature found on many bench power supplies) should be adjusted to the required current.

If the set blows fuses or indicates a high current drain, find the short or overload and repair it *before* continuing with other checks. It may be helpful to note the current drain for each set when the set is first connected to the bench power supply. Refer to the service manual for maximum current drain for a specific set. For a typical CB set, maximum current drains are in the vicinity of:

Standby (receiver unsquelched)	500 mA
Receive (full-rated audio)	1.5 A
Transmit	2.2 A

7.5 DEFINING TROUBLE SYMPTOMS

Although proper test equipment can be used to perform a complete diagnosis of an ailing communications set, without a description of the symptoms, a good description of the problem can often lead you directly to the circuit or part. Service time is reduced by eliminating the time required for a complete checkout of all possible symptoms.

When a service order for a communications set is taken, get a full description of the symptoms, if possible. This description should be defined as precisely as possible. Ask the owner or operator additional questions, if possible, to refine the description.

For example, a symptom of "poor reception" could include an entire range of symptoms from "weak audio" to "short range (poor receiver sensitivity)," "adjacent-channel interference (poor receiver selectivity)," or "garbled voice (distortion)." One of these terms is much more precise in defining the malfunction. More important, the term could further isolate the problem area.

Unless the servicing is to start immediately, jot down a description of the symptoms in correct technical terms. Keep the note with the equipment for reference when the troubleshooting job begins.

The troubleshooting procedures in the remainder of this chapter are grouped by symptoms. The symptoms alone localize the trouble within a portion of the set. The troubleshooting technique that isolates the malfunction in the shortest time does not include checks in circuits that are not related to the symptom.

Using an obvious example, on combination AM/SSB sets, the usual practice is to check the AM mode completely and correct any malfunction before checking the SSB. Of course, this does not apply if the symptom is "failure in the SSB mode only."

7.6 BENCH CHECKS

The following general procedure is used to determine what is wrong with the set if no description of symptoms is available, or if you wish to verify the symptoms and check for additional symptoms (a very practical idea in most cases). The procedure may also be used if you wish to perform a complete performance check of the set and correct any subnormal performance.

1. Connect the set in the basic test configuration as described in Chap. 5 (Fig. 5.1). Use the test connections described in Chap. 5 as necessary to perform the test procedures described in the following steps.

2. Connect an external speaker in parallel with the set's internal speaker.

3. Set the operating controls as follows: Select any channel, set the squelch control to the fully unsquelched position, set the RF-gain control (if any) to maximum, set any accessory mode switches (such as noise limiter or automatic noise control) to off, and select the AM mode.

4. Turn on the set, and turn up the volume until a strong receiver noise is heard on the speakers. If noise is heard on the external speaker but not on the set's speaker, you have already found one of your problems. If no receiver noise is heard on either speaker, adjust the volume control to about three-fourths of the maximum position.

5. Perform as many steps shown in Fig. 7.6, as required to be directed to a troubleshooting procedure. Use the referenced troubleshooting procedure to isolate and correct any malfunction.

For example, the first step shown in Fig. 7.6 is to listen for receiver noise. If there is no receiver noise, the next step is to perform transmitter RF-power check, as described in Sec. 5.1. If the transmitter power check is normal, the next step is to refer to the troubleshooting procedure for radio does not receive, as described in Sec. 7.9.1.

If there is no output on any channel during the transmitter RF-power check, the next step is to refer to the troubleshooting procedure for radio does not transmit or receive, as directed in Sec. 7.9.2. If there is no output on some channels during the transmitter RF-power check, the next step is to refer to the troubleshooting procedure for radio does not transmit or receive on some channels, as described in Sec. 7.9.3.

Now assume that during the first step in Fig. 7.6 receiver noise is low or normal instead of no receiver noise. Then the next step is to perform audio power check, as described in Sec. 5.4.

6. If troubleshooting and repair are required during the steps in Fig. 7.6, recheck operation of the set by repeating the steps in Fig. 7.6, starting at the beginning. Repeat as many times as required until the test results end with "check SSB mode of operation for AM/SSB sets."

7. If an AM/SSB set is being checked, perform as many steps in Fig. 7.7 as required to be directed to a troubleshooting procedure. Use the referenced troubleshooting procedure to isolate and correct any malfunction (as you did during step 5 of this section). If an AM-only set is being checked, skip this step and proceed to step 9.

8. If troubleshooting and repair are required during the steps in Fig. 7.7, recheck SSB operation by repeating the steps in the figure, starting at the beginning. Repeat as many times as required until the test results end with "SSB performance normal."

9. Perform as many steps in Fig. 7.8 as required to be directed to a troubleshooting procedure. Use the referenced troubleshooting procedure to isolate and correct any malfunction (as you did during step 5 of this section).

10. If any troubleshooting and repair are required in step 9, recheck the performance of the set by repeating the steps in Fig. 7.8, starting at the beginning. Repeat the steps as many times as required until the test results end with "all performance normal."

7.7 ANTENNA AND TRANSMISSION-LINE CHECKS WITH LIMITED TEST MEANS

In general, antennas and transmission lines (lead-ins or cables) used with radio communications sets are best tested during troubleshooting by means of commer-

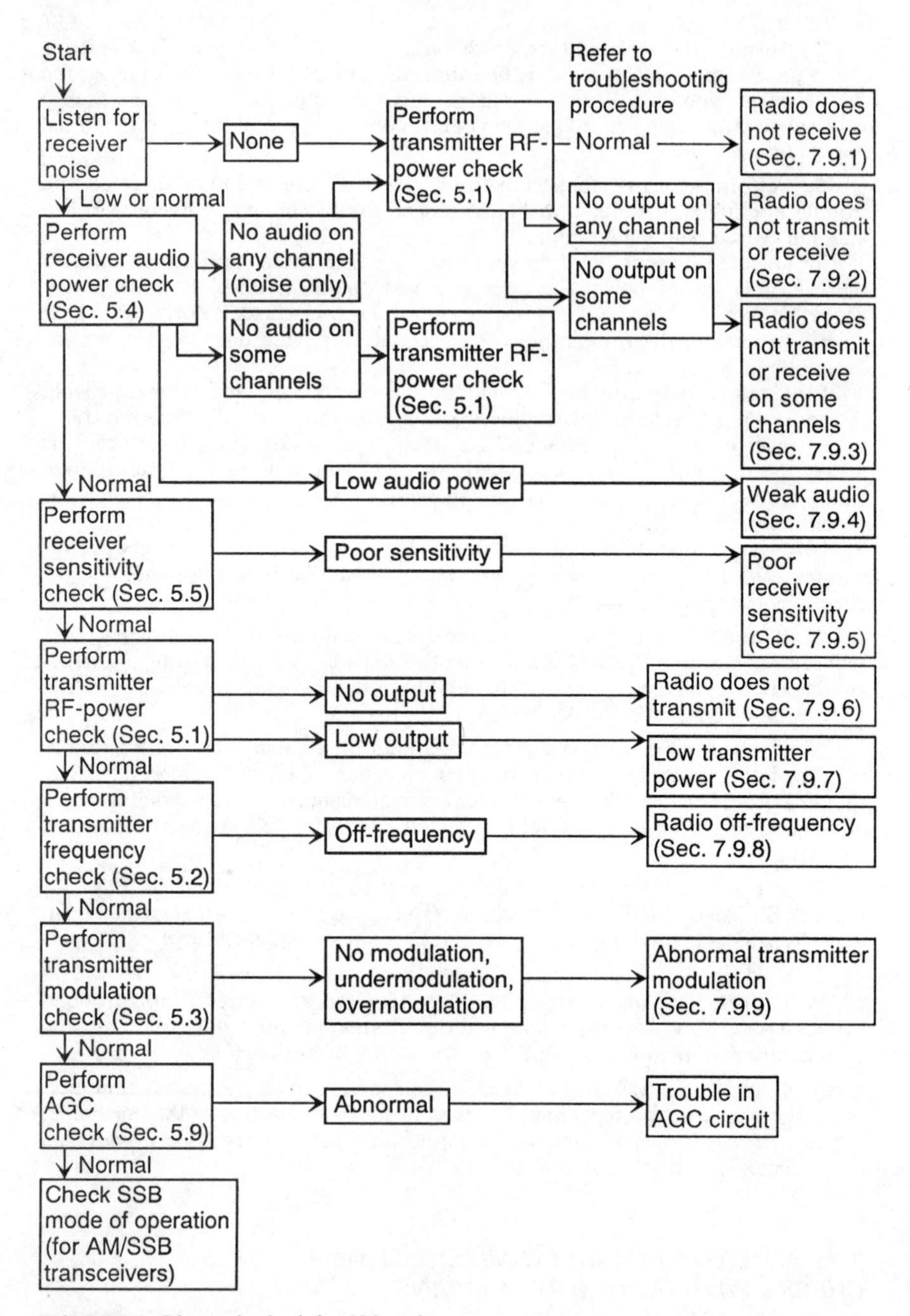

FIGURE 7.6 Diagnostic check for AM mode.

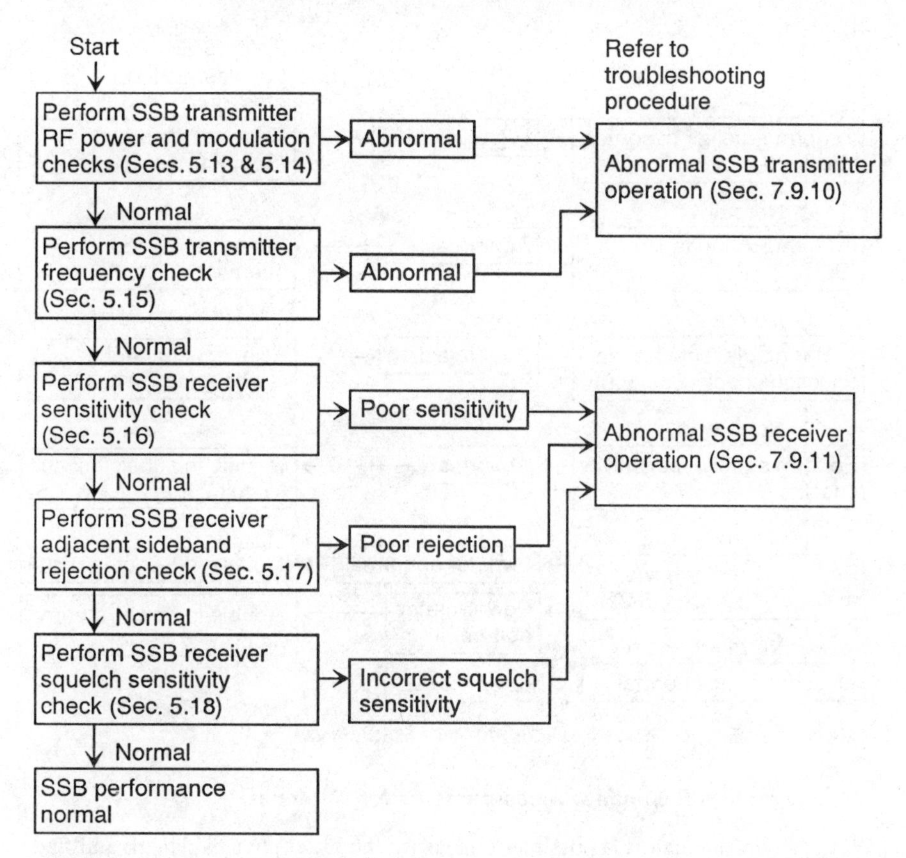

FIGURE 7.7 Diagnostic check for SSB mode.

cial SWR meters, field-strength meters, and the special test sets described in Chap. 3. However, it is possible to make a number of significant tests using basic meters (voltmeters, ohmmeter, and ammeter). The following paragraphs describe how these procedures can be performed when commercial antenna test devices are not available to the troubleshooter. The section concludes with a summary of antenna service problems.

7.7.1 Antenna Length and Resonance Measurements

Most antennas are cut to a length that is related to the wavelength of the signal being transmitted or received. Generally, antennas are cut to one-half wavelength (or one-quarter wavelength) of the center operating frequency.

As a practical matter, the electrical length of an antenna is always greater than the physical length, because of capacitance and end effects. Therefore, two sets of calculations are required: one for electrical length and one for physical length. The calculations for antenna length and resonant frequency are shown in Fig. 7.9.

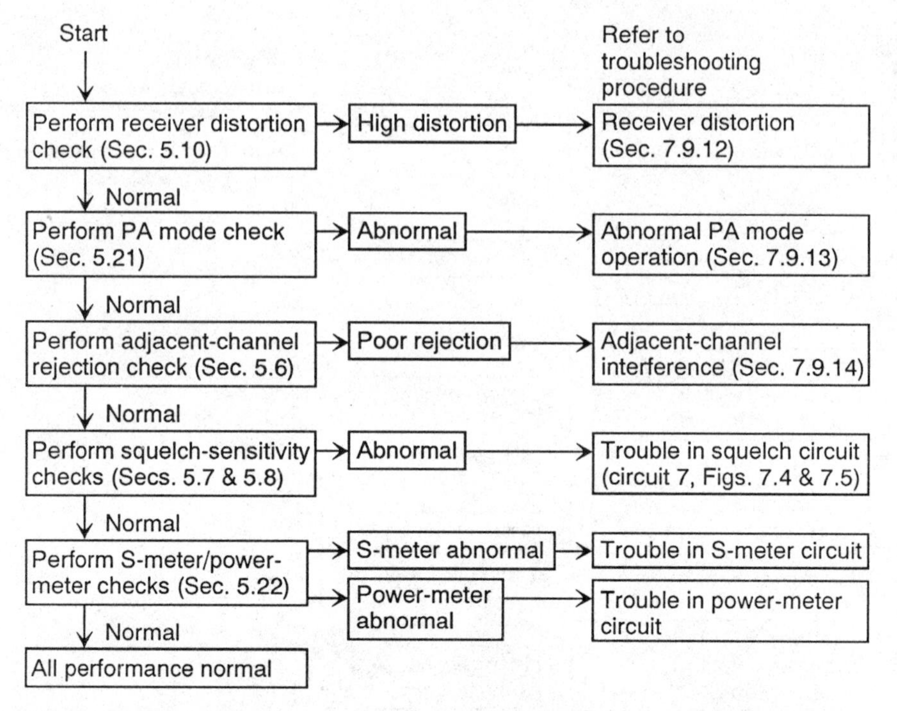

FIGURE 7.8 Final diagnostic check for AM and AM/SSB modes.

7.7.2 Practical Resonance Measurements for Antennas

With a short antenna it is possible to measure the exact physical length and find the electrical length (and the resonant frequency) using the equations in Fig. 7.9. Obviously, this is not practical for long antennas. Also, the exact resonant frequency (electrical length) may still be in doubt for short antennas because of the uncertain K-factor in Fig. 7.9. Therefore, for practical purposes, the electrical length and resonant frequency of an antenna should be determined electrically.

There are three practical methods for determining antenna resonant frequency: dip-adapter circuit, antenna ammeter, and wavemeter.

Dip-Adapter Antenna-Resonance Measurement. The dip-adapter circuit described in Sec. 3.10 can be used to measure antenna resonance. The basic technique is to couple the adapter to the antenna as if the antenna were a resonant circuit, tune for a dip, and read the resonant frequency. However, there are certain precautions to be observed.

The measurement can be made as a conventional resonant circuit provided that the antenna is accessible, allowing the dip adapter to be coupled directly to the antenna elements. If the antenna is a simple grounded element (no matching problems between antenna and transmission line), the adapter can be coupled to the transmission line. However, if the antenna is fed by a coaxial line or any system in which the line is matched to the antenna (which is usually the case for communications equipment), it is best to couple the adapter coil directly to the antenna elements (to get the most accurate results).

|← Half-wave antenna →|

Basic hertz type
(Typical for TV antennas)

Electrical length:

$$\text{Meters} = \frac{150}{\text{Frequency (MHz)}}$$

$$\text{Feet} = \frac{492}{\text{Frequency (MHz)}}$$

$$\text{Inches} = \frac{5906}{\text{Frequency (MHz)}}$$

Physical length (approximate) = electrical length × K-factor

K-factor = 0.96 for frequencies below 3 MHz
= 0.95 for frequencies between 3 and 30 MHz
= 0.94 for frequencies above 30 MHz

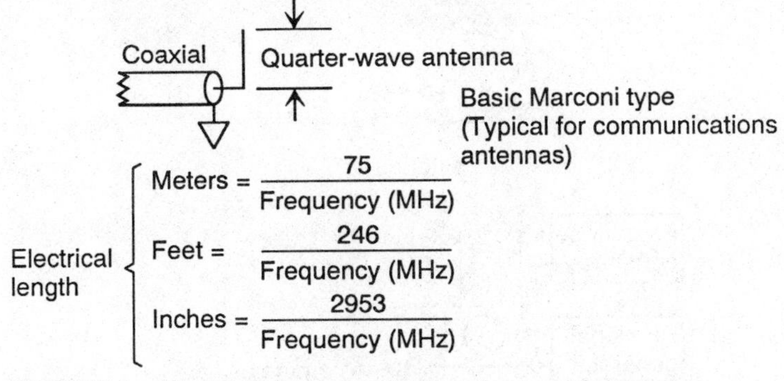

FIGURE 7.9 Calculations for antenna length and resonant frequency.

No matter how carefully the antenna and lead-in are matched, there is some mismatch, at least over a range of frequencies. This means that there are two reactances (or impedances) that interact to produce extra resonances. If resonance is measured under such conditions, a dip is produced at the correct antenna frequency (or at harmonics), and another dip (plus harmonics) is produced at the extra frequency.

It can sometimes be very confusing to tell the dips apart. Also, antenna resonance measurements should be made with the antenna in the actual operating position. The nearness of directive or reflective antenna elements, as well as the height, affects antenna characteristics and possibly changes resonant frequency.

The dip-adapter procedure for grounded antennas is as follows:

1. Couple the dip adapter to the antenna tuner, if the antenna is tuned, as shown in Fig. 7.10*a*.
2. If the antenna is untuned or it is not practical to couple to the antenna tuner, disconnect the antenna lead-in and couple the lead-in to the adapter through a pickup coil, as shown in Fig. 7.10*b*.
3. Set the generator (Fig. 3.9) to the lowest frequency. Adjust the generator output for a convenient reading on the adapter meter.

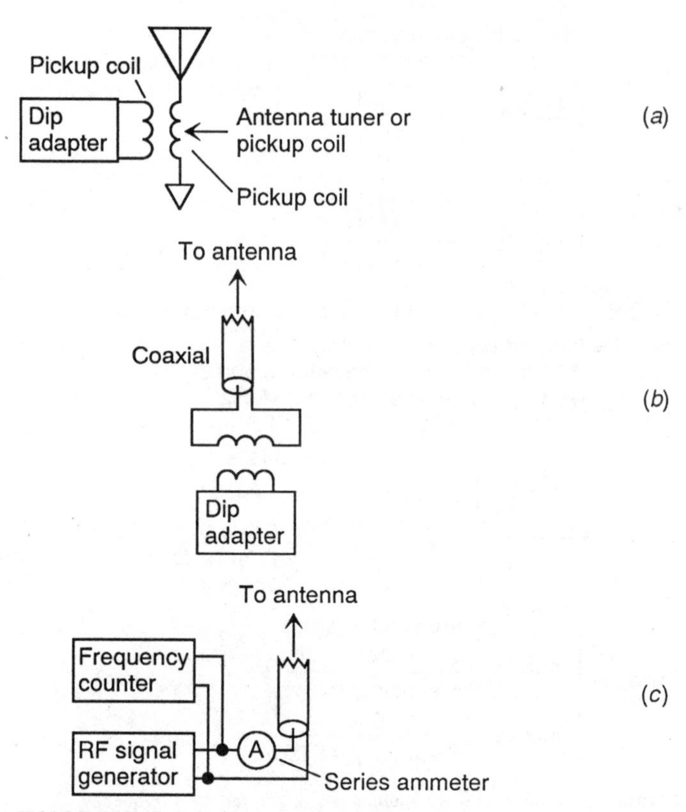

FIGURE 7.10 Practical antenna-resonance measurements.

4. Slowly increase the generator frequency, observing the meter for a dip indication. Tune for the bottom of the dip.
5. Note the frequency at which the first (lowest frequency) dip occurs. This should be the primary resonant frequency of the antenna. As the signal-generator frequency is increased, additional dips should be noted. These are harmonics and should be *exact multiples* of the primary resonant frequency. Check two or three of these frequencies to make sure they are harmonics. Then go back to the lowest-frequency dip to ensure that the lowest frequency is the primary resonant frequency.

The dip-adapter procedure for ungrounded antennas is as follows:

1. Disconnect the antenna lead-in from the antenna.
2. If the antenna is center-fed, short across the feed point with a piece of wire.
3. Couple the dip-adapter coil directly to the antenna. Usually, the best results are obtained by coupling at a maximum current (low-impedance) point. For example, the maximum-current point occurs at the center of a half-wave antenna.
4. Starting at the lowest generator frequency and working upward, tune the generator for a dip on the meter as described for grounded antennas. The lowest dip is the primary resonant frequency.

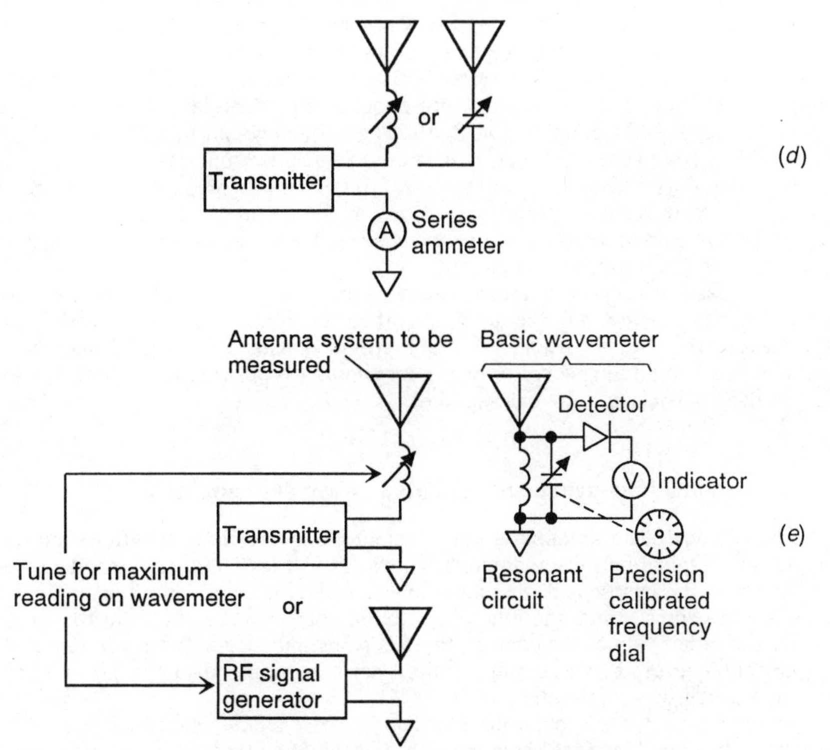

Figure 7.10 (*Continued*) Practical antenna-resonance measurements.

Series-Ammeter Measurement of Antenna Resonant Frequency. A series ammeter can be used to find the resonant frequency of an antenna. The basic circuit is shown in Fig. 7.10*c*. The signal generator is tuned for a maximum reading on the ammeter, indicating a maximum transfer of energy from the generator into the antenna (as a result of both being at the same frequency). The antenna frequency is then read from the generator dial or frequency counter. The series ammeter method has the advantage of measuring the combined resonant frequency of *both the antenna and transmission line.* This is most practical since (in normal communications operation) the antenna is operated with the transmission line.

A version of the series-ammeter method is often used in transmitters as an *indicator for antenna tuning.* Most transmitters are crystal-controlled and operated at a specific frequency with the antenna tuned to that frequency. The electrical length (and consequently the resonant frequency) of the antenna is varied by a reactance in series with the lead-in, as shown in Fig. 7.10*d.* The reactance can be a variable capacitor or variable inductance. With such an arrangement, the transmitter is tuned to the operating frequency, and then the antenna is tuned to the frequency that is indicated by a maximum reading on the series ammeter.

The series-ammeter method has certain drawbacks, one being the frequency limit of the meter itself. Another is that the series ammeter consumes some power in operation. However, the series ammeter has an advantage in that the true antenna power can be calculated (as discussed in Sec. 7.7.3).

Wavemeter Measurement of the Antenna Resonant Frequency. A wavemeter can be used to find the resonant frequency of an antenna. The basic wavemeter circuit, shown in Fig. 7.10*e*, is essentially a tuned resonant circuit, detector, and indicator. Commercial wavemeters (not popular in present-day RF work) have a precision-calibrated tuning dial so that exact frequency can be measured. Such wavemeters have generally been replaced by frequency counters.

When used to measure antenna resonant frequency, the wavemeter is tuned to the approximate resonant frequency of the antenna, and the signal generator is tuned for a maximum reading on the wavemeter. Then the generator frequency is measured on the frequency counter.

When used to tune an antenna, the wavemeter is tuned to the approximate transmitter frequency, and the antenna is then tuned for a maximum reading on the wavemeter. Not all wavemeters are provided with precision tuning; those without precision tuning serve only as a maximum (or peak) readout device, similar to the field-strength meter (Sec. 3.8).

7.7.3 Antenna Impedance and Radiated-Power Measurements

The impedance of an antenna is not constant along the entire length of the antenna. In a typical half-wave antenna, as shown in Fig. 7.11*a*, the impedance is minimum at the center and maximum at the ends. In theory, the impedance is zero at the center. Since the antenna is fed at some point away from the exact center (on either side of the center), there is some impedance for any antenna. A typical antenna used in a communications set has an impedance of 50 to 72 Ω, whereas a half-wave TV antenna has 300 Ω.

Antenna impedance is determined using the basic Ohm's-law equation $Z = E/I$, with voltage and current being measured at the antenna feed point. However, such measurements are not usually made in practical applications.

Radiation resistance is a more meaningful term. When the dc resistance of an antenna is disregarded (antenna dc resistance is usually a few ohms or less, except in very low-frequency, long-wire antennas), the antenna impedance can be considered as the radiation resistance. Radiated power can then be determined using the basic Ohm's-law equation $P = I^2R$.

Practical Impedance and Radiated-Power Measurement for Antennas. On those antennas designed to be used with coaxial transmission lines (as is the case with most communications sets), the antenna and transmission-line impedance must be matched. In this case, the impedance match between transmission line and antenna is of greater importance than actual impedance value (both antenna and transmission line must be 50 Ω, 72 Ω, etc.). The condition of match (or mismatch) between antenna and transmission line can best be tested by the SWR measurement discussed in Chap. 3 and in this chapter.

The following procedure can be used to find the impedance and radiated power of any antenna system. However, it should be noted that the impedance power obtained is for the *complete antenna system* (antenna and transmission line), as seen from the measurement end (which is generally where the transmission line connects to the communications set).

1. Connect the equipment as shown in Fig. 7.11*b*. Disconnect the transmission line from the antenna terminal of the set, and then reconnect the line to the signal generator, with the series ammeter in place as shown. The set transmitter can be

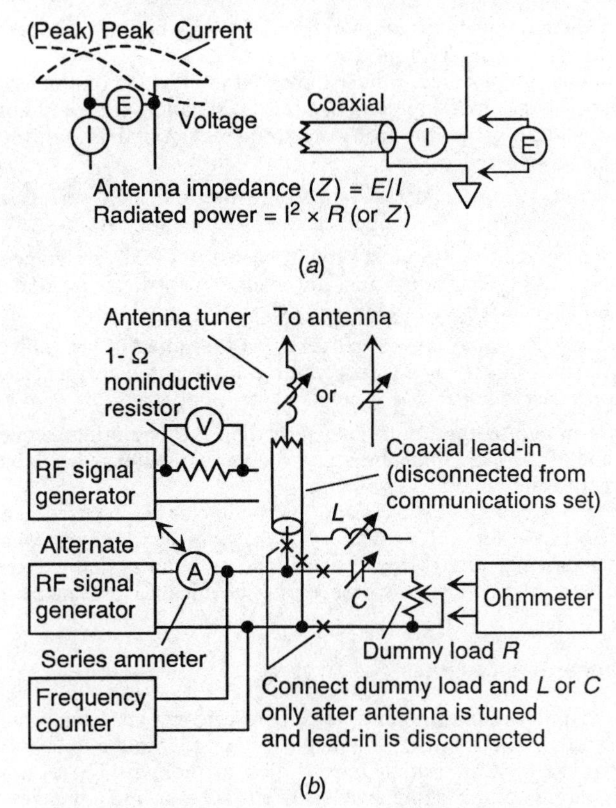

FIGURE 7.11 Antenna impedance and radiated-power measurements.

used in place of the signal generator. However, the transmitters usually operate on fixed frequencies, thus limiting the measurements to those frequencies.

2. Adjust the generator to the center frequency at which the antenna is used (or for any desired operating frequency to which the antenna can be tuned). If the transmitter is used in place of the signal generator, operate the transmitter at (or near) the center frequency.

3. If the antenna can be tuned, tune the antenna for a maximum indication on the ammeter. Record the indicated current. If the antenna cannot be tuned, adjust the signal generator for maximum on the ammeter, and record the current. If the antenna cannot be tuned and you use a fixed-frequency transmitter, simply record the indicated current.

4. If a frequency counter is available, verify that the generator (or transmitter) and antenna are tuned to the correct frequency (after the antenna and/or generator are tuned for maximum indication on the ammeter).

5. Disconnect the transmission line from the ammeter and generator. Connect the generator to the dummy load through the ammeter as shown. If the antenna is tuned in step 3, adjust inductance L or capacitor C for maximum indication on the ammeter. (If the antenna is tuned by means of a variable inductance, use capacitor C to tune the dummy load, and vice versa.)

If the antenna is not tuned, adjust the generator for maximum indication. (If the dummy load is noninductive and is a pure resistance, it should not be necessary to readjust the generator from the frequency found in step 3, when the dummy load is connected.)

6. Adjust the dummy-load resistance until the indicated current on the ammeter is the same as the antenna current recorded in step 3.

7. Remove power from the test circuit. Measure the dc resistance of R with an ohmmeter. This resistance is equal to the antenna-system impedance (or radiation resistance) *at the measurement frequency.*

8. Calculate the actual power delivered to the antenna (or radiated power) using: radiated power $= I^2 \times R$ (or Z) where I = indicated current (amperes) and R = radiation resistance (or Z = antenna-system impedance).

9. An alternative method must be used when the operating frequency is beyond the range of the available ammeter, when no ammeter is available, or when the ammeter presents an excessive load.

A precision 1-Ω, noninductive resistor and voltmeter can be used in place of the ammeter, as shown in Fig. 7.11*c*. With a 1-Ω resistor, the indicated voltage is equal to the current passing through the resistor (and antenna system). Except for the meter and resistor connections, the procedure is identical to that using the ammeter.

7.7.4 Antenna Service Notes

If a communications set shows chronic "poor performance" symptoms, suspect the antenna and/or antenna installation. We do not go into the details of installing a communications antenna (either base-station or mobile). There are dozens of books and hundreds of magazine articles on the subject (including several by the author). Instead, we concentrate on quick-check service methods to determine if the antenna is functioning properly and how to pinpoint the problem in a defective antenna system.

An antenna may not be performing properly because of a defect such as shorts in the coaxial lead-in, poor contact, open contact, broken shielding, etc. If is also possible that the antenna never performed properly because of a mismatch or improper adjustment. Either way, you want to know if the antenna is capable of performing properly now.

SWR-Meter Antenna Checks. The SWR meters described in Sec. 3.9 are the most useful test instruments for checking antennas. You can tell at a glance if the set, lead-in, and antenna are properly matched. If the set has a built-in SWR meter, use it. If not, use an external SWR meter or one of the special test sets (Sec. 3.13) that include an SWR function.

Ideally, the SWR should be checked on all channels. As a minimum, check SWR at the highest, lowest, and middle channels. An SWR reading of 1.1 or 1.2 on all channels indicates that the antenna system is good. *Leave it alone.* If the SWR readings are about 1.2 to 1.5, the antenna system is on the borderline.

Adjustable Antennas. If the SWR reading is good on the lower channels but gets progressively worse (a higher SWR reading) on the higher channels, the antenna is probably too long. If the opposite is true (good SWR on the higher channels), the antenna is probably too short. Some (but not all) communications antennas are adjustable. In some cases, the physical length of the antenna can be changed (you loosen an Allen screw and move the antenna in or out). In other cases, the loading coil is adjustable.

No matter how the antenna is to be adjusted, operate the set on the channel that shows the poorest SWR reading, and make the adjustment to get a good SWR on that channel. Then recheck SWR at the other channels. Always adjust the antenna for the best possible SWR on all channels. (This often involves compromise.)

Antenna-Matching Devices. A number of antenna-matching devices are on the market. These devices are connected between the set and the antenna. In effect, the matching devices change the antenna length electrically. The author has no recommendations on any of the devices.

In general, from a service standpoint, it is better to try correcting the antenna problem rather than to compensate. Of course, if the antenna is good (mismatch at a minimum, proper continuity, no shorts, etc.) but not adjustable and SWR is poor, an external antenna-matching device may be the answer.

RF-Amplifier Tuning. If the antenna is not adjustable, it is sometimes possible to detune (very slightly) the final RF amplifier of the transmitter to get a better SWR. The success of this technique depends on amplifier tuning-circuit design, considerable experimentation, and, even more, luck.

The usual sequence for tuning is to adjust the final amplifier for peak with an RF wattmeter and/or the set S/RF meter, and then check SWR. If there is an SWR problem, try detuning the final RF-amplifier circuit and see if the SWR improves. Do not make any drastic changes in final amplifier tuning (from the peak). If drastic changes are required, a mismatch or other problems are present.

Always make sure that the output frequency is within tolerance and power output is not seriously reduced, if the final amplifier is detuned in any way (even slightly). Verify that the improvement in SWR holds for all channels, high and low. Finally, keep in mind that the technique will not work on all sets. Also, an antenna problem of major proportions cannot be corrected by adjustment of the set, nor should any attempt be made to do so.

Field-Strength Meter Antenna Checks. The FS meters described in Sec. 3.8 may also be used to test an antenna system. However, FS meters do not provide the proof available from an SWR meter. An FS meter shows the *relative field strength* of the transmitted signal. If the FS indication is drastically different for each channel, the antenna system is suspect.

Remember that an FS meter will not produce exactly the same indication on all channels, even in a perfect antenna system. Also, an FS meter indication that is the same for all channels does not prove that the antenna system is good. A severe mismatch in the antenna system may still produce a strong indication on a relative FS meter. Thus, an FS meter indication proves only that the antenna is radiating a signal.

Ohmmeter Antenna Checks. In most cases, a mobile antenna (automobile or shipboard) may be checked with an ohmmeter using a pair of extra-long leads as shown in Fig. 7.12. The procedure is as follows:

Antenna Continuity. Connect one ohmmeter lead to the antenna tip and the other to the inner conductor of the lead-in that plugs into the set. Use alligator

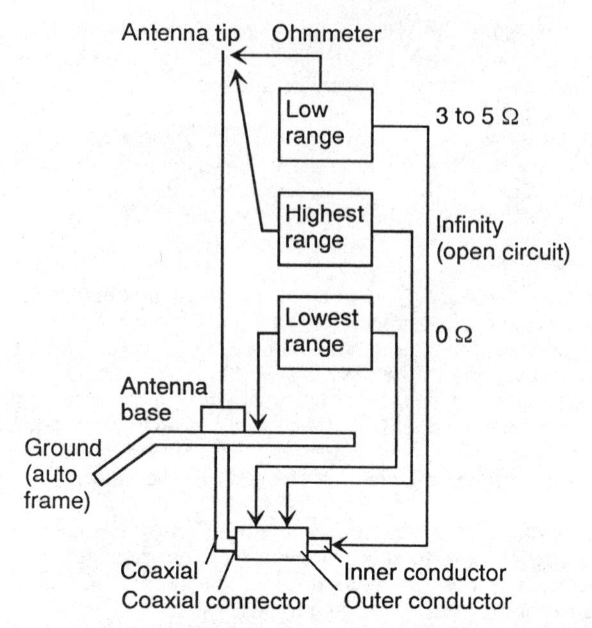

FIGURE 7.12 Antenna checks using an ohmmeter.

clips to make the connection. Set the ohmmeter to the lowest range, and then shake the antenna and lead-in. The resistance indicated by the ohmmeter should be about 3 to 5 Ω for a typical mobile installation. The resistance may be slightly higher for some installations.

If the resistance is considerably higher, or if the resistance varies when either the antenna or lead-in is moved, there is poor contact between the antenna and lead-in. This reduces antenna efficiency and may result in chronic poor performance of the set, both during reception and transmission. An antenna connection that is intermittently making and breaking contact produces a popping or crackling noise in the loudspeaker when the car (or boat) is in motion.

If you suspect high resistance or poor contact, disconnect the lead-in from the antenna and measure the lead-in resistance and the antenna resistance separately. Shake, bend, or twist the individual units to see which one is causing the problem. Look for corrosion or dirt on the coaxial connectors of the lead-in, and make sure that the connectors are tight. A collapsible antenna may have poor contact between segments, and an antenna loading coil may make poor contact with the antenna or base.

Antenna Shorts and Grounds. Once you are satisfied that there is proper continuity from the antenna to the lead-in center conductor, *make certain there is no contact* between the antenna and the outer conductor or shield. Connect one ohmmeter lead to the antenna tip and the other to a ground or the lead-in outer conductor (Fig. 7.12). Set the ohmmeter to the highest range and shake the antenna and lead-in. The ohmmeter should read infinity (open circuit).

If an intermittent short is indicated, a visual check will usually pinpoint the source. The problem is not so easy to locate when the ohmmeter indicates a constant high resistance that does not change when the antenna and lead-in are moved. Such high-resistance shorts reduce antenna efficiency and may also be a noise source.

Look for moisture at the connectors and for frayed braiding on the outer conductor of the lead-in. High resistance shorts may also be caused by an accumulation of dirt at the antenna base, since the dirt can hold enough moisture to form a high resistance between antenna and ground.

Outer-Conductor Continuity. The outer conductor of a coaxial cable is both a conductor and a shield. A broken shield braiding, or braiding that makes bad contact, can also produce noise and reduce antenna efficiency. To check the outer conductor, connect one ohmmeter lead to the outer conductor of the lead-in (at the set end) and the opposite lead to ground (Fig. 7.12). Set the ohmmeter to the lowest range and shake or bend the lead-in. The resistance should be zero or near zero.

If the resistance is not zero, the outer conductor is not making good contact with the ground and is not performing the normal function of shielding the lead-in. In addition to reducing antenna efficiency, poor contact in the lead-in outer conductor offers an easy path for electrical noise to enter the set. This is a particular problem if the lead-in is routed near an engine or other electrical wiring.

If the outer conductor shows poor contact (a high resistance to ground), look for breaks at the point where the shielding enters the connectors. In most cases, a visual check will pinpoint the poor connections.

7.8 RF OSCILLATOR TROUBLESHOOTING

Most communications equipment uses some form of oscillator for signal generation and frequency control. The frequency-control circuits can be quite simple or fairly complex, depending on the set. In the simplest form, such as a hand-held "walkie-talkie," there are two oscillators, one for transmitter frequency control and one for the receiver local oscillator. In other sets, there are two or three oscillators, combined with mixers, to form a frequency synthesizer. In a set with PLL, there is a standard oscillator (probably within the PLL IC but having an external crystal) and one or two other oscillators to form the complete frequency synthesizer.

No matter how complex the circuits appear, they are essentially oscillators and can be treated as such from a practical troubleshooting standpoint. For example, each circuit contains oscillators, which produce signals (probably crystal-controlled). These signals must have a *given amplitude* and must be at a *given frequency* (or must be capable of tuning across a given frequency range) for the set to operate properly. If you measure the signals and find them to be of the correct frequency and amplitude, the oscillators are good from a troubleshooting standpoint.

7.8.1 Oscillator Test and Troubleshooting Procedures

The first step in troubleshooting any oscillator circuit is to measure both the amplitude and frequency of the output signal. Many oscillators have a built-in test point. If yours does not, the signal may be monitored at the collector or emitter as shown in Fig. 7.13*a*. Signal amplitude is monitored with a meter or scope using an RF probe. The simplest way to measure oscillator signal frequency is with a frequency counter.

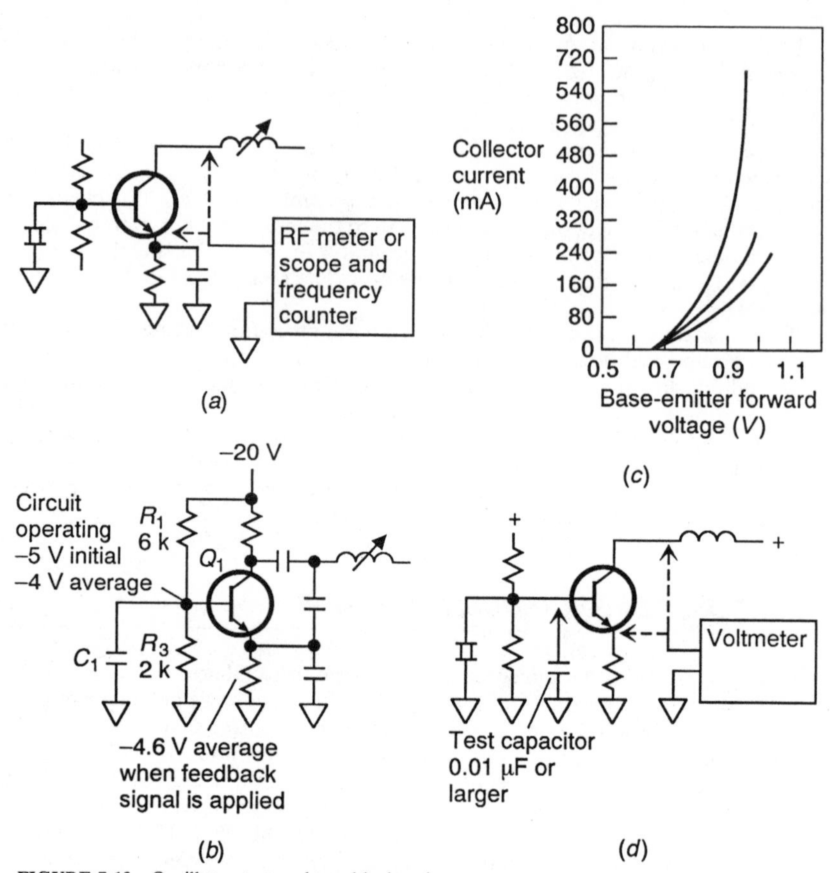

FIGURE 7.13 Oscillator test and troubleshooting.

7.8.2 Oscillator Frequency Problems

When you measure the oscillator signal, the frequency is (1) right on, (2) slightly off, or (3) way off. If the frequency is right on, leave the oscillator alone. If the frequency is slightly off, it is possible to correct the problem with adjustment. Most oscillators are adjustable, even those with crystal control. Usually, the RF coil or transformer is slug-tuned. On many PLL ICs, there is an external adjustable capacitor for each crystal. The most precise adjustment is obtained by monitoring the oscillator signal with a frequency counter and adjusting the circuit for exact frequency. However, it is also possible to adjust an oscillator using a meter or scope.

When the circuit is adjusted for *maximum signal amplitude,* the oscillator is at the crystal frequency. However, it is possible (but not likely) that the oscillator is being tuned for a harmonic (multiple or submultiple) of the crystal frequency. A frequency counter shows this, whereas a meter or scope does not.

If the oscillator frequency is way off, look for a defect rather than improper adjustment. For example, a coil or transformer may have shorted turns, a transistor or capacitor may be leaking badly, or the wrong crystal may be installed in the right socket (this does happen).

7.8.3 Oscillator Signal-Amplitude Problems

When you measure the oscillator signal, the amplitude is (1) right on, (2) slightly low, or (3) very low. If the amplitude is right on, leave the oscillator alone. If the amplitude is slightly low, it is possible to correct the problem with adjustment. Monitor the signal with a meter or scope, and adjust the oscillator for maximum signal amplitude. This also locks the oscillator on the correct frequency. If the adjustment does not correct the problem, look for leakage in the transistor or for a transistor with low gain.

If the amplitude is very low, look for defects such as low power-supply voltages, badly leaking transistors and/or capacitors, and a shorted coil or transformer turns. Usually, when signal output is very low, there are other indications, such as abnormal voltage and resistance values.

7.8.4 Oscillator Bias Problems

One of the problems in troubleshooting many oscillator circuits is the bias arrangement. RF oscillators are often reverse biased so as to conduct on half-cycles. However, the transistor is initially forward biased through bias networks, as shown in Fig. 7.13*b*. This turns the transistor on so that the collector current starts to conduct. Feedback occurs, and the transistor is driven into heavy conduction.

During the time of heavy conduction, a capacitor connected to the transistor base is charged in the forward-bias direction. When saturation is reached, there is no further feedback, and the capacitor discharges. This reverse biases the transistors and maintains the reverse bias until the capacitor has discharged to a point where the fixed forward bias again causes conduction.

This condition presents a problem. If the capacitor is too large, it may not discharge in time for the next half-cycle. In that case, the circuit operates as a blocking oscillator, controlling the frequency by the circuit capacitance and resistance (instead of by capacitance and inductance, as should be the case for RF oscillators). If the capacitor is too small, the oscillator may not start at all. The same is true if the capacitor is leaking badly. From a practical troubleshooting standpoint, the measured condition of bias on an oscillator can provide a clue to operation, if you know how the oscillator is supposed to operate.

The oscillator in Fig. 7.13*b* is initially forward biased through R_1 and R_3. As Q_1 starts to conduct and in-phase feedback is applied to the emitter (to sustain oscillation), C_1 starts to charge. When saturation is reached (or approached) and feedback stops, C_1 discharges in the opposite polarity, reverse biasing Q_1.

The value of C_1 is selected so that C_1 discharges to a voltage less than the fixed forward bias before the next half-cycle. Thus Q_1 conducts on slightly less than the full half-cycle. Typically, an RF oscillator such as shown in Fig. 7.13*b* conducts on about 140° of the 180° half-cycle.

It is commonly assumed that transistor junctions (and diodes) start to conduct as soon as forward voltage is applied. This is not true. Figure 7.13*c* shows characteristic curves for three different types of transistor junctions. All three junctions are silicon, but the same conditions exists for germanium junctions. None of the junctions conduct noticeably at 0.6 V, but current starts to rise at that point. At 0.8 V, one junction draws almost 80 mA. At 1 V, the dc resistance is about 2 or 3 Ω, and the transistor draws almost 1 A. In a germanium transistor, noticeable current flows at about 0.3 V.

For troubleshooting purposes, bias measurements provide a clue to the performance of oscillators, although such measurements do not provide positive proof. The one sure test of an oscillator is to *measure output signal amplitude and frequency.*

7.8.5 Oscillator Quick-Test

It is possible to check to see if an oscillator is oscillating using a voltmeter and a large-value capacitor (typically 0.01 μF or larger). Measure either the collector or emitter voltage with the oscillator operating normally, and then connect the capacitor from base to ground, as shown in Fig. 7.13*d.* This should stop oscillation, and the emitter or collector voltage should change. When the capacitor is removed, the voltage should return to normal.

If there is no change when the capacitor is connected, the oscillator is probably not oscillating. In some oscillators, you get better results by connecting the capacitor from collector to ground. Also, do not expect the voltage to change on an element without a load. For example, if the collector is connected directly to the power-supply voltage, or through a few turns of wire, as shown in Fig. 7.13*d,* this voltage does not change substantially, with or without oscillation.

7.9 TYPICAL TROUBLE SYMPTOMS AND REMEDIES

This section describes typical trouble symptoms and remedies for communications sets such as shown in Figs. 7.4 and 7.5. The diagnostic checks shown in Figs. 7.6 through 7.8 also apply.

7.9.1 Radio Does Not Receive

Use this procedure when there is no received audio and the transmitter RF power is normal. More specifically, use this procedure if there is no receiver noise or if an audio output cannot be obtained when a strong (modulated) signal is applied to the receiver.

The symptom can be caused by failure in almost any portion of the receiver (circuits 1 through 9, 12, and 15 through 19, in Figs. 7.4 and 7.5).

Note that if there is no receiver noise, it is likely that the fault is in the 455-kHz IF or audio sections. If some receiver noise is present, the problem is more likely to be in the RF or first IF sections.

An open circuit in the RF section often does not cause complete failure to the receiver. High-strength modulated RF signals can often be coupled through an open circuit and produce an audio output. Of course, receiver sensitivity is poor. On the other hand, a short in the RF circuits can often give a symptom of no receiver audio.

The following steps isolate the problem to a much smaller are:

1. Perform the PA mode check (Sec. 7.3.5). If the PA mode operates, the audio section is proved good (circuits 15 through 18). Proceed to step 2 if the set is not equipped with an S-meter or to step 3 if the set has an S-meter.

If the PA mode does not operate, the audio section is the problem area (circuits 15 through 18) and can be further isolated as follows.

Operate the set in the PA mode. Connect an audio generator to a loudspeaker. Place the microphone over the speaker. Set the audio generator to 1 kHz. Measure the audio signal with a scope at circuits 15, 16, 17, and 18 (in that order) until the loss of signal is noted. The circuit at which signal loss is noted is probably defective. Check the voltage, resistance, and PC-board wiring in that circuit. Note that this procedure also checks the microphone and microphone connections. As an alternate, you can inject a 1-kHz audio signal at the input to circuit 15 and then check at each of the points in the audio section.

2. If the set is not provided with an S-meter, inject a modulated 455-kHz IF signal into the detector (circuit 6). If audio output is obtained, the detector (circuit 6) and the first audio amp (circuit 12) are probably good. Proceed to step 4.

If no audio output is obtained, measure the audio signal on a scope at each accessible measurement point in the signal path from the detector (circuit 6) through the first audio amplifier (circuit 12), and note the point of signal loss. If the input to circuit 12 is normal but no output is measured, check the voltages of the associated squelch circuit. Certain component failures can cause the circuit to remain in the fully squelched condition at all times.

3. If the set has an S-meter, apply a 1000-μV on-frequency signal from an RF generator to the antenna jack, and note the S-meter reading. If a strong S-meter reading is obtained, the RF and IF sections are operating (circuits 1 through 5). The trouble is isolated to the detector (circuit 6) or first audio amplifier (circuit 12). Inject a modulated 455-kHz IF signal into the detector and troubleshoot as instructed in step 2 of this section.

If no S-meter reading is obtained, the signal is not passing the RF and IF sections. Proceed to step 4.

4. Inject a modulated 455-kHz IF signal at the output of the second mixer (circuit 3). In a single-conversion receiver, the point of injection is the output of the first mixer, and the frequency may not be 455 kHz.

If audio is obtained, the 455-kHz IF section is operating. Proceed to step 5. If no audio output is obtained, the signal is not being passed through the 455-kHz IF section (circuit 4 or 5). Starting at the point of the signal injection and working toward the detector, measure the 455-kHz IF signal on the scope until a loss of signal is noted. This is the defective area.

5. Inject a modulated 7.8-MHz signal at the output of the first mixer (not all sets use 7.8-MHz). If audio output is obtained, the second oscillator (circuit 8) and second mixer (circuit 3) are probably good. Proceed to step 6.

If no audio output is obtained, measure the output of the second oscillator. If no RF is measured, the second oscillator is defective. If RF is present, the second mixer is defective. A defective AGC circuit (circuit 9) could bias the second mixer to cutoff. Measure AGC voltage with no RF signal applied.

6. Inject a modulated signal (at the selected channel frequency) into the first mixer (bypass the RF amplifier). If normal audio is obtained, the trouble is in the RF amplifier (circuit 1) or AGC circuit (circuit 9). If no audio is obtained, the first mixer (circuit 2) or AGC circuit (circuit 9) is suspect.

7.9.2 Radio Does Not Transmit or Receive

This procedure is used when there is no transmitter RF output and no receiver audio on any channel. The symptom can be verified by making transmitter RF-power and receiver audio-power checks on several channels. The symptom is valid if there is no measured power (receiver or transmitter) on any channel.

Other than the power supply, the only circuit common to the receiver and RF portions of the transmitter is the synthesizer (circuit 11). If the synthesizer fails, there is usually low receiver noise present. Of course, there is a remote possibility that there are two simultaneous failures, one in the receiver and one in the transmitter. The following steps help isolate the trouble:

1. If there is any receiver noise, the trouble is probably in the synthesizer. Go to step 2. If no noise is heard, go to step 5.

2. Check the output of the synthesizer with an RF voltmeter or scope and a frequency counter. If no RF is measured or if the RF is at the wrong frequency, the synthesizer is defective. Remember that if one mixer input (say the synthesizer) is missing, there will still be RF output from the mixer, but at the wrong frequency. However, if the synthesizer output is absent or off frequency, the mixer RF output level will be low. See step 4 for an alternative checkout method for the synthesizer.

3. A synthesizer typically contains at least two oscillators and a mixer. If any of these stages is defective, the correct output will not be generated. Measure RF-output voltages at each stage until the defective stage is located; then make voltage and resistance checks to find the bad part in that stage.

4. An alternative to steps 2 and 3 is to use an RF generator, and inject signals of the correct frequency into the synthesizer (to substitute for the signals produced by the synthesizer oscillators). If the injected signal restores operation, the normal signal is missing from that point. Move the point of injection from the output of the synthesizer toward the input, one stage at a time. Each point may require a different frequency (refer to the service literature).

When the point of injection is moved beyond the defect, signal injection no longer restores operation, and you have found the defective area. Remember that this alternate procedure of signal injection requires a *high-impedance probe* to prevent circuit loading. A low-impedance probe used for injection can detune or disable the circuit and render the procedure useless.

5. If no receiver noise is heard during step 1, check the front-panel LEDs or power lamps. If the LEDs are on, input power is getting to the set. Check some of the major distribution voltages (such as the 13.8-V distribution bus). If the LEDs are not on, check input power and fuses.

6. If no problem is found in step 5, perform steps 2, 3, and 4.

7. If no other problem can be found, treat the "radio does not transmit" and "radio does not receive" symptoms as separate symptoms. Refer to the associated troubleshooting procedure for each of these symptoms (Secs. 7.9.1 and 7.9.6).

7.9.3 Radio Does Not Transmit or Receive on Some Channels

This procedure is used when there is no transmitter RF power on certain channels, normal RF output on other channels, no received audio on certain channels, and normal reception on other channels.

This trouble is, in all probability, caused by failure of the synthesizer or PLL, whichever is used in the set. Typically, with either system, the trouble results from a *defective crystal or a faulty channel-selector switch.* These are the only parts involved in switching from one channel to another.

Sets with Synthesizer. Troubleshooting the synthesizer is usually not difficult if you understand the synthesizer scheme (or how the synthesizer crystal-controlled oscillators are combined to produce the various channel frequencies).

For example, Table 7.1 shows the classic synthesizer scheme for a 23-channel communications set. This scheme permits 10 crystals to generate all 23 channel frequencies. The crystals are connected in a 4 by 6 matrix to permit 24 possible frequency combinations. The channel-selector switch is wired to select 23 of the 24 possible combinations. A crystal failure in such a synthesizer usually disables four or six channels.

A defective crystal is easy to identify from the synthesizer crystal-frequency

TABLE 7.1 Synthesizer Scheme for a 23-Channel Communications Set

Channel	Oscillator no. 1 crystal	Oscillator no. 2 crystal
1	A	E
2	B	E
3	C	E
4	D	E
5	A	F
6	B	F
7	C	F
8	D	F
9	A	G
10	B	G
11	C	G
12	D	G
13	A	H
14	B	H
15	C	H
16	D	H
17	A	I
18	B	I
19	C	I
20	D	I
21	A	J
22	B	J
23	C	J

scheme. For example, if the set uses the scheme in Table 7.1 and channels 13, 14, 15, and 16 do not operate, crystal H is defective. If the problem is noted on channels 2, 6, 10, 14, 18, and 22, crystal B is the cause (since all of the channels have only crystal B in common). Remember that crystal frequencies are different with nearly every manufacturer, so you must know the synthesizer scheme (which should be included in the service literature).

Sets with PLL. As discussed, most present-day communications sets use PLL instead of a multicrystal synthesizer (although the PLL may be called a frequency synthesizer in some literature). With PLL, all frequencies are generated by *two or three crystals* and an IC which contains a programmable divider, as shown in Fig. 7.1.

If any of the crystals or the IC fail in a PLL system, *all the channels become inoperative* (in most cases). Therefore, if you have a failure only on certain channels of a PLL system, the trouble is probably in the channel selector or associated wiring. Refer to Sec. 7.1 for additional information on PLL troubleshooting.

7.9.4 Weak Audio

Use this troubleshooting procedure when the receiver audio is below rated power with a strong modulated signal applied to the receiver. The symptom may also be accompanied by high distortion. Start by performing the PA-mode check in Sec. 7.3.5.

If normal audio power output is possible in the PA mode but not through the receiver with a modulated signal applied, the trouble is in the first audio amplifier (circuit 12). Inject a modulated 455-kHz signal into the detector (circuit 6) and measure audio-signal levels in circuit 12. Use a scope for the measurement and look for evidence of signal-amplitude clipping or below-normal signal levels.

If audio power output is below rated power in the PA mode, as well as through the receiver, the trouble is in the audio circuits (circuits 15 through 18). Apply a 1-kHz signal to a loudspeaker, and use the speaker signal to drive the microphone while operating in the PA mode. Make audio-level checks using a scope in circuits 15 through 18. Look for evidence of signal-amplitude clipping or below-normal signal levels.

If one of the push-pull power amplifiers (circuit 17 or 18) is disabled, audio power will drop to about 25 percent of normal, and the sound will be highly distorted.

7.9.5 Poor Receiver Sensitivity

Use this troubleshooting procedure when the receiver provides rated audio output but does not meet the receiver sensitivity specification (Sec. 5.5). The symptom is produced by a malfunction in the RF or IF amplifier section of the receiver (circuits 1 through 5).

1. Set up the test equipment as for the receiver sensitivity check (Sec. 5.5), but increase the RF-generator output until the 10-dB (S + N)/N level is reached.
2. Check AGC voltage for normal values. If it is convenient, disable the AGC and recheck sensitivity. AGC can usually be disabled by shorting the AGC

bus to PC-board ground or connecting a bias power supply to the AGC line and setting the voltage for the normal no-signal AGC level. Consult the service literature regarding the AGC. If normal sensitivity is restored when AGC is disabled, troubleshoot the AGC circuit.

3. Measure the RF-voltage levels of the synthesizer (circuit 11) and second oscillator (circuit 8) outputs. Low signal injection for the synthesizer can cause this symptom.
4. If the service literature includes typical stage gains, make gain measurements and compare them to the specified values.
5. Inject a 30 percent-modulated 455-kHz signal at the detector (circuit 6) and adjust the RF generator output level for a convenient audio-output level (such as 0.5 W) at the speaker.
6. Note the RF generator output level required to produce 0.5 W of audio output.
7. Move the point of injection toward the antenna, one stage at a time, and readjust the RF-generator level for the same audio-reference level (0.5 W). Change from 455 kHz to 7.8 MHz and then to the RF channel frequency as the point of injection moves from the 455-kHz section toward the antenna.
8. Again note the RF-generator output level.
9. The difference in the RF-generator output levels between steps 6 and 8 is the gain (or loss) between injection points. If the RF-generator output scale is in microvolts and the service-literature stage-gain is in decibels (or vice versa), convert the RF-generator output readings as necessary.
10. Troubleshoot any stage with low gain or any passable device (such as a coupling capacitor) with high attenuation, using voltage and resistance checks. If stage gain is low but there is no apparent defect in circuit components, try correcting the problem by alignment of the receiver (using the service-literature procedure).

The procedure described in this section is a good troubleshooting technique even when the service literature does not give typical stage gains. Sometimes, the defective stage is obvious (shorted capacitor, open resistor, etc.).

7.9.6 Radio Does Not Transmit

Use this procedure when there is no transmitter RF power output but receive operation is normal. The procedure also applies to the symptom in which the transmitter cannot be keyed. Start by checking to see if the transmit indicator (if any) turns on when the PTT switch is pressed.

1. If the indicator does not turn on, place a jumper across the PTT pins of the microphone jack (or equivalent point) on the set. If the transmit indicator now turns on (with the jumper), the problem is in the microphone or microphone cord. Try replacing the microphone and cord.

If the transmit indicator does not turn on, the transmit-receive relay is probably the cause. In most present-day sets, the transmit-receive relay is replaced by a solid-state diode-switching circuit (Fig. 5.15). As discussed, these switching circuits are made up of diodes that alternately connect the transmit and receive circuits when alternately forward biased and reverse biased. Solid-state switching circuits are generally more difficult to troubleshoot than is a transmit-receive relay.

2. If the transmit indicator does turn on but there is no RF transmission, the problem can be in almost any of the transmitter RF circuits (circuits 21 through 25). Use an RF voltmeter or high-frequency scope to measure RF voltages. Starting with circuit 21 and working toward circuit 25, measure each accessible point in the signal path until the absence of RF voltage is noted. This is the defective area.

In some sets, circuits 21 and 22 may be part of the synthesizer (or PLL), and the synthesizer output in the transmit mode is a low-level RF signal. If the set uses such a design, start by checking the RF from the synthesizer and then go on to circuits 23 through 25.

7.9.7 Low Transmitter Power

Use this procedure if the transmitter RF output power is below normal and receiver operation is normal. The symptom is caused by low gain or low voltage in the transmitter RF amplifiers (circuits 23 through 25). Measure RF voltages at each of these amplifiers. Follow this with voltage and resistance checks.

If there is no apparent fault but RF power is low, try correcting the problem by alignment of the RF amplifiers, using the procedures described in the service literature. Usually, the RF amplifiers only require "peaking" of the tuned circuits to get proper alignment.

7.9.8 Radio Off-Frequency

Use this procedure if the transmitter frequency is not within specification on any channel or channels. The receiver may also operate off frequency, but it is difficult to detect such a symptom. An off frequency receiver displays symptoms of poor sensitivity (and possibly distortion). Normal results may be obtained during test because the RF generator is usually tuned for maximum receiver output, not necessarily to the channel frequency.

If the set is off frequency, the problem is usually one of the crystal oscillators. If the trouble appears on all channels, the transmit oscillator (circuit 21) is operating off frequency. If the trouble appears only on certain channels, refer to the procedure in Sec. 7.9.3. However, instead of certain channels being inoperative, there will be certain channels that are off frequency. Replace the crystals or tuning components (coils and capacitors) that may detune the operating frequency. Slight off-frequency conditions may be improved by realignment.

7.9.9 Abnormal Transmitter Modulation

Use this procedure if there is no transmitter modulation and receiver audio is normal. Also use the procedure if the transmitter appears undermodulated or if the transmitter is easily overmodulated.

If there is no modulation, the microphone, microphone amplifier (circuit 14), or modulation input to the transmitter RF power amplifier (circuit 25) is defective.

1. Perform the PA-mode check (Sec. 7.3.5).
2. If PA-mode operation is normal, troubleshoot the audio output transformer and audio input to circuit 25.
3. If the PA mode does not produce audio output, inject a 1-kHz signal into the

audio points of the microphone jack (or equivalent points on the set) and key the transmitter.

4. If you get modulation, the microphone or microphone cord is defective.
5. If no modulation is produced, troubleshoot the microphone amplifier (circuit 14).
6. If an unusually high audio level is required for 100 percent modulation (or you cannot get 100 percent modulation), perform the PA-mode check. If rated audio power is possible in the PA mode, troubleshoot the audio-input portion of circuit 25. If low audio power is measured on the PA-mode check, troubleshoot the microphone amplifier (circuit 14) for low gain and the microphone for low output (try a substitute microphone, if one is available).
7. If overmodulation is present (Figs. 3.1 through 3.4), troubleshoot the modulation-limiting circuits, which is usually part of circuit 25.

7.9.10 Abnormal SSB Transmitter Operation

Use this procedure for any abnormal SSB transmit condition (no output, low power, improper modulation, or incorrect frequency), including symptoms in which the SSB transmitter and receiver both show abnormal operation.

1. If there is no SSB transmitter RF-power output, check the SSB-receiver audio power.
2. If both modes are inoperative, the trouble could be in the ring (balanced) -modulator/product-detector (circuit 28) or narrowband 7.8-MHz bandpass filter (circuit 30) shown in Fig. 7.5. Apply a test signal to the microphone and measure the RF output of the ring modulator (circuit 28). If no output is measured, troubleshoot the ring-modulator circuit. If output is measured, troubleshoot the 7.8-MHz bandpass filter.
3. If receiver operation is normal, the SSB transmitter amplifiers (circuits 31 and 32) are suspect. Apply a signal to the microphone and measure RF voltages at circuits 31 and 32. Troubleshoot the area where the RF voltage is first missing.
4. If there is no transmitter and receiver operation on one sideband only, the problem is probably in the 7.8015- to 7.7985-MHz oscillator (circuit 21). This circuit uses one crystal in the AM and one in the sideband modes and another crystal in the opposite sideband mode. Troubleshoot the crystal and mode-selector switch.
5. If transmitter RF power is low, troubleshoot the transmitter RF amplifiers (circuits 31 and 32) for low gain as described in Sec. 7.9.7.

7.9.11 Abnormal SSB-Receiver Operation

Use this procedure for any abnormal SSB-receive condition (no output, poor sensitivity, or poor adjacent-sideband rejection). The AM mode and SSB-transmit modes are normal.

1. If there is no receiver audio or there is poor sensitivity, set up test equipment for an SSB receiver-sensitivity check (Sec. 5.16).

2. Disable the SSB AGC circuit (circuit 27). If normal operation is restored, troubleshoot the AGC circuit.
3. Troubleshoot the 7.8-MHz IF amplifier (circuit 29).
4. If adjacent-sideband rejection does not meet specification, troubleshoot the narrowband 7.8-MHz bandpass filter (circuit 30).

7.9.12 Receiver Distortion

Use this procedure when the receiver audio does not meet the distortion specification or there is a symptom of distorted audio in the AM mode. Start by measuring distortion in the PA mode.

1. If distortion is the same in the PA mode as in the receive mode, measure audio signals in circuits 15 through 18 using a scope. Starting at circuit 15 and working toward the speaker, observe waveforms for a point where a *change in the waveform occurs.*

2. If distortion meets specifications in the PA mode, check distortion with a 1-kHz test signal injected at the detector. If the same amount of distortion is measured, measure audio waveforms in circuit 12, and look for the point where the waveform changes.

If distortion meets specification, check the receiver for an off-frequency condition, lack of AGC action, and IF amplifier distortion.

7.9.13 Abnormal PA-Mode Operation

Use this procedure only when the PA mode is inoperative but all other modes are normal. With a set like the one shown in Figs. 7.4 and 7.5, this symptom can be caused only by faulty PA-jack wiring or by the PA-mode selection switch.

7.9.14 Adjacent-Channel Interference

Use this procedure when the receiver does not meet the adjacent-channel rejection specification (Sec. 5.6) in the AM mode, but all other performance specifications are normal.

1. Recheck receiver alignment as described in the service literature.
2. Check the bandpass filter and all tuned circuits in RF and IF stages as follows:

   Inject a modulated 455-kHz signal at the output of the second mixer. Tune the RF generator to 10 kHz above and below the 455-kHz center frequency. Measure the output of the 455-kHz bandpass filter (circuit 4) as the RF generator is tuned across the band. The *bandpass should be symmetrical.* A defective part usually causes a nonsymmetrical condition (but severe misalignment can also result in nonsymmetrical response in tuned circuits).

   Measure the bandpass at each accessible point in the signal path through the 455-kHz IF amplifier (circuit 5).

   If the 455-kHz circuits show symmetrical bandpass, check the bandpass of the 7.8-MHz IF circuit (circuit 3).

If the receiver meets specifications, little can be done to further improve adjacent-channel rejection.

7.10 NOISE AND INTERFERENCE PROBLEMS

Electrical interference, or noise, particularly interference caused by automobile engines, has long been one of the most annoying factors in RF-communications equipment operation. Manufacturers have taken what steps they can to solve this interference problem by adding shielding and filters to the sensitive portions of the receiver circuits.

Most present-day communications sets include some form of squelch, ANL, and other noise-suppressing circuits. Effective as these circuits are, they only set the noise level and make it bearable. Weak signals that are below this level simply do not come through. The result is a limiting of the operating range of the receiver section and a possible loss of communications.

7.10.1 Causes of Engine Interference

All engine electrical interference or noise originates from the same source—sparking. The gasoline engine is full of spark-noise sources. Spark plugs, alternator slip rings, and any device that has make-and-break contacts all have gaps that electricity must jump in the normal operation of an engine. Modern engines with electronic ignition and solid-state voltage regulation produce less noise than older engines with distributor points, distributor contacts, and relay-type voltage regulators.

Sparks produce undesirable RF signals (radio waves) that can be radiated or conducted into a communications set. Since it is not practical to stop the engine or cut off the alternator each time the set is used, the problem must be attacked at the engine itself. Three basic methods may be used: *arc suppression, filtering,* and *shielding.* A number of *noise-eliminator kits* are available, all based on these three techniques.

7.10.2 Identifying Interference Sources

It is possible (usually) to identify the source of interference by the sound that comes from the set loudspeaker. Here are the characteristics:

Ignition interference (and this is usually about 90 percent of the problem) occurs as a popping sound that is *synchronized with the speed* of the engine.

Alternator noise (the next most common problem) may be identified as a "whine" that starts only when the engine is speeded up. If the noise source is in doubt, temporarily remove the alternator leads (or slip the alternator belt off), and check the noise levels with the engine running. If the noise is unchanged, the alternator is not the source.

A relay-type voltage regulator usually produces a rough, rasping sound as it cuts the alternator in and out of the circuit. Solid-state voltage regulators (found in most present-day engines) do not produce RF noise.

Instrument noise (caused by fuel gauges, etc.) is identified by hissing or crackling sounds. Instruments should be disconnected one at a time until the noisemaker is found.

Tire and wheel noises (usually produced by static electricity) produce an irregular "rushing" sound in the loudspeaker. This sound is present only when the auto is in motion.

Preliminary Tests. The easiest way to run a noise or interference test on any engine is to compare the receiver noise level while the engine is running with the noise level under identical conditions while the engine is off. Try to make the test away from an external noise source, such as other engines, high-voltage lines, neon lights, etc.

Turn off the squelch and any other noise-limiter circuits, if possible. Adjust the receiver volume control until you hear a steady noise level from the speaker. If there are any signals on the air, select a channel with a weak signal and adjust the volume until the signal is barely audible. Then turn off the engine and see if the noise level drops noticeably.

If there is no great change in background noise when the engine is turned off or if there is no change in the weak signal, you do not have an engine-interference problem. If you do note an appreciable change, start looking for the source of the interference.

It is generally assumed that if you cannot hear any noise, there is no interference. This is not always true, since strong radiated interference can desensitize a receiver (particularly the RF and IF amplifiers) so that weak signals are obscured. For that reason, make the noise tests on the basis of background noise level and a weak signal.

Radiated or Conducted Interference. Once you have established the presence of engine interference, your next step is to decide whether the problem is radiated or conducted. With the set operating and the volume set so that you can hear both background noise and signals, disconnect the antenna.

If the noise is completely eliminated, the problem is probably one of radiated interference rather than conducted noise. On the other hand, if the noise remains but the signal is gone when the antenna is disconnected, you have conducted noise.

7.10.3 Arc Suppression at Spark Plugs

Arc suppression is used to combat both conducted and radiated noise. Arc suppression systems or kits are based on three facts: (1) although it takes a high voltage to make a spark jump across the gap of a spark plug, very little current is needed, (2) the amount of interference is proportional to the current, and (3) a high resistance placed in series with the spark plug limits the current to practically nothing without reducing the voltage.

There are some drawbacks in using suppressors even though they are effective. Most present-day autos use suppressor-type ignition wire and suppressor spark plugs. In some cases, the resistance of the external kit suppressors combined with that of the ignition wiring and plugs is sufficient to impair the engine. At the same time, the maximum resistance that can be added may not be enough to eliminate all objectionable interference. It is quite possible that *filtering* (for conducted noise) and *shielding* (for radiated noise) will be required in addition to (or instead of) arc suppressors.

7.10.4 Interference Filters for Ignition Systems

The three basic types of interference filters in common use are filter capacitors, chokes, and wave traps.

The most popular type of *filter capacitor* (for interference elimination) is the feed-through capacitor. Such capacitors eliminate conducted noise in the battery circuit before the noise can enter and degrade or desensitize the receiver section of the set. A *shielded cable* attached to the capacitor eliminates noise pickup in the power cable from nearby spark plugs, alternators, etc. Many noise-elimination systems or kits include several bypass capacitors (combined with suppressors) in various points throughout the ignition system.

Some noise-elimination kits include *choke coils.* These kits are meant primarily for auto stereo and CD players rather than communications sets. The choke coil is connected in the power line between the set and battery.

Wave traps are used primarily to filter alternator interference and are more difficult to install than filters and suppressors (wave traps may require adjustment after installation, unless pretuned at the factory).

A wave trap consists of a coil and capacitor combination tuned to either the frequency of the spark interference or the operating frequency of the set. The wave trap rejects, or traps, interference of this frequency (similar to wave traps in antenna systems). The major drawback to a wave trap is that it can be tuned to only one frequency, and spark noise is often generated on several frequencies at once. However, wave traps can be effective in the case of extreme alternator interference on one specific channel (or a few adjacent channels).

7.10.5 Shield Systems

Even when suppressors and filters are used, it is still possible for an engine to create sufficient interference to affect communications-set operation. As discussed, signals are generated when there is a spark across a gap. These signals are conducted through the lines in which the spark occurs and are radiated from the lines. Just as in the case of conventional radio waves, electrical interference may be prevented from radiating by *surrounding the source of interference with a metal shield.*

Shielding for typical mobile (and marine) installations is available in two basic types of kits: custom and do-it-yourself. (Note that there is generally more of an interference problem in marine communications systems since there is no natural shield provided by the fire wall of an automobile engine.)

The complete custom kits consist of metal shields for the spark plugs, high-voltage wiring with shielded braid, and shields for the electronic distribution unit, all supplied as an assembled harness. A typical do-it-yourself kit consists of spark-plug shields (each with several feet of shielding braid) and a distribution-unit shield (with feed-through filter capacitors as necessary for various points in the electronic ignition system). Installation of the do-it-yourself kits is quite involved and time-consuming (even with the step-by-step instructions).

7.10.6 Bonding Systems

Bonding provides an easy route for radiated interference to reach ground. For example, with proper bonding, a way is provided to route the interference from the hood to the frame of an auto, which acts as a ground.

Direct bonding occurs when two metals are connected directly, surface to surface. Direct bonding is part of auto design and is not the problem of service technicians. However, *strap bonding* can be added to most autos as a noise elimination measure.

Strap bonding is done by connecting the two metals with an electrical conductor, usually a braid with lug fasteners at the ends. (Braid is used because it is flexible and does not break easily with constant movement. Quite often the two metals to be bonded move in relation to each other when the auto is in motion.)

In addition to being more durable, braid produces an additional benefit when used as interference bonding. The rubbing of two metals can produce static electricity, which creates interference under some conditions. When parts are bonded together, the static electricity has a discharge path through the bonding strap.

Typical Bonding Points. Some typical bonding points on an auto interference-elimination installation are:

Corners of the engine to the frame
The exhaust pipe to the frame and the engine
Both sides of the hood
Both sides of the trunk lid
Distribution unit to engine and fire wall
Air cleaner to engine block
Battery ground to frame
Alternator to voltage regulator
Front and rear bumpers to the frames on both sides
Tail pipe to the frame at rear.

7.10.7 Wheel and Tire Static

The interference from static electricity produced by wheels and tires occurs only when the auto is in motion. One way to pin down wheel or tire static is to switch off the engine and let the auto coast down a slight incline. Any interference still present (usually an irregular rushing sound in the speaker) is probably from the wheels or tires.

In most cases, wheel static occurs in the front wheels and is caused by the insulating film produced by the lubricant in the wheel bearings. At one time, there were collector springs, installed in the hub cap, to offset this condition. The springs provide a grounding path from the rotating wheel to the axle. (Note that most present-day communications sets do not require such drastic interference-elimination measures.)

There is also a static discharge between the auto tires and the road surface, particularly on hot, dry days. This may be eliminated by simply wetting the tires, although there were some "antistatic powders" being sold at one time.

7.10.8 Ignition-System Maintenance

Finally, remember that a poorly maintained ignition system causes more interference than an ignition that is in good condition. For example, worn, ragged spark-plug gaps require higher ignition voltage, so both set and engine performance deteriorate. Flat, parallel gaps and factory-maintained settings in spark plugs require less voltage and decrease interference. Careful spark-plug maintenance is therefore the first step toward ignition-noise suppression.

It is usually a waste of time and money to cover up a poorly maintained ignition system with suppression, filtering, and shielding. Also, it is possible that noise suppression applied to a poorly maintained engine will cause the engine to malfunction (rough idle, stall, etc.). Of course, a properly installed noise elimination system should not affect a well-maintained ignition.

INDEX